Technology and Christianity

Technology and Christianity

Essays on the Interface

Egbert Schuurman

WORDBRIDGE PUBLISHING
Aalten, the Netherlands
www.wordbridge.net
info@wordbridge.net
ISBN 978-90-76660-74-5

PAIDEIA PRESS
P. O. Box 1000
Jordan Station, ON
Canada, L0R 1S0
info@paideiapress.ca
ISBN 978-0-88815-351-7

ILLUSTRATIONS: These engravings of swivel trucks built by the Baltimore Car Wheel Company were published in the *Street Railway Review,* vol. V (Chicago: Windsor & Kenfield Publishing Company, 1895), p. 724. As such, they are in the public domain.

TABLE OF CONTENTS

FOREWORD

It has been over 44 years since my *Technology and the Future* was first published by Wedge Publishing Foundation, Toronto, Canada. The second printing appeared in 2009, published by Paideia Press, Grand Rapids, MI. At that time the preface mentioned a prospective Volume II, which "shall be based on the systematic work of the first volume but will focus more immediately on a broad range of technological issues that are present in most societies, such as Ethics of Technology in Industry, Medicine, Bio-engineering and Agriculture." That promise is now fulfilled. I have continued to draw on Reformational philosophy as it has developed at the Free University of Amsterdam for almost a century and later found worldwide acclaim.

My philosophical analysis of technology and ethical reflection cover the range of development of technology and will continue to be relevant to new, coming developments. Unlike others, I continue to call attention to the spiritual-historical background of technological culture. The development of technology is vast and broad. Secularization – culture without God – gives technology a tremendous boost. The Christian philosophical approach is not empty-handed in facing that development. It asks in that technical culture that we obey the call of God and abandon the thirst for technical power and alluring materialism as a driving force. That we foster the pursuit and delimitation of technology in the righteousness of the Kingdom of God, which has come and is coming in Jesus Christ. That perspective offers a healthy delimitation of technology and makes that technology subservient to all that lives: to all that lives in the microcosm, to the plant and animal kingdoms, and to human life and a multi-hued diversity of human societies. That is work in the service of the living God. In our secularized, godless technological culture, that is always the renewing perspective of God's Kingdom. That includes the very latest developments, such as those of artificial intelligence, which comes up here, in passing or in anticipation.

Taken together, this book in my view raises issues and offers perspectives that are in dire need of attention.

I want to warmly thank Ruben Alvarado for translating portions of this book and for his editorial work: a handsome achievement. His perseverance is admirable.

It is also cause for thanks that WordBridge Publishing and Paideia Press, belonging to two different continents, collaborated on the publication of this book. Technology is global. So is a critical reflection on it!!!

Egbert Schuurman, February 2024
Breukelen, The Netherlands

A NOTE ON THE TRANSLATIONS

"Reflections on the Technological Society" trans. Harry van Dyke and Lammert Tenyenhuis, copyright © 1977 Wedge Publishing Foundation, reprinted by permission.

"Christians in Babel" trans. James C. van Oosterom, copyright © 1987 Paideia Press Ltd., Jordan Station, Ontario, CA, reprinted by permission.

"Perspectives on Technology and Culture" trans. John H. Kok, copyright © 1995 Dordt College Press, reprinted by permission.

"The Technological World Picture and an Ethics of Responsibility" trans. John H. Kok.

"The Challenge of the Islamic Criticism of Technology," in *Transformation of the Technological Society,* trans. Kim Batteau.

"The Scientization of Modern Culture," "Systems Theory, Modern Technology, and Cultural Change," "Technology in Christian-Historical Perspective," "Beyond the Empirical Turn: Responsible Technology," and "Information Society: Cultural Impoverishment or Enrichment?" trans. Herbert Donald Morton.

"Christianity and Progress," "Enlightening the Enlightenment," and "Technicism and the Meaning of Technology" trans. Ruben Alvarado.

CHRISTIANITY AND PROGRESS (1974)

There is every reason to speak of Christianity and progress. Although it is not always clear what progress is, it is clear that progress is spoken of within what is called the linear view of history. And this view of history is rightly attributed to Christianity. After all, with the growing influence of Christianity, the cyclical or circular view of history disappeared. This pre-Christian view of history was oriented to the rhythm of nature, to the cycle of day and night and to the passage of the seasons. It is the rhythm of rising, shining and setting.

By contrast, revelation, to which Christian faith is related, speaks clearly of a beginning to history – the creation – and of an end – the completion of creation in the Kingdom of God. It is particularly through this view of history that Christianity has so strongly emphasized cultural development and speaks of progress within that framework. Only – and this is extremely important for the rest of the argument – this linear view of history must not be simply represented in terms of the model of a straight line. For beginning and end and everything in between depend on and focus on God as the origin of everything. God is in a sense vertical to the present and history. In the horizontal course of history, man ought to acknowledge and obey Him as Lord. In the horizontal dimension of history, man's cultural work ought to respond to divine principles and to the ultimate meaning of everything. This responding happens in freedom and in creativity. Just as a fish in water is in its element, so man comes to his proper destiny in following the call of God in listening to the normative power of revelation. However much man may err in this, as long as he recognizes this dimension, there is a transcendental orientation for all cultural work. When this transcendental dimension is discontinued, is no longer recognized, then from the Christian point of view we speak of a *secularized* or *immanent* conception of history. For the most part, Western culture has come to conform to this, and progress takes place purely in the horizontal dimension. Eschatological promises, namely that God will grant man peace, freedom, eternal life and perfection, are translated into perspectives of progress which man himself will come to realize.

This also clarifies the sense in which we use the word *progress*. We speak in culture of advancement and improvements when we compare progress in a subfield with an earlier situation. For example, there are numerous improvements in the development of technology. We can also speak of improvements in social order, etc. But when we speak of *progress*, we do not refer to a sector but to the whole of culture. This concept of progress has been introduced particularly since the time of the Enlightenment. There are mainly two sides to this introduction. The first side is that progress means man(kind) making himself increasingly independent of the Christian religion. From this slavery man must rise as the empowered man who – and this is the other side – knows how to use his intellect, his science and his technology for the progress of culture. This progress becomes the immanent goal of history. The unifying component in this progress is, positively stated, the coherent set of scientific advances and technical improvements.

This indicates, as it were, a resultant of progress thinking. After all, it is not the case that every thought of progress resulting from the Enlightenment constitutes an unambiguous expression of commitment to it. Yet they did contribute, each in its own way, to the fact that in the twentieth century we can say that the development of science and technology has been generally interpreted and in faith accepted as an advance of culture, as an advance in material prosperity.[1]

With Kant we as it were hear of this when by the application of reason he wishes to arrive at happiness and perfection, at a perfect society. He is not so much aiming at general prosperity or well-being but at a moral perfection. But he achieves this moral perfection by conforming to the natural laws of history. These natural laws Kant interprets on the model of the laws of natural science, of which he was quite impressed. In doing so, he also took inspiration from Descartes and Francis Bacon, who are generally considered the forerunners of modern progress thinking, and whose thinking about the development of the natural sciences was inspired by the idea that with and through the natural sciences man could bend reality to his will and solve all problems.

According to Auguste Comte, humanity entered the scientific, the positive stage in the 19th century. One should no longer ask about the origin, about the meaning, about the being, about the destiny of everything, or about metaphysical forces and backgrounds; no, one should pay attention only to the scientific connections between facts. Comte believed that social processes can be formulated and controlled as precisely as physics does with the processes of inanimate matter. He expects man's happiness to increase the more he, with the help of science and technology, succeeds in organizing society. In a sense, Comte's

---

[1] Zie M. Horkheimer and H. Staudinger, *Humanität und Religion* [Humanity and Religion], Verlag Naumann, Würzburg, 1974, pp. 9ff.

vision represents a spirit that would later come upon many. Even when the conception of science changes, as happened in the mid-20th century, the idea that science and technology will be able to solve all problems and move culture forward still remains predominant.

With Karl Marx, the emphases again shift. He recognized that the development of science and technology, especially through the economic relationships within which they are developed, is accompanied by tensions, problems and struggles. But through it all, progress breaks ground through the path of revolutionary action. Marx believed that someday the deepening shadows of alienation and unfreedom in the ongoing development – on the basis of scientific, modern technology – will be erased in and with the inevitable revolution, and that with the dawn of the new day the realm of freedom will dawn in which humanity as a collectivity will be lord and master of the works of its hands.

Descendants of Comte and Marx, for all their differences, still have much in common. Marxists do not assume the freedom of man and a free enterprise-like production within which scientific technology develops. They assume a scientific technology which can only revolutionize and liberate if it is managed in a centralist manner. In this, the use value of the goods produced takes center stage. In positivist-oriented societies, this is the use value or a mixture of exchange and use value. In Marxist society, a centralized technocracy is assumed from the outset, while in the free West politics is democratic, with increasing reinforcement from technocratic ideas.

Yet for all the differences in both conceptions of progress, the development of science and technology is central. On that central development our attention will have to focus, especially since in both systems the problems of the progress of science and technology assert themselves.

Scientific progress as such

The development and progress of science as such does not yet bring the progress of culture. This can only be spoken of when science is used as an instrument of culture; it happens especially under the guidance of economic and political powers. As a result, a large-scale and massive development aimed at economic growth and led by the welfare state sets in. When this development confronts us with major problems and threats, as is the case today, it is natural to subject the economic political powers to criticism. This happens particularly in neo-

Marxism. However, the criticism from that side does not penetrate deeply enough because it pays little or no attention to the problem of science as an instrument of culture. Neo-Marxist critique and action might possibly bring about a changing of the guard even while the *predominant* or *absolutized* place of science is hardly addressed. Nor is this done when one expects the solution from *democratic* politics, which is to guide scientific-technical development in beneficial directions. Nor is a substantive reversal brought about by a change in scientific methodologies, in techniques and in objectives – however much it is to be appreciated in itself. It is only when the *structure* of science as it has developed is critically challenged that one can gain a better understanding of the problems that arise when science is used as an instrument of control or, in other words, *necessarily* passes into application.

For proper understanding, it is first necessary to realize that the scientific question is limited to rules and laws that indicate interrelationships between facts. It does not ask about the origin, meaning or essence of things. This means, for example, that one no longer pays attention to man's own meaning and value; this easily makes man the victim of manipulations and control procedures such as those applied in the scientific control of production processes. Man is then reduced to a thing among things.

A second feature of scientific progress also causes concern. Scientific knowledge is concerned with knowledge of one function or aspect of reality. One forgoes the full reality of experience. In a sense, science alienates itself from that full reality. Consequently, much is lost when scientific knowledge is applied irresponsibly. A simple example makes that clear. In theory, we know how to divide four apples among four children. But in reality it is not at all that easy. One apple is not the other and children's preferences differ and change constantly. In our arithmetical operation we forgo the size and color of apples; but in reality, children are controlled by that. Of course, the application of this simple arithmetic has no earth-shattering effect. It is different when, through modern industrial technology, functional science comes into tension with full reality. In a world of applied science, everything becomes completely artificial, which also makes everything interchangeable, calculable and manipulable. Man in such a "world" is generic man.

The third characteristic follows from the two previous ones. Usually one speaks of scientific truth. Because origin, meaning and coherence are waived, only scientific statements are still correct. After all, correct statements can obscure rather than reveal the essential. Thus – to give an example – it can be said from science that man is a fat-protein compound. That statement is correct, but not true, because it does not express the essence, the inherent nature of man.

When science is applied in the technical mastery of reality, much is indeed accomplished. But when we uncritically label this as progress, we lose sight of the fact that much is also lost. This is so much worse in large-scale applications and in developments the results of which are very attractive. After all, when technocratic ideas gain more influence, more is lost than gained in the name of progress. One then pursues an optimally functioning society for the purpose of promoting general welfare. Meanwhile, in such a perfectly functioning society, humanity may suffer considerable damage. The whole process can be quantified and controlled scientifically and technically. In the end, as far as man is concerned, uniqueness and singularity are lost. Even words like esteem, gratitude, sacrifice, trust and love disappear from the vocabulary. If they are used, the emptying of their essential content can no longer be prevented because what they actually mean is sacrificed to the optimally functioning scientific-technical system. In short, a guaranteed scientific-technical future with more material prosperity and security goes hand in hand with dehumanization and de-naturalization. For even nature is no longer recognized in its own value and meaning but is reduced to a mere object to be controlled.

Of course, this is about a principal trend. Fortunately, humane and Christian values and standards have always offered resistance to this process. Meanwhile, it cannot be denied that the main road of our cultural development has gradually come to be dominated by science and modern technology. This road has brought our culture into crisis.

The seriousness of this cultural situation can be clarified by drawing a comparison between the so-called everyday world of experience and the so-called world of science and technology.[2] The autonomization of the scientific-technical world in particular sheds a clear light on our culture, its tensions, problems and threats.

---

[2] See H. Staudinger and W. Behler, *Chance und Risiko der Gegenwart – Eine kritische Analyse der wissenschaftlich-technischen Welt* [Opportunity and Risk of the Present – A Critical Analysis of the Scientific-Technical World], Schonungh, Paderborn, 1976, pp. 225ff.

"Experience World" and "Scientific-Technical World"

What do we mean by our "experience world"? It is the "world" in which we live, hope, suffer, struggle; it is the "world" in which we see things, in which we feel and love; it is also the "world" of faith and trust; indeed, that faith and trust make up the very core of it. This world of experience is the most original and primary; it is beyond our minds' capacity to reduce it to order; it is complex, concrete, full, richly varied and profoundly unfathomable. Every human activity and its meaning belong to that "world," including science and technology and the meaning of both.

This original, primary world of experience implies a knowing in the sense of involvement and attachment, in the sense of recognition and confession. This knowledge precedes and extends beyond scientific knowledge.

The so-called second "world" is that of philosophy, science and application of science. It is the "world" of scientific-technical mastery thinking. Scientization and technicization is where one passes from the first "world" to the second "world." This second "world" becomes first "world," because one allows oneself to be captured in science and in scientific methodology. The scientific-technical world comes to rule over the everyday world of experience. One lives under the illusion that science gives the one, true, full and concrete knowledge of reality. When since the time of the Enlightenment one goes down this road of the continuing instrumental use of scientific knowledge in the expectation of increasing material progress, much is *finally* lost. After all, through the application of science the abstract, reduced world of science becomes the first "world"; the full world of experience is reduced to scientific abstractions, and that reduction finally results in destruction. Although fortunately it never fully succeeds – the resistance within the real world of experience is too great for that – the tendencies of scientific and technological reduction can be seen in modern urban planning, industrial organization, housing, health care, social work, economics, politics, etc. Perhaps nowhere is this influence more evident than in the increase and concentration of *power* in technocratic society and in the disappearance of *love*, which cannot flourish within the uniform and universal structures. After all, love focuses primarily on the particular, on the unique. That is the reason why so many in the technocratic state, as the center of the scientific-technical "world," complain that no one cares for them and loves them. Talk of the "welfare state" cannot disguise this situation. In the technicized culture, essential community ties are cut and replaced by *artificial* ones. That is why love is cooling, that is why empathy and compassion are disappearing, that is why alienation and loneliness are increasing, and that is why people are crying out for love and compassion in all kinds of protests – or at least the call to do so should be heard in them. Much can be learned from these protests.[3]

If we confine ourselves in this article to a Christian philosophical approach, it will be clear how much that lesson has been taken to heart; but it should be equally clear that Christian philosophical thought makes its own specific and valuable contribution.

Abstractions in Science

To gain an even better view of the problems generated by an unlimited, presumptuous application of science under the guidance of the drive for technical control, it is necessary to examine the abstractions by which scientific knowledge is characterized.[4] For in an irresponsible application of scientific knowledge, these abstractions are projected into full reality on the assumption that this is the true reality.

The path to scientific knowledge is called the path of *analysis* and *abstraction*. The scientist analyzes the multifaceted reality in many aspects or functions. For scientific research, he abstracts from the conjunction of aspects one aspect or function (for example, the physical, the biotic, the economic, etc.). The second abstraction is that the scientist forgoes the concrete, the particular or unique given, but within the framework of the first, functional abstraction pays attention to the general, the universal. The third abstraction is called the abstraction of objectivity, that is, the scientist alienates himself from the visible, observable reality and turns to the laws *valid* for this reality. Finally, a fourth abstraction is that the scientist should forgo his own or others' benefits and interests, that is,

---

[3] For a brief description of those protests, see E. Schuurman, *Techniek: middel of Moloch* [Technology: Means or Moloch], Kok, Kampen, 1980, 2nd ed., p. 68ff.

[4] See H. van Riessen, *Wijsbegeerte* [Philosophy], Kampen: Kok, 1970, pp. 77–109.

science should be altruistic.[5] Within the framework of these abstractions, science is fueled by curiosity in search of *theoretical* knowledge acquisition.

In particular, the abstraction of altruism has come up quite a bit lately. This abstraction, as it were, implies an ethical standard for science which can easily be pushed aside. Political, social or economic interests more often than not direct and stimulate scientific research. When within that framework there is still talk of objective, value-free science, it is a description that no longer corresponds with the content. Incidentally, given the pre- or super-scientific view of scientific knowledge, it is always incorrect to speak of objective, value-free science. This especially comes out when the vision of the application of scientific knowledge is at issue. Broadly speaking, it can be said of the prevailing views that one *presumes* with science to be able to get a grip on concrete reality. That is why little attention is paid in modern philosophical discussions of science to the *problems of abstractions*. In Reformational philosophy, on the other hand, that problem is central because it recognizes that science is increasingly alienating itself from visible, observable, full and concrete reality.

From what we have discussed so far, it follows that the path of science is the path of the four abstractions. Every scientist must learn to walk this path in order to achieve good scientific results. In fact, this means that the scientist walks with blinders on; he excludes a lot from his attention in order to pay assiduous and precise attention to what he brackets out.

It is obvious that the four abstractions mentioned above constitute the characteristics or properties of scientific knowledge. Scientific knowledge is first of all *functional*, because it concerns knowledge of an abstracted function or aspect of coherent reality. Next, scientific knowledge is *universal* because it renounces the particular, unique and individual. Attention in science to the laws prevailing in reality makes scientific knowledge the knowledge of enduring *laws*, in this context – incidentally wrongly – usually spoken of as *objectivity*. Finally, it is (or should be) knowledge *independent* of subjective, social, economic or political interests. When the latter is not the case, it already contains, as it were, an anticipation of the application of this knowledge in order to secure, confirm or strengthen existing interests.

---

[5] See A. G. M. van Melsen, *Wetenschap en verantwoordelijkheid* [Science and Responsibility], Utrecht/Antwerpen: Uitgeverij Het Spectrum, 1969, pp. 174ff.

Before discussing the problems surrounding the application of scientific knowledge, it is still necessary to emphasize that scientific knowledge is interconnected in a *system* of scientific knowledge. This connection comes about through logic. Hence, we can also say that scientific knowledge is *logically connected* knowledge, also called *rational* knowledge.

Application of Science

As mentioned earlier, science in our culture has become an instrument for controlling reality. With science, people have wanted to gain more and more control over reality. They have wanted to use science to solve all the problems that arise. Reality is to be manipulated with science. Motives of economic or political power formation as well as prosperity-fomenting materialism played a major role in this. Precisely because these goals were pursued with the application of science, little or no consideration was given to the path or manner in which these goals were achieved. In fact, however, this meant that reality in the application of science was transformed into the characteristics of scientific knowledge. Science no longer was subservient to praxis but came to dominate it. The abstractions of science became cultural properties, and that under the guidance of powers and interests that utilized abstract science.

This process is still continuing at an increasing rate in many sectors of culture. However, the problems are also becoming more apparent. In particular, this happens when specializations become more abstract and obscure. Not only is resistance to this development currently being offered in all kinds of philosophical movements, but we are also seeing increasing resistance in society itself, through a growing number of protest movements. The multifaceted, complicated, confusing and full reality cannot be squeezed within the abstract frameworks of science. Many – especially young – scientists recognize this problem. Hence, in medicine and the technical sciences, for example, one so often hears the call for alternatives. Nor are the fruits of this lacking. Think of the results of adapted and small-scale techniques and the increasing attention paid to inventions and innovations. Yet such developments – as important as they are – still seem to take place too much on the sidelines of the main trend. That main trend, meanwhile, continues unhindered. And that is where our critical attention should be focused.

Christianity and Progress

Drawing conclusions from the foregoing, it becomes clear how much scientific-technical development offers possibilities to achieve total world control in a functional way. Because this development is essentially foreign to the character of reality and is value-blind – the second conclusion – dehumanization and de-naturalization will increase as scientific-technical development – whether adapted to new social problems or not – becomes stronger. Moreover, development from within will become increasingly unstable and threatening.

With this, the question arises before us in all clarity as to whether Christianity can acquire significance in this development, and whether this influence is such that the development of science and technology can become for man a blessing rather than a curse?

Earlier, I mentioned that the idea of progress originally came from Christianity. But that progress is not contained within human history. We should even say that the expectation of the Kingdom of God is not fulfilled within history, but that this Kingdom comes to man as a gift from outside history. Living in this expectation means that a different attitude is taken regarding the development of science and technology. It distances itself from an immanent sense or purpose of cultural development. Christianity does not expect salvation from science and technology. That salvation is given to the Christian by grace. This is how the Christian stands in the world, and this is how he is co-responsible for shaping and developing society. This means that he must not uncritically conform to the prevailing trend.

The question arises as to whether from the Christian perspective progress is possible without dehumanization and de-naturalization. Indeed, science and technology as non-absolutized cultural possibilities can cooperate in the necessary fulfillment of the basic necessities of life. Without that, no human life and fulfillment is possible. But it is equally true that with increasing material prosperity man is certainly not becoming happier but is experiencing its stifling influence. Apparently we have to accept limits, normative limits to the development of science and technology, which keeps the development of science and technology on track.

The first task, therefore, is to provide the basic necessities of life for *all* human beings through scientific-technical development. This is a demand of love of neighbor, just as it is a demand of love of neighbor to qualify the present development of science and technology through optimal functioning as a development which entails depersonalization. Unhindered scientific-technical development, de-personalization and lovelessness belong together. A different path with science and technology is taken when it is not man that is adapted to that science and technology, but conversely when science and technology are put at the service of man, of all men. Then man and nature are no longer reduced to their functional value. Respect and love for the individual fellow man implies respect for his freedom, for his imperfection as well. This implies a risk that is excluded in a scientific-technical functioning society. But the gain is to see man as a responsible human being from the outset. The ethics of science and technology are then no longer derived from science and technology, no, these ethics should on the contrary correct and put into perspective the current development. On the basis of Christianity, this correction and relativization become possible because the meaning of science and technology lies not in its own progress but because God in His love entrusts us with all of reality, including all of our neighbors, with the call to do our work in love. This is the essence of responsibility, which has consequences for the development of science and technology.

First, we must consider how to free science from the *compulsion* of application. How do we get off the path of trying to use science to achieve an instrumental control of all reality? A first requirement for this is to recognize that science should no longer function as the *main* route in cultural action, but that science must be brought back to the full world of experience, in order in this way to enrich experience and increase the insight of the person responsible for cultural action. Seen in this way, science is a *serviceable supply route* for practice. In its relativity, this is its great value.

On the basis of the Christian faith, which affirms the first "world," the world of experience, as the *world of revelation*, the secondary "world" of abstract thought must be put into perspective. The "world" of science as a secondary "world" must not become primary; it must be relativized from the primary, full world of experience and thus integrated into the world of experience. The depth of the *mystery* of all that exists as creation must not be surrendered to any scientific-technical mastery. After all, on that road one no longer has an eye for a total vision of reality, as it is the road of abstraction. On that road, too, one determines one's own goals based on egocentric motives. One is blind to the origin, perspective and meaning of everything. In science and technology, people forget to learn to live in the path of the divine law of life. People think they are preserving life in science and technology, but in fact they are losing it.

Christian philosophical reflection will therefore also have to emphasize the *necessity* to *continually* return from science and technology to the primary, original "world." We must continually distance ourselves from the absolutized thinking of *instrumental reason*. By this we do not, like cultural pessimists and existentialists, distance ourselves from science and technology as such. No, from the recognition that the world of experience is a world of revelation, the content and direction of which are revealed by divine word revelation, science and technology return to their proper, limited and relative place. They again become meaningful sectors of the full life-praxis. Within the light of revelation, science is given every opportunity, and again becomes a fascinating activity. To renounce the absolutization of science, of faith in science, means to positively recognize a method-openness and a method-pluralism. This recognition stems from the recognition of faith that reality as God's creation is a *mystery* to be investigated, yet profoundly unfathomable and never to be grasped. Living from this mystery implies first of all *listening, receiving, thanking, joy, rest, love.*

It is possible, given the openness of man's heart to the fullness of revelation, to continuously incorporate (new) scientific knowledge into direct, concrete human knowledge and experience, which have at their center the *fundamental knowledge of trust*. In this process of integration, the one-sidedness and abstractions of scientific knowledge are removed. In this way, scientific enterprise is practiced from faith and to faith, and thus believing insight is strengthened and enriched, pre- or super-scientific knowledge deepens, and responsibility and a sense of stewardship increase. Scientific enterprises then no longer aim at the expansion of scientific knowledge as such, expecting with this expanding science to exercise greater *power* over reality as the "first" world from the so-called "second" world.

Scientific knowledge serves *growth in wisdom* and thus it leads to increasing comprehensive insight and *creative circumspect action*; the responsibility to act-in-love is increased, and we can better orient ourselves to the proper, normative direction of cultural development. This *wise*, restrained responsibility precedes and should therefore never be dominated by philosophical or scientific thinking and by technical mastery. The reverse should be true: responsibility should accompany science and technology. Only in this way can the *demanding, entitlement-driven, joyless, restless* and finally *deadly pursuit of the future* be cured and prevented.[6]

A Different Vision

In modern thinking, praxis is usually seen as an extension of theory; praxis is then applied theory or science. In view of the reductionism and the other problems this entails, it must be recognized that we are dealing here with a fundamental fallacy. We encounter this fallacy when technology is said to be *applied* science or when – really modern – computer technology is considered to be applied systems thinking.

A prerequisite for the renewal of culture is to forgo an appeal to a science, new or otherwise, to solve problems, at least in the first instance. After all, the problems, as we saw, are mostly a consequence of the predominant *use* of abstract scientific thinking as an instrument. One stays within the framework of this "abstract world" if, given the problems in praxis, one wishes to solve them from theory. This intractable automatism of attempting to bridge the gap between science and praxis through the application of science must be broken. In this respect we might well speak of a necessary detoxification, in view of the fact that almost all of us have become addicted to abstract theories under the influence of the spirit of the West. Ethical action has increasingly degenerated into slavishly following newly (or not) constructed systems of thought. In this we have renounced the order of creation and redemption given to us.

To (once again) recognize this order means to return to the original and normative relationship between science and praxis. The bridge between theory and praxis is *given* with the *creative responsibility of man* as the image-bearer of God. That responsibility is not to chart one's own chosen path, but, standing in the truth and living out of faith, hope and love, to respond to the Creator and Redeemer's word. Within this vision we distance ourselves from what is usually

---

[6] For elaboration see E. Schuurman, *Kultuurkritische evaluatie van het systeemdenken en de computertechniek* [Cultural-Critical Evaluation of Systems Thinking and Computer Technology], in: *Samenlevingsproblematiek en Systeemdenken* [Society Problems and Systems Thinking], VU publication, Amsterdam, 1981, pp. 123–143.

understood by the *application of science.*[7] We want to emphasize the *subservience* of science to practice.

Such subservience requires, first of all, an understanding that the four abstractions mentioned earlier be restored in growing wisdom. Interpretations of scientific knowledge must take into account that which in science is placed outside brackets. This concerns first of all the abstraction of the functional in relation to the coherence of reality. All of reality must come into its own, and that reality has more aspects than the single aspect examined by the one particular science. A single scientific approach to solving a problem is therefore insufficient, especially now that in our time problems have become so complex. A multidisciplinary approach to problems is therefore rightly called for. Yet this multidisciplinary cooperation as a sum of many forms of professional scientific knowledge is not yet a sufficient condition for justice to be done to the coherence, versatility, and concreteness of reality. That realization only dawns when it is realized that scientific knowledge is characterized by several abstractions. There is the abstraction of the general, the universal. This abstraction is often misunderstood, i.e., universal scientific knowledge is identified with knowledge of concrete, individual reality. But that reality consists of unique entities. The restoration of the abstractness of the universal contributes to an enriched knowledge of that unique reality, rather than that reality being surrendered to the abstraction, or science pretending to encompass and master that uniqueness.

In connection with the previous abstractions, the abstraction of objectivity should still be mentioned separately. Since under the influence of dominant positivism there is a general opposition to any form of metaphysics, this abstraction is often misunderstood, and scientific knowledge is often presented as the only truth. Indeed, positivism misunderstands scientific knowledge as an expression of the laws *valid* for visible, observable reality. Then it is not recognized that scientific knowledge is an *estrangement* from concrete reality. This does come out in the view of Toulmin,[8] who therefore compares scientific knowledge to a

---

[7] To what extent we can still speak of this, see E. Schuurman, *Techniek en toekomst – Confrontatie met wijsgerige beschouwingen* [Technology and the Future – Confronting Philosophical Reflections]. V. Gorcum, Assen, 1972, pp. 359–370: "Fundamentele onderscheidingen" [Fundamental distinctions].

[8] See S. Toulmin, *The Philosophy of Science*, Hutchinson University Library, London, 1965, 7th edition, pp. 105ff.

floor plan, a map, that applies to reality. According to this view, one can get a better orientation in reality through science. Not so with Popper[9] when he compares scientific knowledge to a safety net, the meshes of which, as scientific development progresses, become smaller and smaller and with which one gains more and more grip on visible reality. Within such a vision it is obvious to say that culture advances with the development of science. Objectivity is then equated with rationality and the "application" of this rationality leads to an ever more rational and thus – in the philosopher's eyes – a better reality. In fact, however, reality increasingly comes under the grip of a logically coherent inexorability. If consistently implemented and applied, reality becomes completely scienticized and technicized in its functionalization, atomization, fixation and isolation.

Neo-Marxism, as already mentioned, has mainly drawn attention to the fact that the practice of science is more often than not conducted in the service of economic, political and social interests. This is correct. In the *fundamental* practice of science these interests should be renounced. In a great deal of scientific research this does not happen, because this research is a priori employed by certain powers. If one wishes to use the results of this scientific practice, one must be aware of the influence of these powers and subject them to criticism. Incidentally, it should also be remembered that the criticism exercised from the neo-Marxist side often does not go beyond criticism of the powers themselves. And yet, if existing powers are replaced by new – possibly democratically elected – powers, the other abstractions will still exert their disastrous influence if science is applied uncritically.

Responsibility

Integrating scientific knowledge into the full reality of experience is a necessary condition for making science subservient to cultural or practical action. This subservience then revolves around responsibility. In that responsibility, we might say, lies the bridge between science and praxis. In order for the word "responsibility" not to become a kind of magic word, it is necessary to know what its content is.

---

[9] See K. R. Popper, *The Logic of Scientific Discovery*, Harper Torchbook, New York, 1965, pp. 59ff.

In cultural or practical action, there is great diversity in responsibilities. The cohesion and unity in that diversity are formed by the cultural motive as the central driving force in cultural action. With respect to that motive, however, there is far from unanimity in our culture, if there ever was any. Increasingly we are dealing with a culture that is spiritually inwardly torn, as a result of which there is no longer a unified vision of history, the future, mankind, mankind's cultural task, and so on. This greatly complicates responsibility while making its effectuation increasingly difficult.

Within the framework of this narrative, we refrain from addressing the difficulty in the pluralistic society posed by the acceptance of different norms and values caused by the spiritual disintegration of culture. We deliberately limit ourselves to the Christian cultural motif. That motif is the command to build in and preserve creation, the motif to bring creation to openness, so that everything in creation comes into its own. As such, this motif implies the perspective of justice, the summary of which is given in the double commandment of love. Its elaboration is our responsibility. This means that in cultural action justice must be done to a number of supra-subjective or supra-arbitrary norms, each of which should receive its concrete form in the rich diversity of cultural activities. This gives an opportunity for new cultural paths in the midst of the process of scientization and technicization.

First of all, in cultural development one should comply with the *cultural norm* of differentiation and integration, of continuity and discontinuity, of small scale and large scale, of decentralization and centralization, of uniformity and pluriformity.

Cultural disclosure is equally facilitated by meeting standards *expressed in words*. This means that any cultural action must provide information in a clear and open manner about the norms used. This openness is at odds with the secrecy pursued in many modern developments. In that case, justice can also no longer be done to the *social norm*, the norm of communication. Elaboration of this norm means that all those involved in their diversity in the development of culture must be able to live out their different responsibilities.

Until now, the *economic norm* has often been concerned only with (industrial) production processes. This has led to undervaluation of human labor and to large-scale environmental pollution and destruction. Responding fully to the economic norm means taking into account that people and nature also have their own specific economic value, which cannot be ignored with impunity. In this respect, we cannot be economic enough!

Normative cultural development is also promoted by responding to the *norm of harmony*. This norm must also be worked out in all aspects, in the relationship of people, culture and nature, and all on the basis of the aforementioned norms and in the light of those yet to follow; especially also in the light of the (mostly political) elaboration of the *juridical norm*. Resistance is thereby offered against every possible injustice; at the same time, justice is pursued with regard to everything, to nature as well as to man and culture in their rich diversity and variety. And this, in turn, is possible by obeying the ethical norm of love; love for all kinds of cultural activities and in them love for the *many* neighbors and care for the *many* "natural" fellow creatures. However, if we confine love to love of science alone, scientists, without otherwise wanting it themselves, will easily fall into the hands of those who wish to use or even abuse science in the service of selfish motives for deliberately evil practices.

The perspective outlined so far is at odds with the predominant process of technicization. For it is the perspective of cultural vocation and cultural awareness, in which *service in responsibility* is put front and center. Against the pretension to create reality and to give meaning to everything in orientation to the kingdom of man, there stands the obedient acceptance of the order of creation and redemption in which the central thing is the unlocking of creation, of the meaning of everything in the hope of the Kingdom of God. Responding to this *norm of faith* implies resistance to any pretense of using science to regain the lost paradise, or to realize a salvific state, and to any misuse of scientific knowledge. At the same time, it means making science subservient to the unfolding of all reality.

REFLECTIONS ON THE TECHNOLOGICAL SOCIETY (1977)

Preface

Western culture, characterized by science and by modern technology, is in a crisis. This is shown not only by the philosophical reflection on our culture, but also by the many problems facing our culture.

In this brief volume I have collected a number of lectures that attempt to show the spiritual and historical backgrounds of that crisis. These backgrounds are generally ignored, yet they must be examined if we are to find a meaningful perspective for the future.

The first essay is a slightly revised and expanded version of my inaugural lecture at the Technological University of Eindhoven in the fall of 1973; it analyzes the tension in our culture between technocracy and revolution. The second essay deals particularly with the spiritual roots of our environmental crisis. The third essay is a revised version of my inaugural lecture at the Technological University of Delft, Fall 1975; this essay consists of a critical analysis of the relationship between science and culture.

My aim throughout these essays has been to reduce the dilemma facing our civilization to its basic elements. This will prevent rushing into matters of detail and put us in a position to pose the right questions. It will also stimulate us to seek the only sure basis on which we can begin to solve a major issue that plagues our culture.

The many footnotes documenting my arguments have been reduced drastically. The works I relied on chiefly are listed in a Select Bibliography.

I thank Harry van Dyke and Lammert Tenyenhuis for the care with which they have translated chapter one and chapters two and three respectively.

Egbert Schuurman
Breukelen, The Netherlands
Spring, 1976

Foreword

Dr. Egbert Schuurman is professor of Christian philosophy at the Delft and Eindhoven Institutes of Technology. He was appointed to these two positions by the Foundation for Calvinist Philosophical Education, which avails itself of the legal privilege of private associations to appoint professors and lecturers of their choice in special chairs at state universities and comparable centers of higher education in The Netherlands. Dr. Schuurman also teaches philosophy of culture at the Free University of Amsterdam. Prior to pursuing his doctoral program in philosophy, he completed a graduate program in mechanical engineering. His major publication is *Techniek en Toekomst: Confrontatie met wijsgerige beschouwingen* [Technology and the Future: A Confrontation with Philosophical Views] (1972) which is being translated[10] and forms the basis for the chapters in this booklet.

Schuurman discusses technology in the context of the Christian philosophy that finds its roots especially in the thought of Herman Dooyeweerd (1894–1977). He thus continues for our generation what his mentor, Professor Hendrik van Riessen, began immediately after the second world war, namely, an analysis of the place of technique in Western civilization in the light of Christian assumptions. This analysis thus runs parallel with the work of two fellow Christians – Jacques Ellul, the French thinker who published *The Technological Society* in 1954, and George Grant, the Canadian philosopher whose *Technology and Empire* appeared in 1969. One way of introducing Schuurman to the English-speaking world is by comparing him briefly with Ellul and Grant.

With Ellul, Schuurman rejects the positivist's and pragmatist's adoration of technology as the neutral tool by which we can create the great and global society of the future, in which all of mankind's needs supposedly will be met by the inexhaustible resources of an industrial-technological apparatus. Further, Schuurman accepts Ellul's diagnosis of the cause of the destructive derailment of technology in contemporary culture. Both single out the notion of the autonomy of technology – as if it were a law unto itself, not subject to a norm outside

---

[10] *Technology and the Future: A Philosophical Challenge* (Toronto: Wedge Publishing Foundation, 1980).

itself – as the immediate cause of that derailment. But at this point Ellul and Schuurman part ways. Ellul seems to say that the very nature of modern technology implies its autonomy. For him technology is basically evil and inhuman. Moreover, he does not search for a deeper source of the notion of autonomy. Schuurman, in contrast, argues that autonomy is not inherent in technology but that it is the religio-spiritual assumption of post-medieval modern man.

Here Schuurman and Grant present parallel diagnoses. Ever since his rejection of the religion of progress in the early sixties, Grant has forcefully argued that Western technology is rooted in the main heresy of the modern age, namely, the belief that man's essence is his freedom, his autonomy. In view of this both Grant and Schuurman point to the weakness of the counter-cultural critique of technology because that critique is still founded on the ideal of absolute human freedom. After all, that ideal gave rise to Western man's mastery of nature by science, technology, and industrial production. In modernity, freedom means mastery. Thus both Grant and Schuurman realize that within modernity's dialectical swing from freedom to mastery and from mastery to freedom it is impossible to overcome the autonomy of technology.

If there is then so much similarity between the diagnoses of technology that Grant and Schuurman present, why should one bother to translate and read Schuurman? Because, while Grant suggests that in our coming to grips with the error of modernity we should painstakingly recall the Greek view of nature and man's place within it, Schuurman suggests that we listen to the biblical revelation of reality as creation and man's place within it. While for Grant and other representatives of the neoclassical school of cultural reflection – like Eric Voegelin – classical philosophy and biblical revelation are correlative and complementary, for Schuurman they are to be carefully distinguished. For Grant the autonomy of modern science and technology presupposes the rejection of the Greek view of nature; for Schuurman it presupposes the rejection of the biblical view of creation.

This means that Schuurman not only can present a critique of an *autonomous* unfolding of science and technology but also can point to the foundation of their *responsible* disclosure, even at this late hour in the disintegration of Western society. For the revelation of creation, known to us in Jesus Christ, presents the dynamic, normative guidelines for the execution of man's task in history, including the structurally limited but nonetheless positive service required of science and technology. In his critical appraisal of the role of autonomous science and technology, Schuurman therefore calls for the recovery of the consciousness of created reality as the proper foundation for science and technology. That call is imperative in our era of gnostic futurisms, in which the *alpha* of creation has been repudiated to make possible the realization of a secular *omega,* an immanentized eschaton.

Bernard Zylstra
Institute for Christian Studies Toronto

Between Technocracy and Revolution

It has been estimated that perhaps eighty percent of all the scientists and engineers who ever lived are living today. A figure like that makes one realize the exceptional character of our time. A growing number of people have jobs related in one way or another to the development of science and technology. In addition they find themselves surrounded more and more by the products of technology, both on the job and at home. We live in a technological civilization that is growing all the time in strength and scope and that is spreading across the globe.

Until recently this development was generally applauded as the surest way to progress. Today, however, as man turns his mind to the future, he sees enormous problems ahead. These problems seem to have been produced by the close cooperation between science, technology, economics, and politics. Confronted by the growing perplexities and conflicts in society and by the threat of ever greater catastrophes, some people have lost their enthusiasm for technological advancement. The destruction that man has brought about and continues to bring about with the aid of technology is so terrifying and seems so irreparable that it looks as if even man himself may eventually become a victim of technology.

A tension has thus arisen between, on the one hand, the seemingly anonymous, impersonal, and objective development of technological-scientific power and potential and, on the other hand, the subjective, personal decisions of human beings who contribute to this development. It is a tension that plagues those who are involved in this process; yet this same tension has an even greater effect on people who do not directly contribute to science and technology, but who may well be swept along by the avalanche of their cumulative applications as they give way to a spirit of apathy or anxiety, unbearable tension, alarm, or even panic.

These growing misgivings notwithstanding, there are still many technologists, engineers, and technicians who champion the unremitting growth of technology. Today's problems, they say, are the problems of a technology in its infancy; they can be solved by exploiting more fully the possibilities of technology. The method of technology should be extended to other areas, such as economics and politics. What is good for technology is good for all culture. As technology advances, so will our society as a whole. The people who argue this way are called technocrats. Looking at the possibilities of technology, they are as optimistic about the future as ever.

Meanwhile the view of the technocrats only intensifies the reaction of those who take a radically different stance: the revolutionary Utopians. Their assessment of our situation today is very somber. Looking at present trends in technology and at the growing power of science over our lives, they are pessimistic about the future and prefer to stake their hopes on alternative plans for tomorrow's society. They dream of a utopia in which man lives a free, happy, and carefree life. This utopia is to be realized by overturning our technocratic society through revolution. Obviously the realization of these ideas would have far-reaching consequences for the further development of technology and for its possible role in economics and politics.

The field of tension in present culture is governed by the two poles of technocracy and revolutionary utopianism. These constitute the contrasting orientations or twin foci of the struggle that is being waged in modern culture. In this essay I cannot discuss all the intermediate positions that have been or could be proposed.[11] Therefore when I refer to specific philosophers and futurologists to illustrate the spirit of technocracy or of revolutionary utopianism, I do not

---

[11] Another recent current in Western culture that I shall not discuss in this essay is the passive counterculture. This movement suffered from much internal disagreement, divided as it was over a number of short-lived, competing subcultures (cf. Alvin Toffler, *Future Shock*, chap. 1). Together they formed a persistent and intriguing phenomenon on the fringe of Western civilization. Reacting against the lack of freedom of technocratic society — in this respect they agreed with the revolutionary Utopians — the members of the counterculture chose for the impossible: a culture with bare hands and on bare feet (cf. Theodore Roszak, *The Making of a Counter Culture*, pp. xi, 50ff.). Worse still, they had a tendency to switch at any time to the camp of the revolutionaries (ibid., pp. 61, 63). Their terror of technocracy was expressed by their escape from culture into romanticism, pantheism, and naturalism — as evidenced by their fascination with hallucinogenic drugs, their enthusiasm for astrology and occultism, their frantic search for truth in the expanded consciousness and the ecstatic experience, and their radical rejection of science and technology (their continued dependence on them notwithstanding). For a critical discussion, see Os Guinness, *The Dust of Death* (Downers Grove, Illinois: InterVarsity Press, 1973).

mean to imply that their thought coincides nicely with either one of these cultural currents; my point each time is that the pattern of thought in question contains elements that confirm the contrary tendencies in our culture. Precisely because the issue is one of basic "orientations," one rarely runs into a thinker who is a strict technocrat or a strict revolutionary utopian. In most cases, however, a man's thinking clearly gravitates towards either the one pole or the other and unmistakably reflects its specific temper.

In this essay I shall investigate in what respects technocrats and revolutionary utopians differ, what kind of future each group pursues, which methods and which criteria each uses, and how each judges the other.

To gain a deeper insight into the cultural conflict I have outlined, I shall deal briefly with the historical background of the intellectual and spiritual mainsprings behind the rise and development of science and technology in Western civilization. Such a historical analysis, highly desirable in itself yet woefully absent from most treatments of the topic, should demonstrate the contemporary relevance of a Christian philosophical framework. The relevance of a Christian vision must not, however, stop once the critical evaluation has been made; it should also indicate concretely the direction in which we might seek a solution. That is what I propose to do; however, the scope of the present essay leaves me room for a perspective only. I hope this perspective will show in broad outline how we can break through the cultural tension between technocracy and revolution and rise above the dilemma with which the current debate seems to saddle us.

The Technocrats

The technocrats – also known as the ideologists of planning – include such men as Herman Kahn, Antony Wiener, Olaf Helmer, Karl Steinbuch, Erich Jantsch, and the Marxian philosopher Georg Klaus. To a man, this motley group whole-heartedly approves of modern science and technology. Technology is for them the motor of progress, and scientific discoveries are the fuel. These people have set their hopes in particular on the development of the computer in close connection with systems theory and cybernetics. The computer is regarded as a mighty tool with which to investigate, guide, and control the future. The problems of the present, it is claimed, even if they have been called into existence by technology itself, can be solved and overcome by applying the latest discoveries in the field.

In the past, the so-called technological-scientific method has proved very successful in asserting man's sovereignty over "inanimate nature." The technocrats argue that if we want to progress still further in our mastery over nature, we should apply the same method to areas not specifically technical. The method should be used to analyze man himself, to dissect society, and from there to reconstruct the future. Thus the technology of production can be duplicated in "organization technology" and in "human techniques."

It is especially the modern planners that are under the spell of this technicistic type of thinking. They promote, sometimes unwittingly, the imperialism of technology and its scientific method. This is particularly clear in the planning procedure they follow.

On the basis of empirical findings, modern planners first draw up universal laws for the way society develops. They look at the past, observe a certain trend there, and then carry this trend into the future. In this way they arrive at certain forecasts for all sorts of areas. The procedure is based on the assumption that knowledge that is useful for explaining the past is equally useful for predicting the future. In cases where it is impossible to speak with absolute certainty, the planners resort to the language of statistical probability.

In a certain sense what the planners are doing is simply extrapolating. Lines of possible (or probable) trends are drawn from the past, through the present, and into the future. To put it differently, a picture of the past is faithfully projected onto a larger screen called the future. Thus the future is a mere projection of the past. Or rather, it is a projection of one's picture of the past, that any picture of the past remains an abstraction of the past is compensated for, presumably, by calling in the aid of multi-disciplinary research and a variety of research methods. Nevertheless, for all that effort, the final step in the whole procedure of the planners is a projection, pure and simple. This is true even when the "circular model" is used: that is, past phenomena that revealed a circular or cyclical structure are assumed to repeat themselves as such in the future (e.g., the human life cycle, the seasons, solar and stellar revolutions, and until very recently, economic booms and recessions).

For studying the interrelatedness of the more salient phenomena, such as rising energy demands and diminishing fuel supplies, futurologists resort to modern systems analysis. By means of this type of analysis, they are inclined more and more to try to capture the entire development of our growing world community in one great system with many sub-systems. The activities of the Club of Rome in particular have caught the imagination of many scientists and technologists for a world model of this sort. Simulating systems like these in computers is supposed to provide us with reliable knowledge of what is ahead of us.

These studies by futurologists lead to a second stage: planning or "assembling" the future. Plans are drawn up on the basis of the collected data relevant to the future (or to possible futures). The different sub-plans, such as the technical, the economic, the social, and the political plans, are mutually adjusted so that they harmonize, and are then integrated into a single total plan. By using computers to simulate such models of the future, planners can presumably discover whether today's problems will thus be solved and catastrophes averted. The planning phase can be drawn to a close by deciding upon various alternative plans for the future.

In the third phase, a choice is made from among the drawn-up plans, and the plan chosen is then executed. The choice can be left to, say, the economists or to those in the government bureaus, who are the present wielders of power; or it can be left to the people, through the usual parliamentary channels or by direct plebiscites, prepared for by information sent out over the communications media. However, in either case it is a matter of choosing among the plans that are drawn up and worked out by the experts who serve the present ruling powers in society, i.e., by the scientists and the technologists, who will continue to direct the show as the future is being realized, a future that others have opted for but that they have designed. Thus no truly democratic process, direct or indirect, influences the making of the plans, nor is any democratic influence likely to be realized during the phase of executing the plans.[12] The deterministic nature of the plans makes compulsion a real threat, for it will in practice restrict, if not

---

[12] Direct democracy, as it is called, even facilitates research of the future since one can make use of the laws of large numbers to reduce the surprise element in the future by basing one's forecasts on the verdicts of the people. I have worked this out further in my forthcoming book, *Technology and the Future* [published in 1980 by Wedge Publishing Foundation].

eliminate, the role of free and responsible people. Accordingly, many people recognize that in developing, as well as in executing, such plans we are dealing with a trans-political phenomenon no longer open to discussion. The expertise of an elite serving the present cultural powers is moulding and mastering the future on behalf of all of us.

This picture of a future society does not pose much of a problem for Marxist-socialist countries. They have long since merged the economic, technological, and political forces in the power of the State. Besides, Marxists claim that the people's representatives and the party's experts automatically promote the objective interests of the people; this is guaranteed by the very principle of their system of "democratic centralism." Thus, if anything, they reinforce the development towards a totalitarian, collectivist technocracy.

For the "free" west, however, the case is different. If the ideas of the technocrats are to be realized here, the trend of Western society will consciously have to be turned, voluntarily or by force, in the direction of a totalitarian technocracy. Naturally, a voluntary transformation is to be preferred and, according to the technocrats, is in any event the most likely to succeed. The criteria they use, after all, aim at an increase in production and, what is more, at sufficient productive employment for all to be able to share in the material fruits of our technological progress. This will assure everyone of survival and well-being. We must therefore simply carry on with the present situation and even reinforce it, and over and above that solve the problems and avert the menaces hanging over us today.

Is this a real future in the sense of a time to come, an open door to new and surprising challenges and opportunities? No, at bottom the technocrats de-futurize the future. Their tomorrow is a day without surprises. They try to bring the future within their grasp by turning it into an extension of the past and the present. Their projection of the future reinforces the present and reaffirms the controlling economic and political powers. On their screen of the future the existing order of things is simply magnified. For this reason these men are also known as the futurologists of order. By reinforcing the technological-scientific order, which is already characteristic of today, they only contribute to the enlargement of the power of science and technology, a power under which people groan or to which they willingly adjust for purely materialistic reasons.

In short, the future of the technocrats, as ideologists of planning, is a technological future, a future in which diverse cultural pursuits and societal relationships are progressively levelled to parts of one gigantic, all-embracing totalitarian system. In that system man will be reduced to a cog in a wheel, to a standardized part of a machine – interchangeable and replaceable. This all-devouring collectivism will culminate in a technocratic world state that will be in complete control of the future.

The Revolutionary Utopians

Among the revolutionary utopians may be numbered such men as Herbert Marcuse, Arthur Waskow, Claus Koch, Robert Jungk, Ernst Bloch, and in a certain sense also Jürgen Habermas. These revolutionary or critical futurologists, as they are also called, are opposed to a rigid, highly deterministic future under the leadership of a technocratic elite. They are opposed, first of all, because in such a future all present misery, suffering, evil, injustice, and oppression will be intensified, not abolished. For instance, even the moral impossibility of a nuclear war becomes a logical possibility in the thinking of a technocrat like Kahn. The reason for this – and this is the second major objection of the revolutionary utopians – is that the technocrats equate history with the advancement and aggrandizement of science and technology, with the result that the present cultural powers are quantitatively strengthened.

The revolutionary utopians oppose the evolution of culture under the leadership of the technocrats by rejecting order and harmony in the development of culture. Instead, they champion protest, conflict, struggle, action – in a word, the *revolution* of culture. For them, revolution is the locomotive of history, propelled by utopian imagination. They bitterly oppose the development of society that leads toward a totalitarian technocracy that gradually imprisons man (even though he will be granted some freedom of movement by the grace of the possibilities of cybernetics). The revolutionaries realize that history and its future will be frustrated if technological progress gains the ascendancy and if work is reduced to purely productive labour, the burdens of which have to be compensated for by the sop of increased consumption. Seeing modern man already weighed down by growing technocracy, they take up his cause, even though he may be so sedated by compulsory production and consumption that he is not even conscious of his plight. The revolutionaries also perceive clearly that the technocrats' solutions to today's problems are solutions that will soon saddle us with even greater problems and threats.

Clearly, the position of the revolutionary utopians is diametrically opposed to that of the ideologists of planning. Rather than defend the inherited situation, the revolutionaries stress the discontinuity of history, the new and surprising elements that may set men free. A pre-determined future is not their ideal. On the contrary, they urge an open attitude toward the future. Yesterday must not be carried into tomorrow; the existing powers must not be strengthened by any increment in knowledge; it is *imagination* that must come to power. In their sketch of utopia, the revolutionaries try to negate the present dehumanizing tendencies and to rough out a future that is open and free. But utopia is not to remain a mere dream. The revolution – especially the political revolution, since it can have the most far-reaching effects – is to be a radical leap out of present reality into the realm of possibilities. In and through the revolution, utopia must abolish itself in the very process of being realized. Utopia thus functions as a catalyst for a radical humanizing process of change.

The formula of the revolutionary utopians, in other words, consists of a categorical negation of the past and the present. The establishment is to be resisted, defied, challenged, and brought low through a conflict culminating in a revolution. Only then will the road lie open to a new future with new possibilities for the coming of the reign of autonomy, the kingdom of freedom, and the rule of peace in what Marcuse calls "pacified existence."

They are possibilities and no more than that. Developments in Marxist countries have taught the neo-Marxist revolutionaries that the revolution does not come with scientific certainty and will not necessarily bring freedom; whoever believes that fails to recognize that the revolution can turn into its opposite, namely, a comprehensive, technocratic dictatorship. Thus warned, they do not plead for a scientifically predictable revolution but for a utopian revolution, which is to be as permanent a revolution as possible. Only in a society in which the revolutionary inspiration replaces acceptance of the existing order and in which revolutionary action takes the place of order, harmony, and technological progress will things be restructured to allow everyone to realize his private and group objectives. The aim should not be to develop science and technology further as instruments of power, but rather to give all possible encouragement to revolutionary creativity and the creative revolution. Over against the logical thought of the technocrats, which they call "mad rationality," the revolutionaries champion a fully rational thought and even more a historical and ethical consciousness. The revolutionary mind addresses itself to the free man, turning against the ideas and thought patterns of vested-interest groups that only magnify the catastrophic present and seal off the future. This revolt is to issue in destruction, for creativity has a chance only in the chaos that comes after the passion of destruction has abated.

Whereas with Marx the revolutionary dialectic ended with the coming of the communist society, with the neo-Marxists the revolutionary dialectic never comes to a standstill. It must not or history would once more become a continuous process, new institutions of oppression would sprout up, and our troubles would begin all over again. Marxism has become questionable to the neo-Marxists because in the transition from private enterprise capitalism to state capitalism the power structures, instead of being conquered, are actually strengthened.

Neo-Marxists therefore espouse a critical dialectic. They are much less confident about the success of the revolution than are the classical Marxists. Marx – at least the older Marx – predicted the success of the revolution with scientific certainty. For the neo-Marxists there is no such certainty. Utopia may well remain utopia, because the establishment will leave nothing undone to restrain the revolutionary forces and prevent them from ever realizing their dream. And it is likely to succeed in this effort if for no other reason than that the revolutionary élan both within and without the technocratic system is still far too weak. The majority of people still fail to see the perils of unbridled industrial technology and still have their eyes closed to the social and political threats connected with more fully exploiting the potentials of the computer and of cybernetics.

The revolutionaries' alternative vision for the future implies a less important role for technology and still has little appeal, because it militates against the current idolization of economic growth, with its promise of higher profits and higher standards of living. The revolutionaries, in particular Marcuse, have been quick to recognize that man has become alienated from nature because of the abstract artificiality of modern technology. They also recognize that man has become alienated from all sorts of things, from his fellowman, and even from himself because of technology's universality, dynamism, and absolutism.

Yet here again it becomes evident that Marx erred. He thought that human alienation could be abolished through communal labour and collective ownership. Nothing could be further from the truth. For while Western man has an unprecedented variety of work and jobs to choose from and has little to complain about in the way of material things, alienation has never been so great. Man's self-alienation, it turns out, is more than social and economic. Marx did seek to eliminate the external symptoms of alienation, but he was insufficiently aware of the real needs of man. Moreover he failed to recognize the power of modern technology and the effect it would have on man. Modern man has been reduced to a statistic, he is lonely, he has lost his identity, and he has no real sense of belonging. Today's production-consumption complex conceals only too well that man in truth has become the captive of the constant stream of scientific and technological change. Man no longer has technology under control; technology has gotten out of hand and now controls him.

The method of the revolutionary utopians is no backward-looking method, such as that of the technocrats, who project the past via the present into the future; it is, rather, a forward-looking method, in which the future has to guide and rule the present. Rather than take their starting point in the science and technology of yesterday and today and thus serve the establishment, they take their cue from a future in which man will be liberated from growing constraint and in which utopia will act as a catalyst for human creativity to set in motion a permanent revolution that will once and for all abolish every form of alienation.

In contrast to the *quantitative* criteria of the technocrats – technological perfection, efficiency, performance, universal order, productive work, consumption, abundance, and progress – the revolutionaries choose what they call the true needs of man: peace, freedom, joy, fun, love, happiness, individuality, simplicity, play, sex. These are needs that come to expression in creative spontaneity and experimentation, in short in *qualitative* changes. They reject the shallow, reductionist image of man current among both capitalists and old-style Marxists – *homo faber,* and *homo economicus;* they choose instead *homo ludens.* Man is not first of all the creature who works, cultivates, creates, and produces, but rather the creature who plays, relaxes, has a good time, and develops himself.

A Family Quarrel

The revolutionaries claim to offer a future that is open, yet at the same time they voice great uncertainty. And this could hardly be otherwise, for against their better judgement they cherish the illusion that critical destruction of the old order means *ipso facto* the birth of a new and better one. Moreover, their belief that revolutionary changes in society will also bring about a change in man himself and in human aspirations cannot conceal the fact that their passionate yearning for the new order disregards the realities of life. This element, too, contributes to the uncertainty of their future. But the more they realize this themselves, the louder they proclaim that the revolution is almost here. The revolutionary élan in many cases approximates the hurry of desperation.

The major difference between the two categories of thinkers is this: whereas the technocrats want to reduce history to the continuity of technological progress in order to gain control of the historical process, the revolutionaries concentrate history in the present as the decisive moment in which subjective man shakes off the burden of the seemingly objective chain of time and converts the present into a function of the utopia. The present is placed in the light of a new and more human future. Each moment must be charged with creativity and freedom. Only in this way is there hope for a better world tomorrow.

The society of the technocrats, argue the revolutionaries, is an inhuman society of power-hungry men, a society of alienation. In the name of reason, it even entertains nuclear genocide as a possibility, thus insisting that the unthinkable become thinkable and the intolerable become tolerable. The threat of self-annihilation through atomic weapons epitomizes the diseased state of a civilization that prides itself on its unprecedented progress.

The planners, in turn, see the propagated revolution as a clear symptom of decadence, as wanton sabotage of our culture. A revolution, they warn us, will lead to chaos and ultimately to poverty, the very evils science and technology have so magnificently conquered.

There is much that makes sense in the analyses offered by both the technocrats and the revolutionaries and in their criticisms of each other. The important question, however, is from what vantage point we can gain a better insight into the nature of their contest – how it arose and why it developed into the conflict we witness today. In one respect – a crucial one – there is fundamental agreement between the two rivals. At bottom, their quarrel is a family quarrel, a feud between factions of humanism, for neither yields to the other in endorsing from first to last the humanist view of man. Both factions view man as an autonomous being. Both tacitly assume that man is a self-sufficient being who has outgrown the need for God. And in keeping with that assumption, both factions alike take for granted that the world is a closed world and that history is exclusively the affair of man.

But how is it possible then, given this basic unity, for Western society to drift into such a crisis? Does man's claim to absolute self-reliance and self-determination entail a civilization divided against itself?

The Spiritual Roots of the Present Conflict

We will not really understand the crisis our culture is in unless we first find out how we got into that crisis to begin with. To this end we shall have to go back in the intellectual history of Western culture. What is it that has engendered this curious dichotomy in our civilization, in which, on the one hand, we see men with infinite arrogance enlisting the services of science and technology in their cause, but in which, on the other hand, we see men haunted by deep-seated uncertainty, stricken by doubt and despair, and filled, in the face of our present technological culture, with revolutionary desperation?

It is commonly agreed that the development of science and modern technology was possible in Western culture only because at the beginning of the modern era man was introduced to a new sense of history and historical development and to a new view of nature and human freedom.[13] Attention was called to *this* world as a world to be cultivated by man. This new outlook, however, lacked internal unity. From the outset two spiritual forces were at work that were moving in fundamentally opposite directions. I am referring, of course, to the Renaissance and the Reformation. These movements gave different answers to the question of the origin of all things.

Basic to the Renaissance was the idea of human autonomy: man discovered himself and proceeded to affirm himself; and before long he came to see himself as the lord of creation. The Reformation likewise saw man as a being called to freedom and mastery, but always in dependence upon God, to whom he remains

[13] Cf. Herman Dooyeweerd, *Reconstruction and Reformation*, chap. vii. See also R. Hooykaas, *Religion and the Rise of Modern Science*, esp. pp. xiff., 98ff.

accountable for all his actions. In Reformation thought, man is less the lord than he is the steward of creation.

At first this basic cleavage in the spiritual foundation of Western civilization was not easily detected. For one thing, the key figures often used identical concepts. In the Renaissance, however, these concepts were invested with a meaning that placed man at the pivot of the universe without any transcendental point of orientation. Terms like freedom, responsibility, nature, and control over nature thus acquired a secular meaning. Another reason is that, initially, the practical influence of the Reformation was immense, almost eclipsing that of the Renaissance. Gradually, however, it was the Renaissance view of human autonomy that conquered the world of philosophy and learning. And it was precisely from the intellectual sphere that the idea of human autonomy eventually arrogated to itself the leadership of the whole of Western culture.

Hesitantly at the outset but gradually with greater boldness, people denied that man's rule over nature is a gift from the Creator – a concept that necessarily implies a limit to man's self-declared independence and freedom. The Western philosopher (and increasingly also the Western scientist, under the influence of philosophy) wanted to achieve lordship over nature on his own steam, in his own right, and to his own credit. He tried to base his lordship on the idea of a self-glorifying autonomy, attempting to realize it in his science and to confirm it in technology. And so man's proud place in creation was secularized. Instead of listening to God, he would henceforth listen only to his reason.

Scientific knowledge became the weapon with which man cleared his path toward the future. And this path he could travel with increasing ease as his technological prowess achieved greater heights. People began to believe that man and world could come to self-fulfilment and consummation through the use of modern technology. And so Christian eschatology retired from the field in favour of a technological utopia: the hope of a new heaven and a new earth was crowded out by the expectation of a man-made heaven on earth. When, in the nineteenth century, the material fruits of the alliance between science and technology began to mature, the secular belief in progress extended its influence to include the masses. In the meantime, belief in progress has spread so widely that expressions like "the wonders of science" and "the age of technology" have become part of our everyday language.

As the consequences of the philosophy of thinkers like Descartes worked themselves out, man began to occupy the central place in Western culture. Nietzsche, and after him Jaspers and Heidegger, has shown how the human ego thus comes to be governed more and more by the will-to-power. Western "egology," which has spawned such terms as self-expression, self-affirmation, self-realization, self-preservation, and self-sufficiency, has found its historic expression in modern technology; its egocentrism has called up powers that have magnified the tensions in the world beyond imagination.

In the twentieth century, it has become apparent that the attempt to raise the world to a state of technical perfection carries with it enormous drawbacks. The ideal of "peace for all time," which could presumably be attained through modern science and technology, has been rudely shaken by two world wars and continues to be in grave jeopardy. The ideal of unprecedented material prosperity may have been partially realized, yet at the same time it has become clear that gains are often made at the expense of our environment, and that with all our welfare we are sitting on top of a volcano that may be about to erupt.

The belief in progress, with its avowed commitment to unlimited production and consumption, is now threatened by the fact that creation is finite after all, that its resources can indeed be exhausted, and that therefore there is a limit to exploiting it. To add to the grim picture, instead of rendering himself and the world more "real" and more "human," man is discovering that he is actually becoming more and more estranged and alienated from one portion of reality after another. Hence the rise of the neo-Marxist-tainted revolt against the stifling domination of scientific technology.

If we would better understand the source and the development of the conflict between technocracy and revolution, we will have to look more closely at the compelling lines of thought along which the tension has built up. Both the technocrats and the revolutionaries, as I suggested earlier, stand in the tradition of Cartesian philosophy. Both are guided by the idea of human autonomy. Man wishes to be a god unto himself. The supremely self-confident ego is made the centre of the universe and at the same time the source and origin of all reality.

In this tradition, the technocrats represent the line which declares rational thought to be the immanent origin and meaning-giver of all that is. Man's reason is absolutized on the basis of his claim to autonomous, absolute freedom. Scientific, logical thinking is pried loose, a priori, from its integrated place in life and set apart in a sovereign position of its own. Consequently, the products of rational thinking, too, are set apart in an unassailable realm of their own. The inevitable result of this arbitrary isolation of reason from the total life context is that the absolutized, universally valid mathematical and mathematical-physical laws for reality ultimately threaten to subject man's reason itself to an unyielding determinism – and that would surely spell the end of freedom, the very freedom that reason was anchored in at the outset. However, in reaction to this threat, human freedom puts up strong resistance and asserts itself as the counterpole to rational determinism.

Such, in brief, is the dialectic of modern philosophic thought. And every new attempt at reconciling the polarity is doomed to failure if it is pursued on the same old basis of the autonomy of human thought. When men break the created coherence of reality apart at the outset, by virtue of their pretended autonomy, then so long as they maintain their stance, there is nothing that can put reality back together again.

With the Enlightenment, the dialectical tension sketched above began to burst out of the confinements of mere philosophical theory and to pervade the whole of culture. The people of the Enlightenment aspired not only to understand the world by the light of reason but also to reshape it according to the dictates of reason. Choosing autonomous reason for their instrument, they aimed at developing a society in which freedom could be realized at last.

However, the objective structures, which were first contrived on the basis of reason and then projected into practice, have since turned into powers in their own right, powers which, as autonomous structures, have turned against cultural freedom. And so, on the basis of autonomous, free reason, a technological-scientific society has been constructed. It is a society characterized by virtually autonomous forces that pose a real threat to man's freedom in shaping his culture. And the more dynamically these forces develop themselves, the greater the threat will be, until man is no longer able to oversee the whole, let alone introduce any changes.

It is especially in our day that human freedom is imperiled. Since the time when science opened up the possibility of industrial technology, the new industrial forces have been allied with the political powers, and science and technology have been pressed into service to master the future. Yet people assert their subjective human freedom and resist the technocrats' passion for control. Their resistance is reinforced by the signs that the tension inherent in technocracy – namely, between the desire for infinite expansion and the hard fact of a finite creation – will sooner or later erupt in disasters and catastrophes. The pollution of the environment, the energy "crisis," the risks surrounding nuclear energy, and the growing shortage of raw materials already seem to point in the direction of a collapsing civilization.

For all these reasons, it is understandable that in our day the voices of the revolutionaries find a strong echo. Neo-Marxists in particular speak for those who revolt against the growing technocracy. They turn against the powers of the "establishment," which are used to control history objectively, disregarding the human subject. Their reaction finds its outlet by recognizing man as a free cultural agent and making him the key and cornerstone of all their thought and action. That is why they plead for imagination and creativity and why they work to overthrow the existing order.

Caught in the polar dialectic between absolute determinism and absolute freedom, between the continuum of control and permanent revolution, they choose for revolution and apotheosize man's role in the shaping of culture.

The Future

Is the apotheosis of freedom a solution? Do the revolutionaries indeed present a viable alternative to the bleak future of the technocrats? To answer these questions, we must take another look at the nature of the quarrel between technocracy and revolution. The polarity, as I said, derives from a common root: human freedom, declared autonomous and made absolute. It is this root that first generates the passion for control, a passion that then pushes us in the direction of technocracy. Soon technocracy begins to lead a life of its own, threatening to obliterate human freedom. Men thereupon take up positions at the freedom pole and declare war upon all controls. Obviously, the two poles simultaneously presuppose, penetrate, and repulse one another.

As a result of this mutual penetration and repulsion, the technocrats continually run into the "irrational factor" called human freedom, in spite of all their efforts to exorcise it. The progress of technology, of which they boast, is only possible if human beings continue to be free to use their ingenuity and inventiveness; the very extension of technocracy is only possible if human beings continue to make decisions in its favour.

Meanwhile we also observe the opposite: the revolutionaries, for all their desire to be absolutely free, can in fact never leap free of continuity, control, or power. As a historical or cultural agent, man is bound to objective cultural means and cultural power. Without these he cannot express himself culturally. It is for this reason, too, that we often see revolutionaries either going over to the camp of the "establishment" or making their revolutionary programmes still more radical. In the latter case, they may either promote total chaos or make a lunge for power themselves. Either way, the revolutionary ideals are given up, the revolution's own children are consumed, and a dictatorship sets in that is more powerful and despotic than the preceding one. From this point on, the cultural dialectic from establishment to establishment and from revolution to revolution can only increase in magnitude and intensity.

In this conflict over the direction of culture, the technocrats appear at a decided advantage over the revolutionaries because they need not rely on men as free historical agents. Instead, by sheer economic power they can capitalize on the objective cultural potential as it becomes available in the most recent scientific and technological possibilities, such as systems theory, cybernetics, and the computer. The technocrats enjoy an additional advantage because the masses either depend so completely on them that they are impotent or they surrender body and soul to them in the confident hope of receiving still more of the good gifts of science and technology.

Moreover, when the technocrats are confronted with embarrassing problems or imminent perils, they simply respond by changing their strategy. And as they do so, they do not spare human freedom; in fact, if need be they restrict it even further. This tendency has been evident, for example, at recent world conferences devoted to such pressing problems as the population explosion, environmental pollution, and the economic development of third world countries. In the face of all the clamour for more rigid controls, it should come as no surprise that the revolutionaries become even more radical; to achieve their goal of overthrowing the existing order, they may resort to greater instruments of destruction.

If my analysis is correct, the tension in our culture can only grow to more disastrous proportions. Technocracy is becoming increasingly centralized, and it threatens to encompass the entire globe; its tight control will turn our world into one great prison-house for free and responsible people. Revolution, on the other hand, if consistently realized, will inevitably result in violence and destruction. Therefore, if a third way is not found, the only choice open to mankind is between a technologically streamlined society of perfected standardization and mass culture and a cultural self-annihilation and suicide.

There is a real danger that science and technology *as such* will be blamed for our present dilemma. In many quarters, in fact, people have already come to this conclusion. But then the nature of our crisis has been woefully misunderstood. It is not science or technology but man that bears the blame. Western man has chosen to accept this world and himself as his first and his last point of reference. He has gradually closed his eyes to any transcendent reality. The purpose of history and the meaning of life have been restricted to this world; they have been made immanent. And man, no longer open to God, is now thrown back upon a purely this-worldly reality.

All the same, the Western mind suffers from the fact that divine revelation once instilled into it notions of perfection and consummation, notions that retain their appeal and that refuse to be silenced. However, since Western man no longer looks to God for the fulfilment of these promises, he is obliged to arrogate to himself the task of realizing them. Thus, as he moves further and further away from God, man secularizes God's promises and begins to think that he can realize these for himself through science and technology. Having placed his faith and confidence in technological progress, he appears to have thrown himself and his future at the feet of technological-scientific development.

In this light, one can well understand that where true history is fundamentally closed off and a purely man-made history is advocated in its place, modern technology in its many applications must grow to inordinate dimensions. In fact, it is already assuming monstrous proportions and is beginning to betray features that are actually demonic.

As the process of secularization widens and deepens, as man's sense of responsibility diminishes, as his hopes are increasingly pinned on science and technology, and as the possibilities of the latter become more frightful, the thinking of technocrats and revolutionaries actually approaches two rival forms of nihilism. The house of the closed worldview is a house divided against itself. It is the natural home of cultural tensions and catastrophes. The rule of technocratic control is self-willed, man-made, power hungry, and hence normless. The reaction it elicits is an equally lawless revolutionary freedom, in which every last cultural achievement must be annihilated again. The nihilism of the lifeless mechanical order of the technocrats has its obverse in the nihilism of revolutionary turmoil and chaos.

Is there hope for the future? Unlimited technological-scientific development will lead to loss of freedom, the exhaustion of nature, and possibly the destruction of the world. The unleashed revolution, intended to liberate man, will only lead to greater slavery. This nihilistic dialectic, which is growing in scope and intensity as history moves forward, marks the way a culture advances toward its dissolution.[14]

Estranged from God, our civilization carries about within itself the seeds of its own death. Already it is experiencing the blight of decay, which will not stop spreading until all lies in ruins. In all of this, however, God himself is clearly making us feel that apart from him there is neither life nor survival.

The Way Out

As I indicated at the beginning, I have restricted my analysis to the extremes of the opposing tendencies that divide our culture. Fortunately there are still sufficient counterforces to keep the hopeless dialectic from breaking our society apart. However, no counterforce can change the fact that the two tendencies are working in opposite directions and are causing serious cracks in the foundations of our society. There is not a person who does not experience daily something of the increasing tension between technocracy and revolution. We are left with the pressing question: is there a way out?

In order to point a way out of the predicament, I must first focus on the problem of human autonomy and the attending secularization of culture, for it is autonomy – or rather the pretension to autonomy together with the closed worldview of the secular mind – that constitutes *the* problem of our time.

---

[14] This nihilistic dialectic is apparent in Theodor Wiesengrund, *Adorno's Negative Dialektik* (Frankfort, 1966), p. 338: "One should try to live so that one may believe one has been a good animal." Cf. Günter Rohrmoser, *Das Elend der kritischen Theorie* [The Misery of Critical Theory] (Freiburg, 1970), pp. 31, 34: "According to the school of negative dialectics, in a world impounded in its own inverted order the extent to which one can still realize something of his humanity is to be a good animal." For this reason it is more appropriate, in my opinion, to speak of nihilistic rather than negative dialectics.

For example, man's autonomy cannot possibly be absolute. Man is and remains, in technology as elsewhere, fully dependent upon *given* materials and structures. And, likewise, he is and remains dependent upon himself, in the sense that he cannot ground his own existence in his *belief* in autonomous, independent existence. His very mortality shows up the emptiness of his claim to being his own origin, his own source of life, and his own god. Where people do indeed perceive this much, however, they often end up resigning themselves to human existence as to being-unto-death. But the realization that death is the end then infuses the whole of life with meaninglessness. The closed worldview of secularism ultimately allows no other vista than that of nihilism: a prospect without hope.

A different perspective will require a different mentality. Man must recognize that he is incapable of autonomously determining the direction in which culture should develop. He must wake up to the fact that he cannot pretend to be the lamp that lights the road he should go. So long as he keeps up this pretension, his world – as history since the days of the Enlightenment clearly shows – will continue to grow darker and more menacing all around him.

I said earlier that modern man labours under alienation from self, from his fellowman, from nature, from culture, and from history. The deepest cause of this ailment is that man has become alienated from the origin of all things, God. Alienation from God always brings with it the other forms of alienation and must finally issue in the utter meaninglessness of everything.

Anyone who has seen and acknowledged the deepest cause of our cultural crisis knows that there is a better way. It is the way in which man is not the measure of all things. It is the way in which man is conscious of being carried and guided by a Creator God, the God who has given him life and who has crowned him with honour and dominion for the sake of responsible stewardship. It is the way which requires that man be open to the meaning of history and to the meaning of his historical existence. The meaning of human life on earth cannot be found in visible, temporal reality itself, nor in any part of it. To restrict meaning to created reality is to constrict it. It is to close meaning off and ultimately to choke it by isolating created reality from its life-giving Origin and by deifying the creature.

Meaning transcends reality. When we recognize that, a horizon opens up that stretches beyond the horizon of this world, with its tensions and its distress, with its sin and evil and death. Before that opened horizon, life has a future again.

I would emphasize that this way is shown us not by philosophy but by God himself. We know from the divine revelation of Scripture that God in Jesus Christ is Lord of history, that he alone rules and governs the world, that he holds all things together and brings them to fulfilment in the final consummation. With Christ's coming into history, the kingdom of God has broken into the world of man, to conquer and to heal it. Consequently, the quest for the kingdom must also be expressed through responsible activity in science and technology.

Only after confessing belief in God and his sovereign rule over the whole of human life can we find a philosophy that orientates itself to that rule and kingdom and that can be of service in indicating the way of deliverance. From that point on, a Christian philosophy of culture and a Christian philosophy of technology have their work cut out for them. For they must point out solutions and avenues of escape, especially at those junctures where the development of modern culture has tied itself into knots.

Some Implications

The implications of what I have been saying so far are many. These cannot all be worked out here, since my chief concern has been to make a proper diagnosis. Still, I should like to examine briefly what a Christian philosophy can say of significance about science, planning, and the meaning of technology.

(i) Science

The predicament of our "scientized" culture calls for a re-evaluation of science. It is especially important that we critically re-examine the absoluteness with which scientific "truth" is presented. First, people need to recognize that "science" as such does not exist. Only flesh-and-blood scientists exist, and they hold certain scientific theories. These theories are not, as is so often believed, objective and neutral; rather, they rest on believed assumptions and are based on hypotheses. This is the reason scientific theories are always conditioned, coloured theories. Further, scientific theories are *relative,* relating as they do only to the knowledge of certain aspects of reality, such as the physical, the biotic, the social, the economic, among others. Therefore, scientific knowledge is necessarily a knowledge of limited scope, abstracted from the fullness of reality, which is itself far more complex. In addition, since reality's complexity even penetrates each of its aspects, scientific knowledge as knowledge of a certain aspect must always be a knowledge for the moment, never finished, never complete; in short, it is *limited* knowledge.

The view that scientific knowledge is conditioned, relative, and limited was never more needed than it is today. For example, more controls on people's activities are inevitable because of the uncritical use of scientific knowledge in systems theories and quantification and the facile recourse to computers in conjunction with these methods. To conduct science naively and to apply it indiscriminately is to cast out human responsibility. If scientists admit the qualifications and limitations of science, however, the knowledge that science yields is taken up into a more comprehensive, responsible knowing. Instead of making human responsibility unnecessary, the growth of scientific knowledge heightens the need for it. Moreover, mankind's heightened responsibility is becoming increasingly a *communal* responsibility, and this development requires a much stronger sense of community among people than has been experienced up until now. It is a condition that can really be met only if there is a common sensitivity to norms and a common religious consciousness. But as everyone knows, these are the very things our secular culture lacks most.

(ii) Planning

The method so frequently applied in planning the future is the so-called technological-scientific method, the same method that is used in manufacturing lathes, computers, wireless receivers, and soon. Using this method in planning for the future entails the grave danger that people are going to be manipulated as though they were mere things or simple parts of a machine. Planning is intended to banish evil, but its practical effect is to abolish man. Planning cancels out man as a free and responsible creature made in the image and likeness of God. If the finished plans are ever carried out, man will increasingly be enslaved to powers beyond his control.

This outcome is so certain because throughout the phases of planning – doing the research, drafting the plans, executing the plans – it is science and the scientific method that men allow to have the first word and the final say. In the first stage of research, moreover, scientists often simply accept the facts at face value without questioning their normativity. In the second phase, too, this question is rarely raised. Also, the planners are presumptuous as they "assemble a model for the future"; they forget that it is not given to mortal man to draw a picture of the future that takes into account all the factors that will then be at work. What the planners lack is enough humility to halt before the unknown, before the unknowable, unexpected, and surprising aspects of the future.[15] The adverse effects of their presumption are reinforced twice over when the planners integrate their various sub-plans into the total plan: man is levelled, society is collectivized, and power is concentrated in the hands of a few. In the third phase, accordingly, when the plans are being executed, there is little if any room left for man; he can no longer play a free and responsible role in shaping the world of tomorrow.

Most people would agree that our society has become so complex that it is no longer possible to make intelligent decisions about the future without using scientific knowledge and resorting to scientific analysis. But science must not lay down the law for man or dictate to him what to do. It should elevate his personal and communal responsibility, not eviscerate it. When science is thus subordinated to human responsibility, men will also be free to explore any new developments that could open up unexpected opportunities for solving problems that now seem insoluble. Today's planners often force their way to a solution in a high-handed, technocratic manner, blocking the road to genuine solutions.

As a rule, planning involves integration and collectivization. This tendency, too, must be resisted. If we wish to give priority to human responsibility, we need to provide for differentiation alongside the tendency towards integration. Where so many things are done on ever larger scales, there ought to be more room for activities organized on smaller scales. Promoting only the super-size ventures leads to power concentrations, collectivism, and drab uniformity. Smaller, differentiated ventures not only create excitingly new opportunities for human involvement and inventiveness, but also help develop a pluriform culture by spreading authority and competence, in both the economic and the political spheres, over a greater number of people and a greater variety of institutions.

One other matter should be clear from my analysis: the many different human communities and relationships must not be denatured into subordinate parts of a technological world state. Rather, they should be respected according to their own normative structures.

---

[15] In this connection, compare Deuteronomy 29: 29 and Revelation 10: 4.

The cardinal question in any human relationship is where the line between authority and freedom lies. To decide that question requires much wisdom and reflection. But whatever the answer, two evils must be avoided. On the one hand we must avoid any absolute authority of persons and rule books, of anonymous powers and the exigencies of technocracy. On the other hand, we must avoid unbridled revolutionary individualism.

Within any given relationship, be it a family or a school, a business or an engineering firm, the normative structure for leadership on the one side and subordination on the other is meant to serve a healthy unfolding of life. When the normative structure is not acknowledged, it will inevitably assert itself in a corrupted fashion, either as dictatorship or as anarchy. In either case it will then be impossible for the members of the relationship to serve each other in love – which is the very purpose of the normative structure for leadership and subordination. Those who are in authority and must give leadership have a responsibility to create conditions that will enable everyone to work at his task in optimum freedom; they ought also to prevent any disturbance of the community by intrusions from outside or by internal disruption caused when members misuse their freedom. Those who are in a subordinate position, meanwhile, have a responsibility to work together to achieve the purpose of the relationship. To underscore the communal task and to forestall misguided or biased decisions, members need to help, encourage, and correct one another. Fruitful interaction of this sort is realized best if policy and goals are discussed in an open atmosphere and if those who carry out policy are required to give an account of their doings at regular intervals.

I admit that these guidelines are anything but easy to work out in practice. Nevertheless, they offer a sound framework for forming genuine communities of men.[16]

(iii) The meaning of technology

The meaningful functions of technology are many. They include emancipating the body and the mind from toil and from drudgery, repelling the onslaughts of nature, providing for man's material needs, and conquering diseases. They also include eliminating unnecessary burdens, freeing time, promoting rest and peace, and evolving new ways and means for advancing the disclosure of culture.

---

[16] Cf. Hendrik van Riessen, *The Society of the Future*, pp. 290 ff.

The meaning of technology concerns the elevation of culture – through fostering reflection, through stimulating internal communication, and through making possible a wide variety of rewarding jobs and tasks.

All this is a far cry from what technology actually is today, for the meaning of technology has been perverted. Partly under the influence of economic power structures, technology has produced superfluity, waste, and pollution. Work has been reduced to "productive" work, to work that is "economically justifiable"; and the resulting emptiness in the workers is compensated for by more consumption. Instead of freeing men to devote themselves to works of assistance and service, of care and mercy, of creativity and beauty, technology has all but banished these forms of meaningful labour.

This impoverishment, as I noted earlier, developed because Western man began to *believe* that technological know-how, assisted by economic efficiency, would bring cultural progress. Technology was expected to deliver what it never could: the redemption of life. People produced whatever could be produced as materialism, pragmatism, and positivism reinforced their faith in a technology that was absolute and intrinsically anormative. Technology has buried life under the yoke of a terrifying power.

If we want to experience the true meaning of technology again, we will have to abandon our mad pursuit of the future, obsessed as we are by our technological prowess, by our will to power, and by our mania for consumption. We will have to take up the battle against superfluity and absurd luxury. To answer to the meaning of technology we will have to devote all our powers in love to the development of our fellowman, of our natural environment, and of ourselves, in happy accordance with the ultimate purpose of each. This focus will guide us is making conscious, responsible choices about *what* will be produced, instead of being governed in such choices by our insatiable thirst for continuously expanding our technological development. The autonomous dynamics of technology will have to be decelerated, so that we may have time once again to reflect upon the meaning of it all.

We will have to determine how much of the damage we have done can be set right. Perhaps we need to recover some forgotten traditions in the history of technology and adjust these to our needs. The so-called alternative technologies will have to be given more than the usual attention; I think, for instance, of durable technical products that can be made with little capital investment and low energy expenditure. That approach would involve using as much as possible the natural materials and sources of energy available to us. It would result, moreover, in individual products that give satisfaction to those who make them, that are geared to specific cultural patterns, and that cause a minimum of pollution. Such scaled-down forms of technology seem almost unreal in the overpowering presence of our gigantic forms; indeed, they are not easy to realize, and they will certainly make heavy demands on our technological imagination.

Alternative technologies offer one way to attack the problems that present technology has fostered in the workplace: the rift between work and leisure and a growing separation between the worker's living quarters and place of work. These obvious problems could be reduced by developing new trades and crafts that use modern technological equipment and also by using modern information and communication techniques on a larger scale to include private users.

In short, to probe the meaning of technology does not mean to throw technology aside wherever possible. Rather, it means to appreciate technology's proper and meaningful place within culture and to develop technology intensively and responsibly.

We must not allow technological-scientific possibilities and economic forces to dominate our culture. Rather, spiritual and cultural values must make technology serviceable to life. That means technology must be opened up socially, so that all who are involved in it are entrusted with responsibilities. Disclosing technology implies, further, that the economic development of technology should not be limited to maximizing profits or catering to consumers; we should prevent waste and strive for a frugal use of nature, even though that will inevitably mean cutting back "economic growth" and tempering consumption. In addition, when we eliminate harmful waste products and when we refuse to use nature as our rubbish heap, we are showing that we have an eye for the aesthetic dimension of technological development; technology must not mar nature but rather be developed in harmony with it. Further, justice demands that we preserve nature, not imperil it, by keeping it clean. Where nature is being destroyed, our laws and courts ought to intervene and punish offenders, both to protect nature against mutilation and pollution and to protect people from the dangers inherent in a deteriorating environment.

The disruptions we are experiencing in our culture only arise when the ethos of men is wrongly directed, so that in all their technological doings they myopically focus on something within created reality, absolutizing and thus asphyxiating it, instead of offering themselves and all their deeds as living sacrifices in the service of God, who in Christ rules over creation. Man needs to be delivered from his short-sightedness and from his satisfaction with short-run measures of relative success. Only the wider perspective of the kingdom of God can accomplish that. Only through openness to this transcendent reality can Western culture look forward once again to a meaningful future.

If we try to implement these guidelines in our practical affairs, we will find that what stretches before us is not a dead-end road but a highway of liberation. This highway alone avoids the stagnation and the entropy of technocratic culture. This highway avoids the kind of development that kills human initiative and that consolidates an elite in citadels of collectivized and centralized power. This highway, too, keeps our culture from being pulled into the vortex of the ideology of revolution.

The Environmental Problem: Its Religious-Philosophical Backgrounds

Environmental pollution is a subject widely discussed today. Daily the media confront us with the looming disaster of a polluted, deteriorated environment in which life's suffocates and man's outlook for the future is hopeless. Higher temperatures, oxygen consumption outstripping its production, increasing carbon dioxide in the atmosphere, menacing radioactivity, poisoning of soil, water, and air – together these symptoms are the ingredients of a very pessimistic prognosis for the survival of mankind.

The report of the Club of Rome, however, pointed out that there are more factors than these involved in the ruinous course our world has taken. These factors, such as the diminishing supply of natural resources and energy, the population increase, and the concomitant problem of food supply, have a bearing on the environmental problem. Nevertheless, I am going to focus our attention particularly on the problem of pollution itself. And even then, I must restrict myself to one side of pollution.

The restriction is indicated by the subtitle of this essay: the religious-philosophical backgrounds of the problem. This may be surprising because it is far from customary to associate man's religion with the pollution of the environment. Usually this problem is approached from the standpoint of a special science, such as biology, technology, or economics. Even the answers proposed are generally limited to scientific or technological ones. Yet I think such approaches tend to touch only the surface of the problem, both in the discussions and in the solutions offered. Faced as we are with the enormous drawbacks that have attended the development of science and technology (such as the destruction of the environment), it is essential that we not flee from the crisis in order to seek refuge with science and technology themselves. Instead, we must enter into the religious heart and source of our scientific activities. For it is in this religious background of our technologized culture that we shall discover the origins of environmental pollution.

This kind of investigation does not occur very frequently because we are too impatient to await the results of new reflections on actual problems. What's more, the scientific and technological enterprise has occupied us to such a degree that the deeper currents the problems generally do not receive any attention. Man no longer seems able to perceive that science and technology are based on philosophical views and that these views, in turn, are rooted in religious convictions. However our blindness does not mean that these religious-philosophical influences do not exist.

Man expresses his fundamental convictions concerning reality and its questions in his religion. This basic religious attitude is the foundation for his living and hoping. It also inspires his philosophical thinking. Philosophical thought is not neutral, objective, or devoid of any values, but is inspired religiously. This religiously inspired philosophy, in turn, exerts its influence on the special sciences, which are related to that philosophy. Furthermore, modern technology, based as it is on modern science, especially modern natural science, also experiences the influence of religiously inspired philosophy.

Therefore, philosophy, inspired by basic religious convictions, plays a decisive role in the development of science and technology. The religion of Western man leaves its sediment in philosophy, and this religiously inspired philosophy influences the views of the special sciences. These views, in turn, assume a tangible form, as it were, because they are projected into daily life through their application in the development of modern technology.

Thus, in the face of the massive problems and inevitable calamities brought on by modern technology, we feel driven to investigate the content of the religion operating beneath the surface. Considering the enormous technological dislocations in the human environment, we must ask whether or not the error ought to be located especially at the spiritual level.

If my interpretation is correct, the causes of pollution lie in the religious origin of Western science and technology. Therefore, we risk giving myopic and irresponsible guidance if we concentrate all our attention on the startling and, at times, overwhelming character of pollution. For we need to realize that what we see surfacing on a large scale today has determined the content of spiritual attitudes for centuries. Long contained by man's view of nature and of himself, it is now violently emerging. Only when we discern the spiritual-historical background of the environmental problem will it be possible to go beyond a simple appeal to science and technology to avert the imminent dangers to an appeal that involves modern man's religious convictions, which affect his attitude towards nature and culture.

God, Man, and Nature

It is difficult to understand my analysis unless we go back in the intellectual history of mankind. Initially, man lived in complete harmony with nature in the Garden of Eden. He did not need to plow or sow, for trees, "planted" by God himself, produced fruits and sustained him. This harmonious relationship between man and nature was changed drastically, however, by man's fall, in which he rejected God and wished to be a god himself. From that moment on, man had to earn his bread by hard work; he was constantly threatened by his natural environment, all the more intensely as his apostasy continued. In fact, men came to believe that nature was determined by mysterious powers. They conformed to the demands of nature and did not often have the courage to intervene. But when they did, they also carried out magical rites to propitiate the gods of nature.

It is true that a few changes took place in the ancient Greek world, but none of these was fundamental. Reality was seen as one organic unity in which everything had a fixed position. The ancient Greeks thought of man as inserted into a hostile nature. Its capriciousness could be averted only by concentrating on the stability and immutability of the supernatural, the world of the stars.

During the Middle Ages, men came in contact with the divine Word-revelation and realized that nature is not divine but is instead a created reality. Nevertheless, the predominant view of nature was similar to the Greeks' view. Greek ontology as a philosophy which deals with the totality of beings was simply adapted to fit the understood meaning of Christian revelation. Men were not yet motivated to examine the various functions of nature, since the worldview of the day contended that everything, including nature, had its fixed and immutable place. Because nature formed a given and static unity that did not tolerate any intervention, men resisted a dynamic development of science and technology. They feared changes because they could disrupt and disturb what had been given to man. This cautious approach to nature also characterized the development of the crafts.

Basically, the medieval attitude toward nature rested on a dualistic view of reality. A firm faith in the hereafter for the soul prevented people from paying adequate attention to this world and to nature. As a result, man's investigation of nature did not go beyond a hierarchical ordering of natural data. Nature was interpreted in terms of what was alive. This organistic view of nature implied that whatever in nature was judged inorganic or "dead" had to be regarded as less than alive, and therefore not worthy of full attention.

At the beginning of the modern age, this situation changed. During both the Renaissance and the Reformation, people began to pay attention to this world on the grounds that it needed cultivation. The people of the Reformation did not regard nature as divine, and they denied the existence of a divine and static order for nature that man had to leave untouched. While they recognized the legitimacy of science and technology, they emphasized that science and technology had to be normed by God's Law, which held for the whole of created reality. The people of the Reformation could not accept science and technology operating autonomously, independent of the creator of all things. Instead, they said scientific and technological activity were performed in obedience and praise to God rather than to man.

A starkly contrasting view emerged from the religious attitude of the Renaissance and the later humanism that dominated the development of science and technology. During the Renaissance man was proclaimed completely autonomous, independent, and self-reliant. Aided by science and later by modern scientific technology, man attempted to make his life secure by developing science and technology in absolute independence. Initially, the influences of the Reformation, the Renaissance, and the later humanism were largely the same. But since the Enlightenment, the development of science has been inspired chiefly by the religion of humanism, which regards man as sufficient in himself. That does not imply, however, that Christians no longer participated in developing science and technology, only that their Christian religion became increasingly isolated from their scientific and technological endeavours. They seemed to hold a dualistic view of reality and human activity; their Christian view was adapted to the humanistic tradition, while the authentic character of the Christian religion disappeared.

For a proper understanding of the religious backgrounds of the problems surrounding the environment, therefore, we must examine the tenets of the humanistic religion, particularly its view of nature. Cartesian philosophy illustrates these tenets quite well. Descartes doubted all tradition and belief. His search for religious certainty was limited to man himself; he found it in scientific thought, which was considered to be the ultimate certainty and truth. Descartes's philosophy was determined primarily by two substances: *res cogitans* and *res extensa,* thinking man and space. They were treated as two opposite poles within reality, but thinking man took the primary place. That is to say, everything that exists beyond and outside of man was interpreted from the vantage point of thinking man.

While Descartes reached this conclusion with the help of mathematics, Galileo reduced everything to the object of natural science. The earlier stress on the being of beings was replaced by an emphasis on the function of beings. Previously, men had regarded God as the origin and therefore also as the unity of all beings, but this unity was increasingly bestowed on thinking man. The human subject became the central point of reference for everything that exists. In modern philosophy, man became the predominant centre of the world. The human subject became pivotal, while nature was relegated to the status of a mere object of natural science, which dealt with reality in terms of cause and effect.

As a result, nature was interpreted mechanistically. It was looked on as a whole made up of interacting forces that could be calculated and expressed in formulas. In fact, reality was thought to be an ingenious mechanism. Animate and inanimate nature were related to this mechanism in degrees of complexity.

From its outset, this mode of thinking housed the possibility of reducing nature to a mere object of technological manipulation. Man was no longer seen as a child of Mother Earth or as part of the whole of natural reality. On the contrary, henceforward man was to be the autonomous lord and master over nature, a power he had to demonstrate. This meant for technology that man was no longer limited to using the things nature offered and that he was able to take from nature as he saw fit. The so-called organistic view of nature was replaced by the mechanistic view, in which man, aided by technology, shaped nature to match his wishes. The natural process was pictured as inanimate but ordered, and its irrevocable and binding laws could be revealed by mathematically inspired natural science. The greatest danger inherent in this view is the absolutization of the method that it gave rise to. The natural-scientific method was thought to be superior to all other scientific methods, and everything that did not come within the scope of this method was written off as unreal, hence negligible. The abstract knowledge peculiar to natural science was thought to be a full, concrete knowledge of reality.

Therefore in humanism man no longer saw nature as created and upheld by God. On the contrary, he denied that nature is intrinsically related to God and has a sacred character. Moreover, humanism rejected the existence of a divine order of things to which nature is subjected and which does not tolerate human interference. I must emphasize that I only partially agree with the second interpretation. For we need to remind ourselves continually that the humanist view was completely secularized. It was a view that regarded nature as a mere workshop at the arbitrary disposal of sovereign and autonomous man, the pretended creator of nature, who could shape the order of nature as he pleased. Of course, this absolutely independent sway over nature was packaged with many promises: for instance, that there was no need to continue living under the threat of a capricious and hostile environment. Instead, doors were opened to a world that was *man's,* a world that he designed, controlled, and subjected to himself.

This, in short, was the radical shift brought about in man's relationship to nature. Modern natural science removed the ethical and religious restraints of earlier times, hesitantly at first but with increasing boldness as confidence and momentum grew. Those views of reality that did not concur with the dictates of natural science lost ground until they were denied altogether and finally vanished.

As a result, only one attitude towards man and his relationship to his surroundings appeared to be valid and worthy of respect; this, of course, was the one prescribed by modern natural science and its method of quantification. This method was used to interpret everything exclusively in terms of numbers and quantifiability. That is to say, everything was *reduced* to numerability, measurability, and estimability and was thus prepared for calculation and subsequent technological control.

The mechanistic view of nature monopolized man's thinking and changed nature's rich diversity into a frightful, monotonous homogeneity. As measurement, estimation, and enumeration became the only valid ways of approaching nature, its inherent coherence disintegrated.

Initially, this novel attitude remained restricted to the area of science. Its practical consequences, therefore, could hardly be detected. At the beginning of the eighteenth century, however, this situation changed because of the advance of modern technology based on modern science. The economic and material circumstances of the day also contributed to the activity of modern technology, while the population increases so stimulated its further development that it seemed imperative. Fueled by the additional available manpower, only the development of modern technology could satisfy the tremendous growth in human needs.

Modern technology did not develop on its own steam. For the spirit of the Renaissance and of humanism, which involved a firm faith in the progress promised by technological development, was also at work. Previously, this faith had been embraced by scientists only, but now the masses accepted it as well; thus its influence was extended to everyone who did not object to the new and infectious prospects of riches and liberty instead of obligatory poverty and suppression. Modern technology became a liberating force.

It is very important, therefore, that we keep sight of both the influence of natural-scientific thought on modern technology and the absolutization of this thought, which inescapably leads to the objectification and constriction of the meaning of nature. For men interpret nature primarily in terms of mathematical categories, such as the numerical aspect of space and the physical categories of time and causality, under the illusion that nature is composed entirely of these primary characteristics; meanwhile, others, such as colour, smell, and sound, are under-valued. To be sure, these characteristics are still called secondary qualities at first, but gradually they are ignored altogether. Inevitably, man misjudges the intrinsic value and meaning of nature because he no longer recognizes nature's essential and normative meaning. This attitude is illustrated in the works of Karl

Marx, for instance, when he wrote in his Paris manuscripts that "nature, when seen abstractly and in itself, has no meaning for man when alienated from man."

We should not conclude from this that objectifying and constricting the meaning of nature brought about immediate large-scale dislocations the moment modern technology began to develop. Although various people pointed out the inevitable disastrous effects of allowing technology to dominate human life without any resistance at all, such warnings were not heeded. And as a result, nature today, strangled as it is through the advances of technology, harbours potential disasters. Instead of enabling man to live, nature has ceased to fully support life. Mankind's consistent application of the mechanistic view of nature produced the threatening results that confront us even now.

The application of natural-scientific knowledge was not a matter of coincidence. Descartes had already pointed out the tremendous usefulness of the new view for meeting man's needs. With an expanding knowledge at his disposal, man would master all forces and movements in nature until they conformed to his wishes and worked to his advantage, simply because man was the possessor of nature.

Thus nature functioned as the object of human knowledge, while at the same time it served as the object of human mastery, satisfying man's arbitrary cravings for utility. This is clearly demonstrated by a statement from Marx's *Outline of a Critique of Political Economy:* "The material of nature, insofar as it has remained untouched by human labour, has no value, for value is but substantiated labour...."

Since the time of Descartes and Galileo, therefore, the modern scientist can be described as a rational engineer, who autonomously directs and controls the forces and movements inherent in nature, rendering them useful as he sees fit, for he is responsible to himself only. In other words, the compulsion to dominate the world by means of technology precedes modern technology itself.

Francis Bacon was particularly appreciative of the potentially positive effects of the new view of nature. He anticipated the achievements of today's technology, but he failed to pay sufficient attention to its possible negative aspects. Bacon combined the persistent desire to apply natural-scientific knowledge with the idea of unprecedented material progress, making him the first to believe in progress brought about by means of science and technology. According to him, man had the duty to pursue new scientific knowledge to establish his unprecedented power.

The dominant view in modern technology, therefore, is that man is in a position to command the world as he wishes. With technology as his tool, man sets out to create a world in which he alone is lord and master. He is motivated and stimulated to do this because of his need to safeguard his autonomous position, so that he may continue to enjoy and consume the fruits of his own labour.

Such ambition will result in a fixed and constricted technological world, far removed from the rich fullness of the whole of reality. The man-made technological world is but a fragment of the whole world and reality. Yet people persistently view the fullness of the whole of reality as if it coincided with the world of technology. In other words, modern man has blown up his desire to dominate by means of technology until it fills all of reality.

In so doing, man reduces reality to one of its aspects. Reality, however, will not tolerate such reduction, because it consists of a diversity of aspects. Everything in reality exists in a coherence of meaning given with creation itself; man cannot reject this coherence without suffering the consequences. Concealing this coherence or failing to reckon with it will inescapably result in serious derailments and dislocations. As technological control becomes absolute, ruinous side-effects will begin to appear. To be sure, this tendency can be ignored for a short time. But with industry growing, with agriculture almost completely industrialized, and with traffic increasing continually, this trend will assume such proportions that it is safe to speak of the qualitative doubling of disruptive side-effects. People will be unable to ignore such a situation for very long. It is these side-effects that together constitute the problem of environmental pollution today. The nature that we control and dominate threatens to turn on us. Destroyed and polluted, it has become a definite threat to the survival of mankind. The religious faith in progress has combined with technological progress itself to bring mankind to a critical stage.

Secularized Motives

For a better understanding of the above, let us briefly describe the motives that have inspired man in his technological endeavours. They indicate the religious nature of the major influences affecting technological development, since man has placed his fundamental trust in technology.

Ever since Francis Bacon said that knowledge itself is power, man has considered scientific knowledge to be the very gateway to universal progress. Of course, this idea gave him a tremendous incentive to increase his knowledge, for science and technology became the means of concentrating on an assured future. Attention shifted from mankind's immediate anxiety, deficiency, distress, and suffering to the achievement of general prosperity for the whole human race. The limitations in this attitude are reflected in Bacon's statement, "To dominate nature, we must obey her." The striking element in this statement is that nature was seen only through the spectacles of natural science. Moreover, obedience in this context implied no more than subjection to the laws of nature. Thus the task of technology was seen as shaping the force and movements of nature to gain ultimate control over it. Thus responsible obedience was reduced to shaping nature technologically by using the laws of nature. The peculiar, intrinsic value of nature itself did not enter the picture, with the result that indispensable ecosystems have been destroyed.

Bacon's second remark also indicates that technology will not be unfolded normatively; instead, its character will be stifled and it will become destructive, not only of the environment but also of life on the social, economic, aesthetic, juridical, and ethical levels.

Inherent in Bacon's view is the idea of man thinking entirely and exclusively in technological-scientific terms, an idea that implies absolutized theoretical domination and control of our world and its future. Modern technology must assist in realizing this ambition.

Engineers, in particular, have often been inspired by the idea of technology for its own sake. It is an idea that leads them to contend, among other things, that whatever can be made should be made. Any adverse side-effects produced in the process, such as noise from supersonic aircraft, will need to be taken in stride. Only when side-effects assume unacceptable proportions are technicians prepared to intervene and cure such ailments.

Such a concept of scientific knowledge, however, leads to a one-dimensional, technologically streamlined society. The quest for universal progress is frequently accompanied by an unchecked greed for power, if not openly, at least covertly. Man is eager to break down all barriers in time and space. He is so occupied by this ambition that he becomes a reckless devotee of the development of science and technology. Infatuated with it, he finally surrenders and throws himself and his future on the mercy of technological development.

These various motives imply a common view of nature, namely, that nature is inanimate, insensitive, and open to unchecked interference. It follows that wherever nature proves too stubborn to comply willingly, it is neglected, as for example, in plant and animal species that have become extinct. The quest for progress and the desire for power leave room only for the utterly pragmatic and utilitarian desires of man. The technological *utility* that nature offers is the one thing that deserves attention regardless of the cost to fellow creatures. Sovereign and autonomous man has become conscious of his unique position in creation; at the same time, he has perverted this position because he fails to observe the normative restrictions it entails. Thus he chooses to abuse nature rather than to manage it according to his original mandate. The normative relationship between technology and nature is broken. Man uses technology so that nature is exhausted prematurely, while everything that does not fit into the scheme of technological control is wiped out.

Instead of promoting harmony between technology and nature and thereby unfolding nature according to its meaning, man interferes in nature in such a way that he devastates it. This process will have sweeping results, especially if technology is so caught in the grip of economic powers that profit-making becomes the only valid criterion in the development of technology. Man will then have to discard anything that increases costs; he will exploit nature carelessly, turning it into a refuse heap.

The pursuit of progress and power has produced relatively short-term successes. Nonetheless, there are serious drawbacks, for with neither foresight nor responsibility, man has developed a technology that threatens his survival; he has become the victim of technology instead of its master. Man's inspiration was his vision of total happiness and welfare achieved by his autonomous efforts and aided by cooperation between technology and economic powers. In many ways, he realized just the opposite of such dreams by shaping a world where threats continue to grow in intensity and proportion.

At bottom, the quest for domination, progress, and power has been *secularized.* That is, Western man increasingly has attempted to subdue the world even as he denied that God exists or remained oblivious to God's works in this world. Western man acts on his own imagined strength, according to his own insight, and at his own discretion. The only decisive criteria he accepts are material welfare, avarice, and his compulsion to produce and accomplish. He is, as it were, possessed by a nearly demonic passion to dominate and control, urged on by his dreams of the ultimate fruit of his labours: a paradise, a utopia in which he will surely find complete happiness.

Meanwhile, technology has become an idol. The Swiss author Donald Brinkmann says that faith in technological salvation has replaced Christian eschatology. Philosopher Oswald Spengler agrees and says that the dominant motive in technology is "the desire for a small, self-created world that reflects the large one because it moves on it own power and obeys the human hand only. The past and present Faustian dream of the inventors is to be God, a dream that has generated the design of all machines." Spengler further states that "technology is eternal and everlasting like God the Father, it redeems mankind like the Son, and it illuminates us like the Holy Spirit."

When men lose their "vertical" orientation, they leave themselves without a proper "horizontal" perspective for this world as well. When men autonomously determine the laws for technological development, they fail to respect the meaning of created reality, i.e., that its meaning cannot be found within created and temporal reality itself, but only in God, its creator and life-giving upholder. Hence they assume that all things exist solely to satisfy the desires of self-centred, technological man and ultimately turn technological strength into a destructive power, as today's environmental problems clearly demonstrate. The ancient legend of King Midas illustrates the point. When Midas was allowed to beg a favour from the gods, he asked that all the things he touched be changed into gold. Like modern economic-technological and materialistic man, King Midas fell victim to greed. Within a few days, he turned to the gods again, asking them to reverse their favour, because literally everything he touched, including his child and his food, had turned into lumps of gold. Even with all that gold at his disposal, Midas was not happy because he could not satisfy his deep personal hunger for love, and he was doomed to starve physically and spiritually.

The moral of this story is that the idols of modern superstition indeed behave as proper idols. They are merciless in their willingness to grant man's every request instead of only what seems to be best for him. Nothing man asks will be denied, even if it costs him something else that he may need. This kind of idolatry ruins life, including the life of cultures. Our technological world amply demonstrates man's basic desires and his deep-seated motivation. The imminent destruction of the environment is a woeful reflection of man's egoism and materialism. Thus today's environmental crisis complements a fundamental spiritual crisis. The Spanish philosopher of culture Ortega y Gasset anticipated this long ago when he said that a man who is forced to live believing in technology and nothing else will lose his meaning. "It is for that reason that these years are the most intensely technological years as well as the most depleted years the history of mankind has ever seen."

Alternative Approaches

The implication of what I have been saying so far is that the environmental crisis is basically neither a technological nor an economic problem. It is first of all a *religious* problem. Therefore, a solution must be sought in that direction. But first I will take up two views that cannot offer any help because of their reactionary origins.

The first is the message proclaimed by the true counter-culturalists: that man needs to wake up to his natural origins and adopt a life style accordingly. It is a call for a return to nature, back to the nurture and care of Mother Earth. We are told that we must learn to conduct ourselves as true sons and daughters of nature.

Secondly, some people have also suggested that the pollution of the environment can be halted by drawing attention to the divine character of nature. This is basically a pantheistic idea. Intervention in divine nature (with modern technology, for instance) is a sin that will not be left unpunished and will bring about inevitable and enormous calamities. This view betrays the influence of eastern thought.

Both views are opposed to modern technology as it has developed today. They also express a romanticized view of nature. Man is not considered the master and controller of nature, but an inseparable part of nature (although he has been inserted into it). If these views should ever be worked out in society, the consequences for mankind would at least match those of an unchecked technological development.

For without the possibilities offered by modern technology, life would become impossible for many. The weak could become extremely vulnerable since they would be unequipped to deal with a hostile and unyielding nature. No meaningful solutions to our problems can be found by taking a position that simultaneously favours nature and opposes culture. Such an approach lacks sensitivity to nature's two faces – the benign and the hostile – and to the unique position of man; nor does it recognize the meaning of history.

In the face of our idolatry and misjudgement of technology, it is important to recall man's original mandate concerning technology. However we cannot fully appreciate this mandate unless we first recognize and confess the true Authority, who gives life to man and who crowns him with honour and dominion. Thus man is accountable to God for responsible stewardship that stresses service instead of autonomous control. His technological endeavours are subject to norms that are supra-human and, therefore, not subject to whim or caprice. God calls upon man to cooperate in the progress of history and in the disclosure and development of creation.

The Gospel – the good news – offers a timely perspective here. Christians especially need to begin with self-criticism, for too often they have simply adapted themselves to the dominant materialistic tendency within Western culture, at times even defending their weakness with texts from the Bible, thus compromising themselves in a frightful way. We know from the Bible that through the work of Jesus Christ, God has reconciled himself with his creation, unsettled as it was by the effects of human sin. This reconciliation implies that man must resist a self-centred and avaricious exploitation of nature. The abuse of nature cannot be in harmony with responsible stewardship in God's creation. Any human claim to the ownership of nature must be denied, for the materials that constitute life can never be anyone's property. This is amply demonstrated by the functioning of ecosystems. "Building materials" have been used by a chain of succeeding generations of living beings throughout the centuries. This should also be possible in the future, unless this generation of mankind is to disrupt this chain by idolizing a technology based on an absolutized scientific method.

To avoid disrupting the chain, man needs to recognize that scientific knowledge will always be of limited scope, since it is abstracted from the fullness of reality. Reality cannot be fully understood through scientific knowledge alone. By refusing to admit the limitations to scientific knowledge, we have dislocated the complex coherence within reality. Multi-disciplinary cooperation may bring about a partial improvement in our efforts – partial, because we will not gain a comprehensive knowledge of the fullness of reality simply by adding the abstract and limited knowledge of various aspects. We will be able to restrain the devastating powers unleashed by modern technology and to restore its constructive and benevolent properties only after we acknowledge both man's individual and communal responsibility to his creator and, therefore, for his fellowman, for his fellow-creatures, and for nature. Accepting this responsibility clears the way for recognizing the limits of the human capacity to command our complicated, varied, and dynamic reality. Such responsibility should also lead us gratefully to employ the various kinds of abstract knowledge. Science cannot be expected to integrate these forms of knowledge, since such integration is part of the responsibility of man himself.

This is by no means an easy task. Anyone aware of the difficulties it raises will proceed cautiously and with restraint, especially where we are concerned with altering reality, as in technology. But this insight will lead us to handle nature prudently and carefully. It will also help us maintain the delicate balance between the natural environment and living organism. On that basis every suggestion that leads to solving the problem of environmental pollution should be regarded positively. The discussions of a production process that involves recycling need to be changed into action as soon as possible. Moreover, we need to take seriously the calls for austerity, for new forms of asceticism, for decentralization of industry, for an end to unbridled consumption and the technological compulsion to produce and accomplish. We also need to pay attention to those who suggest that we change societal structures, work for more efficiency in political decision-making, break up elite groups, expand the effort to democratize decision-making processes, reinforce community consciousness, and place restrictions and normative guidance on our dynamic technological development. All of these are positive suggestions that deserve our full support.

If, however, the inspiration for these proposals is again the subjective desires of autonomous man, they will have only limited and short-term effects. For subjective religious attitudes as a source of inspiration will always be one-dimensional.

Let us grant for a moment that the pollution of the environment can be almost entirely eliminated technologically, assuming the cooperation of economic forces. If the basic changes are limited to using technology merely to deal with symptoms, then technology, bureaucracy, and existing organizations will probably continue to grow in power at the expense of individual freedom and communal responsibility. Man will then become the prisoner of his own technologically streamlined society, which will become a universal prison-house, in which historically, socially, and ethically meaningful life is constricted and finally choked. Life will then hold little more purpose than it does for a fly caught in a bottle; mankind will be doomed to die. Our quest for solutions will only produce more problems as long as we pursue them within a strictly horizontal framework and take no account of man's essential vertical orientation to God, who has indeed issued clear and normative guidelines for a development of technology that harmonizes with nature.

That means, concretely, that technological development should also unfold in a social sense and that everyone involved in such development bears a measure of social responsibility. Further, we should not close off the economic side of technological development, but we must stem any development that turns profit-making or the compulsion to consume into absolutes. We need to prevent squandering of resources and goods and to promote a frugal management of nature, even if such a move should be accompanied by a lowered economic growth and by restricted consumption.

In addition, if we halt the accumulation of garbage and industrial wastes, we will be demonstrating our sensitivity to the aesthetic disclosure of technology. Technology should not be permitted to spoil nature but must be developed in harmony with it. A clean and protected environment is also a judicial concern. Wherever nature is abused, the courts should use their power to intervene and dole out suitable punishments to guard against further pollution and mutilation of nature. In so doing they will ward off the threat that a ruined nature poses.

Nature is dislocated wherever man's ethos assumes an erroneous content because those involved in the development of technology arrive at their point of reference by making something within created reality an absolute. In that situation, they do not consider their actions as part of their service to God who rules over creation in Jesus Christ. They exchange the perspective of the kingdom of God for shortsighted and relatively short-term success because, being unreceptive to the Transcendent One, they also lack vision. It is precisely such receptivity and vision that make a meaningful perspective for our culture possible.

The most fundamental basis for a different attitude towards nature must come from the recognition that we are living in God's creation, a coherent whole in which every part has its peculiar value and place. With that recognition, the "vertical" relationship between God and man will be restored. Only then will we be able to establish the kind of perspective that will allow us meaningfully to cultivate the "horizontal" relationship. We will be able to participate communally in the normative meaning of everything, including the meaning of science and technology, rather than communally to pervert life until man himself is driven to curse. We will be able once again to enjoy nature and to see that all creatures great and small are the letters that together form one great book telling us about the power of God the Creator, of Jesus Christ the Redeemer, and of the Holy Spirit the Restorer.

Reflections on the Technological-Scientific Culture

It is well-known that the study programmes designed for prospective engineers leave little or no room for reflecting on technology and its possible consequences for culture. The emphasis in technological education is on the continuing development of modern technology; only marginal room is left for reflecting on that development. Even then, it frequently amounts to no more than an option for engineering students.

As a result, engineers emerge from their training naively engrossed with the idea of permanent progress brought about by technology. Thus they think that their efforts will contribute to a future in which technology will eliminate many maladies and in which material welfare will continue to grow. Whenever engineers are confronted with problems in shaping such a future, they usually lack the time to reflect on their causes and automatically summon the help of science and technology to solve them. In his daily work the engineer is often so preoccupied with the impressive results of his work that he has little interest in dealing with philosophical questions concerning technology, even if he had the time.

Furthermore, the engineer's role in society is increasingly limited to promoting the development of technology from his position on a team of colleagues. Being a specialist himself, he grows more and more dependent on other specialists because the making of a technological design has become an immensely complex project. This trend is reinforced by advances in the methods and tools of technology, especially the computer. As a result, the individual engineer does not have the insight or ability to help guide the development of technology along a particular course. Given the meaning of modern technology for culture, the responsibility of the engineer has greatly increased; yet his training and his work situation make it difficult to him to carry it out. Unfortunately, the engineer often does not see this himself.

During the last few years, however, there have been a number of substantial changes in society and in engineering circles. For a long time, technology was thought to be the only means of moving towards a good future. But today people who live in the midst of the technological-scientific culture are troubled by persistent doubts. Many engineers refuse to ignore the problems of this technological age any longer. Among them there is a keenly sensed and growing pessimism concerning not only technological advancement but also the heart of technological development itself.

An atmosphere of crisis prevails in engineering circles. Although engineers realize that technological development has meant tremendous material enrichment to many, especially in the west, they also recognize that technology has developed into a power dominating the human situation. Technology has so thoroughly unsettled our society that we can no longer evade an important question: is this development of technology moving in a right direction? Engineers are now asking questions which were not posed until recently. Reflection on those questions should be part of the prescribed programme of the engineering student.

I would like to contribute my share, taking reformational philosophical thought as my point of departure and focusing especially on the influence of science on technological development. Although this influence is extensive, it is not often examined critically. Instead, people assume too often that incessant specialization and a high intellectual level are the prerequisites for coping with impending problems. I would like to challenge this view and to examine critically the prevailing view of the relationship between science and technology, its deepest foundation, and the motives at work behind it.

The Problems of the Technological-Scientific Culture

The central significance of science is demonstrated from the outset by the various problems surrounding the technological-scientific culture. By way of explanation, let me briefly summarize the most serious problems of that culture.

The first problem area concerns the position of man in modern technology compared with his position in technology of earlier days. In the technology of the trades, man's productive activity was determined by his eyes, his hands, his feeling for materials, his fantasy, and his ability to shape. His physical and intellectual capacities were involved directly. Quality, particularity, and uniqueness characterized his work and the results of that work. This situation is entirely different in modern technology, where quantity, rate increase, and volume characterize man's work. Work has become quantified labour which must continually grow more intense.

At the same time, working man has become part of a process which is objective, impersonal, and insensitive, a process in which cold business gains the upper hand. The result is hardened and indifferent workers. Their attitude is quite understandable, since man feels himself locked within an overwhelming, gigantic mechanical process that drains his resources. Even man himself is set aside when complete automation is introduced, and he remains at the sidelines as an unemployed spectator.

An existential fear often takes hold of him in this situation, a fear that understanding cannot relieve. In fact, man grows more afraid the moment he realizes what is really going on: the labour process, in which he is but a small cogwheel, requires unchecked and gluttonous drains on the sources of energy and on the reserves of natural resources. The drain goes on as if they could never be depleted, whereas these sources and reserves are, in fact, finite. Therefore as technology grows to even greater proportions, it becomes clear that it must also come to an end; and the more the dynamics of the process are stepped up, the sooner that end will come. This will inevitably lead to enormous disasters in a culture that has become highly dependent on technological development. Even at an early stage in technological development, Romano Guardini saw that unlimited technological development harbours chaos as the final outcome. The dictates of technological perfection disguise an enormous abyss. A catastrophe of infinite proportion and swiftness threatens to put an end to finite physical reality.

Moreover new frontiers of scientific technological advancement are already inspiring fear. The development of nuclear energy, for instance, has been sold to the public with promises of a guaranteed energy supply to meet increased demand. Of course, we will also have to endure the many unsolved technological and political risks, whose magnitude is difficult to estimate. The question of whether or not man can effectively control the elementary forces released by matter continues to plague us.

The case of computer technology is quite similar. A sober analysis indicates that the computer works fast and accurately and that its results will never go beyond the programmed instructions. Yet people's fear of growing more dependent on the computer remains real because the computer operates independently of man himself, because its results contain *a* limited element of surprise, and because the user is not necessarily the programmer. Moreover since the user changes again and again, he cannot know by what set of criteria the computer works; he is forced to surrender himself in trust to the dictates of the computer. This problem will be aggravated when the self-adapting and self-reproductive machines, predicted by computer specialists, are introduced in the future. A computerocracy is imminent when these machines are expected to find their place not only in production processes but also in economics, politics, and government. Manipulation of data, especially as derived from data banks, and manipulation of people could well assume dangerous proportions.

The threats attending modern technology are particularly striking when we observe the latest military techniques. The most deplorable possibility is nuclear suicide. Many dangers are also associated with bio-technology or eugenics because it is based on biochemistry and biophysics and therefore possesses the built-in possibility of manipulating genetic materials.

Another important problem area in our technological-scientific culture concerns the relationship between man and nature. The small-scale technology of earlier days was integrated with nature, but today a breach exists between man and nature. Aided by modern technology, man continually interferes with nature on a large scale, until the given coherence of nature is disjointed or even broken up entirely. In addition, man interferes so frequently that nature is not given a chance to repair the damage. Add to this the combined adverse side-effects of modern industrial technology, of traffic, and of industrial agriculture and we may conclude that the danger of environmental collapse is even more real. We draw from nature more than it is able to produce, and we discard more waste products than it is able to break down, not to mention those that cannot be broken down. The possibility of completely destroying the environment is becoming very real indeed.

In summary, the threat modern technology poses can be looked at from two angles. From one angle we see the gigantic, massive structure of technology and the frontiers it advances, while from the other we see the technological disruption of the coherence in nature and the alienation of man and nature.

All told, it is understandable that many people look on the explosive and ambitious development of technology as an autonomous development, particularly when they take into account its socioeconomic context. The research needed for the development of technology occurs in mammoth organizations and within the so-called military-industrial complex, often in secret. Should the state ever take over this task, we would experience technological development as an inescapable and unattractive fate. Our specialized and massive technology no longer meets actual needs and desires, which have given way to goals systematically fixed by a system-technology founded on growth-ideals. The unscrupulous use of technocracy and the dehumanization it brings continue to grow, both in proportion and in complexity.

When we look at the influence science exerts on technology, we can better understand the problem that has arisen in man's relationship to technology and to nature and its socioeconomic coherence. Because modern technology and the technological-scientific method of designing rest on a scientific basis, the characteristics of natural science project themselves into the production process and the results of technological activities. These characteristics emerge wherever technological products are introduced, leaving a decisive mark on various cultural sectors. As some cultural sectors dominate others, the characteristics of science also become cultural characteristics.[17]

The characteristics of natural scientific knowledge are these: it is *universally valid, abstract,* and *remote.* That is, it is a reduced and limited knowledge of reality because it is a knowledge of certain aspects of reality only, abstracted from the fullness of reality. Further it is *enduring* and *perpetual,* and it shows a *logical coherence.* These features of scientific knowledge are frequently mistaken for full and complete knowledge of reality. However integrated reality is marked by the uniqueness, the coherence, and the interchangeability of everything. Therefore a tension arises between scientific knowledge and reality as a whole.[18]

---

[17] Cf. Schuurman, *Technology and the Future,* sections 1.4.2, 2.4.4, 2.4.5, and 2.4.9. Cf. also Chapter 1, section 2 of this book.

[18] Cf. H. van Riessen, *Wijsbegeerte* (Kampen: Kok, 1970), section 4.7.2.

The features of scientific knowledge – universality, abstractness, and perpetuity – will also become technological features because science and technology are so closely associated, even joined. Since scientific knowledge is based on rational, logical abstraction and exhibits a purely logical coherence, the products derived from it will also feature this logical coherence. As a result, the absolutization of scientific, logical knowledge by technology is attended by an unprecedented determinism. The mathematical and mathematical-physical laws for reality are then held to be the true and full picture of reality.

Developing a technology based on such absolutization will inevitably lead to enormous control over people. There will be little room for man to use technology creatively to shape his historical situation. Technological activity becomes regulated and constricted; hence man is robbed of his inventiveness, of his responsibility, and even of his freedom. The upshot of this is the rise of a monotonously standardized technological mass-culture, in which people are equalized and levelled out to be average. That is to say, their work situation and their consumption patterns can be interchanged, statistically managed, and manipulated.

Modern technology has become part of all world cultures, moulding these cultures into its own uniform and monotonous mass pattern. The characteristics of technology are the same everywhere. Building construction, for example, is identical everywhere, whether this be in the Netherlands, the United States, Japan, or a developing country. Meanwhile, this process has dislocated and even eradicated many ancient cultures.

Moreover, the technological-scientific culture itself is in danger of being internally torn and segmented. However surprising this may seem, it is the inevitable result of scientists in each specialized area projecting entirely different views onto reality. The problem is the result of indiscriminately applying scientific knowledge.

Consequently, no genuine cultural integration takes place, unless we mean technological-scientific integration, which implies a tremendous reduction of cultural life.

We can already see the impact of such reduction as work grows monotonous, as nearly everything becomes standardized, as human life is broken into distinct and even isolated units, such as home, work, and recreation, and, finally, as we disrupt the coherence in nature and break the intimate bond between man and nature. We can be sure that in the future technology will expand still further, for it displays a tremendous dynamic. But this expanding grip on life will stifle and restrict life. Ultimately, it will eliminate the rich variety, complexity, and stability inherent in human life.

Already our culture has become a very unstable mono-culture with tyrannical tendencies because of the bias and the dynamics of the technological-scientific culture. An obvious indicator is the emergency created by a sudden "energy crisis" and the socioeconomic difficulties it caused. This danger is still with us. But the totalitarian character of the technological dynamic makes it very difficult to alter this established pattern, in part, because people do not agree on how the culture should be changed and also because the force propelling today's culture is so strong. While this force has brought material relief and progress for humankind, we do not seem able to harness it without risking immense catastrophes.

I can only conclude that the problems of the technological-scientific culture can no longer be described as incidental or acute; rather, they are structural and chronic. They are an expression of an emergency situation that is growing.

Initially, man seemed to use technology to subdue natural forces and to deal with emergencies – however resistant and unpredictable; still, he found this task difficult. With natural forces subdued, man is now faced with the problems accompanying cultural forces. The road to mastery over natural forces, for all its hopeful beginning, seems to end in the morass of a cultural crisis, one that is serious enough to threaten both man and nature.

A Look at Some Philosophical Views of Technology[19]

The basic problem in the technological-scientific culture rests in the view that correlates science and technology. However, this problem is not often stressed in today's philosophical assessments of culture. Surprisingly, it is the pragmatists and the positivists – the philosophers who communicate readily with engineers – who typically ignore this problem. They take their bearings from the existing coalition between science and technology, evaluating its

[19] Cf. Schuurman, op. cit., chs. 2, 3. Discussed there are pragmatists and positivists Norbert Wiener and Karl Steinbuch, the Marxist Georg Klaus, the transcendentalists Friedrich Georg Jünger, Martin Heidegger, Jacques Ellul, Hermann J. Meyer, and the neo-Marxists Habermas and Marcuse. Cf. also ch. 1, section 3 of this book.

development positively and interpreting it as a confirmation of human strength and ability, a great stride forward on the road to general welfare and prosperity. Whenever obstructions occur on this road, these philosophers frequently issue an uncritical appeal to science for a solution. When the problems assume the proportions we know today, they do not hesitate to advocate a total domination of culture, aided by modern systems theory and cybernetics.

These philosophers are confident about the future, assuming that they will enjoy the continued support of the current development of science and technology. Infatuated with the dream of complete mastery of the future, they resist religion (Christianity in particular) and those philosophers who demonstrate their dependence on and guidance from the *Hinterwelt* (meta-world, transcendent reality); they must resist them because they would prevent them from realizing their ideas. The closed worldview of the technological-scientific culture is a *conditio sine qua non* for pragmatists and positivists.

The orthodox Marxist also accepts the importance of the development of modern, scientific technology, stressing man's technological achievements and contending that man can reach his ultimate goal of total freedom by developing the potentialities of technology. Unlike positivists and pragmatists, the Marxist is aware of the many problems and dangers involved in this development, including the possibility of alienation and loss of freedom. But he remains confident and argues that continued socioeconomic development based on technological expansion, together with the inevitable revolution, will ultimately eradicate the growing shadows of alienation and restricted freedom. They will open up the kingdom of total freedom at the dawn of a new day, a kingdom in which humankind will reign collectively as lord and master over the work of his own hands.

Positivists, pragmatists, and orthodox Marxists all have great respect for science as a means to control. However their assessment of its effect varies because of underlying differences in their views of society.

The Marxists do not begin with either man's freedom or the production by free enterprise within which technology is to develop. Instead, they emphasize that technology can benefit the revolution and the liberation of humankind only if it is guided centrally and if goods produced are judged by their value for consumption instead of exchange. In other words, they propose a centralist technocracy.

In passing, it should be noted that, in view of the growing gravity of the problems, positivist and pragmatist philosophers have had to reassess their views of increased state intervention. Thus many have correctly concluded that Marxist society and non-Marxist society are beginning to resemble each other.

The existentialists and the personalists, who are keenly aware of the problems facing the technological-scientific culture, take a transcendentalist approach in their analysis. They emphasize the transcendental influence on all experience, interpreting the ongoing development of the technological culture as a menace to the human subject, especially to his personal uniqueness, to his freedom, and to his individuality. They oppose science and technology as forces that are autonomous and anonymous. To be sure, they oppose not the big problems of the technocratic society but science itself, arguing that the scientific method leads to oppression and the suppression of human freedoms.

The existentialists and the personalists make quite an impressive protest, but, in view of their presupposition that science is an autonomous original force, we have to ask whether they are capable of pointing out a meaningful perspective for the future. They submit themselves to the present development by sheer necessity, look to the past nostalgically, or attempt to escape from the culture in a transcending retreat towards freedom; thus they rise above the problems looming over science and technology. But even this freedom is threatened continually. While they reject the urge to dominate, which is so common to technological-scientific thought, they rely on their own thought to help them rise above that urge and to enter a space of spiritualized freedom. However, they must often repeat their spiritualized flight as their concern for external aspects continually grows.

In their evaluation of man's position in society, the neo-Marxist revolutionaries agree with the transcendentalists. Although they do not consider science and technology to be autonomous powers, they fail to submit science and its relationship to technology to a basic and critical examination. Their criticism amounts to attacking economists and politicians, charging them with elitist tendencies and a willingness to enlist the service of science and technology to their own advantage. Inspired by the vision of a future utopia in which everyone will be free and happy, they want to transform present society until it conforms to their model. If their revolutionary ideas are realized, they will have sweeping consequences for the development of science and technology, including their place in economics and politics. For the neo-Marxists would give priority not to the solution of practical *problems,* but to the realization of practical *purposes,* a process that may not involve restraint (v., Habermas).

The neo-Marxists offer revolutionary resistance to the ideology of technocracy so as to give human authenticity a chance in everyday practice. In their view, man is not primarily the creature who works and produces, but the *homo ludens* who plays, relaxes, and has a good time. Thus they hope to fulfil human existence by meeting man's need to enjoy life and to express his passions fully, goals that were previously nearly impossible.

The Countercultural Critique

No matter how important and penetrating the critique of the transcendentalists and the neo-Marxists may be for problems of the technological culture, in this study I would like to devote special attention to those people who have spoken for the so-called counterculture, particularly Theodore Roszak. The ideas advanced by these people and their followers have served as a marginal undercurrent within Western culture since the days of Romanticism. Recently these ideas broke through to the surface and occupied the minds of many, and not just the young.

The representatives of the counterculture formed a front with the transcendentalists and the neo-Marxists in their serious objections to the position of science and technology in modern culture. With the transcendentalists they pleaded for a return to authentic thought and for its renewal. However, when they advocated ways of applying this renewed reflection in the culture, their proposals did not correspond with those of the revolutionaries.

For one thing, the revolutionaries and the transcendentalists differed in their concept of revolution. The neo-Marxists proposed a revolution of society, a transformation of societal structures, and therefore a change in the function of science and technology. The spokesmen for the counterculture, however, advocated a *spiritual* revolution, an internal revolution of the mind. This amounted to an attack on science itself, for they critically examined the structure of science and the peculiar course science and technology have taken in Western culture. Roszak stated that the rise of the counterculture must be seen in the light of the meaninglessness and utter despair produced by the technological culture and the mad pursuit of the excessively specialized sciences, which tended to dissolve the integrated character of human knowledge. Countercultural critics called for communal awareness, authenticity, and meaning in the midst of prevailing social upheaval, alienation, and meaninglessness.

If the human mind is to be cured and healed, critics said, it first needs to be aware of the cause of its illness. Roszak thought that this illness originated with the rise of modern science, particularly stimulated by the thought of René Descartes and Francis Bacon. From its start in the seventeenth century, science has developed into the final and highest court of appeal. The technological-scientific culture is based on such science, used as an instrument to gain control. Therefore this culture now looms large and looks like a totalitarian, artificial, and abstract culture having a linear, horizontal dynamic. Roszak emphasized that the groundwork for this development had been worked out by Judaism and Christianity, since they used a linear view of history, which was advanced especially by later Calvinism. With the rise and development of science, the knowledge of the *natural* sciences has been accorded a superior status. All other forms of knowledge are considered inferior to the lucidity of objective scientific knowledge.

As I pointed out earlier, scientific knowledge is limited in its scope; it is the result of the human mind observing strict and objective limits and leads to a partial, reduced view of reality. It is precisely this constriction of the mind that provides scientific knowledge with its apparent strength, for within this uncomplex, uniform, and single view, all of reality is reduced to the categories of mathematical and physical laws. Thus some can pretend that such reduction clears the way to a true and adequate picture of reality that can be manipulated at will. Roszak called this the *religion of science.* This religious veneration of science helps explain the scientist's willingness to sacrifice his energy and his triumphant claims that he has grasped reality through the eye of reduced knowledge. Thus reductionism is on the increase, and more is lost than gained in the name of progress. The end is a shrunken technological-scientific view of the world, that can only lead to a meaningless and nihilistic culture.

The countercultural message was addressed to this culture and advocated a salutary return to the mystical and visionary sources of authentic culture. The apologists of the counterculture wanted to keep the objective, constricted, and alienated mind at a distance and return to a dedicated, visionary consciousness based on feeling rather than on reason. This, they suggested, is the first step towards liberation from the total alienation caused by the technological-scientific culture.

They hailed other forms of knowledge, such as those provided by imagination, intuition, wisdom, mystery, inspiration, ecstasy, contemplation, meditation, myth, gnosis, passion, the unspeakable, the mysterious, and the holy. Their aim was not to know as much as possible, but to know as profoundly as possible, to move away from continual abstraction towards more deeply meaningful, "transcendental" knowledge.

They suggested that people appreciate the non-intellectual faculties of the human personality, that are nourished by visionary sparks and by wholesome human experience. Only by returning from expertise to wisdom, they said, could men and women participate in the good, the true, and the beautiful. These newly enlightened thinkers called for a mind that participates in the divine; they sought the natural mind, perfect in itself, as opposed to veneration of science and technology. They preferred child-like receptivity and simplicity to the mania for self-esteem common to men of intellect and culture. They pled for union of heart and feeling as opposed to the gap between heart and intellect that troubled technological-scientific man. Modern man, they said, not only controls his technological culture but is also controlled by it at the expense of his multidimensionality, of his wisdom, and of his independence.

The values and norms of the counterculture have formed a striking contrast to those of Western culture since the scientific revolution of the seventeenth century. In a sense, countercultural apologists advocated a return to a far earlier time. In effect, they replaced a linear view of history with a circular or cyclical view, never mentioning cultural continuity and progress. Of course, they did not reject science and technology altogether.

Science and technology had their place insofar as they were required for the survival of mankind, but the countercultural apologists insisted they must be cut back to size.

The distinguishing feature of the counterculture, therefore, is that it offered a limited, human scale, that was varied even in its societal forms, rather than monotonously identical at every level. It stressed the organic above the mechanical, simplicity and frugality above abundance, and meaningful and delightful labour above production-oriented labour.

Thus countercultural advocates called attention to many important issues often neglected in the past. Much of their analysis of the technological-scientific culture was correct, especially their critical observations concerning the structure of science and their emphasis on the religious backgrounds of the issues they raised.

However when the apologists of the counterculture reacted against the adulation of science and technology, they reduced them to a mere requirement for the survival of mankind. Since they did not recognize a cultural mandate, they have been unable to shape an alternative direction for present culture or even to deflect its current direction. In effect, the ideas they introduced into public discussions could only sponge on a culture that is internally torn and segmented.

The philosophy of counterculture took an extreme position on the continuum between rationalistic technological-scientific culture and the irrationalistic reactions it evokes. Since the counterculture offered no alternative for cultural development, it could offer little or no actual resistance against increasing cultural dislocation. It may well have had a completely opposite effect because no real resistance could be provided. If the spirit of the counterculture were to dominate the heart of the technological-scientific culture, the latter would be undermined from within, for it would lack the courage to accept cultural freedom and responsibility. Our culture would then be confronted with imminent collapse, which would have far more serious consequences than the hazards it has been shown to harbour today.

The Motives Behind Technological Development[20]

Present cultural development must be slowed down, modified, and redirected. But how is this to be done? To answer this question, we need a clear view of the motives underlying and driving this development. We also need to understand the fundamental religious background of these motives and of the cultural development they spawn. I agree with Roszak that the major motive in Western culture today is man's will to master and control, combined with the idea of technology as applied science. After all, the dictatorial character of science can be seen most clearly in its application, that is, in modern technology.

As man attributes absolute power to science and to the technological-scientific method, technological development appears to reflect scientific knowledge.

---

[20] Schuurman, op. cit., section 4.8.

This leads to an extremely scientistic technology, particularly on the design level. As a result, human creativity recedes or even disappears altogether. Thus all invention is stifled, and the possibilities for redirecting our technological efforts are reduced. As I said earlier, scientific knowledge is enduring and perpetual because it is knowledge of a fixed and limited subject matter. In turn, this continuity is projected into technology, which becomes fixed and limited also.

Thus domination by rationalism has made the development of technology not only dynamic and immense but constricted as well. Innovative work receives only a very small place because people influenced by rationalism singled out the prevailing technological development as the only proper one. Thus the development of technology becomes stifled and rigid in its rational determinism and relentless logic. However, our obsession with technological accomplishments blinds most people to the narrowness of our current technological development.

However the human will to control by technology is not the only motive operating in Western culture. In fact, it is intertwined with various other motives, some of which are corrective, but most of which buttress the dominant trend. There is, for instance, the motive of *technology for the sake of technology.* Let us call it the imperative of technological perfection, demonstrated most notably by the exclusiveness of the ideas proposed by engineers. Whatever can be made and perfected must be made and perfected. This leads to unchecked and meaningless technological power, which engineers claim they master and control but which in fact victimizes them. Although the motive of technology for its own sake often anticipates progress, it actually does just the opposite: technology dominates man with absolute power. Even nature and culture cannot escape the menace posed by this technological power.

Frequently, rationalism is also allied with the motive of technology as the servant of economic powers. These powers dominate the development of technology, often turning profit-making into an absolute good. Admittedly, this makes it possible to interrupt technology's current development, but it also means that technology is developed without the norms which ought to be applied in the first place, such as responsible management of the environment. Under the influence of this economism, technology ceases to answer to its essential meaning. It no longer serves actual needs but creates artificial needs and superfluous products. The result is an authoritarian technological development that leaves behind a trail of waste, pollution, and destruction.

These motives are especially influential on those who are actively engaged in the development and social direction of technology. Those active elsewhere may be influenced by different motives: for example, the belief that technology is a neutral and autonomous power. Yet they, too, frequently close their eyes to the dangers of technological development. Blinded by an insatiable desire for welfare, they willingly adapt to the prevailing development in order to derive private benefit from it.

Thus our search has taken us to the fundamental basis for these various motives. Could it be that the large-scale problems and threats of a large-scale technology arose because these motives are based on large-scale human pretensions?

The Spiritual or Religious Background of the Technological-Scientific Culture

Roszak was correct when he went back to thinkers like Descartes and Bacon for his evaluation of the technological culture. For it was Descartes who pleaded for the autonomous and self-sufficient position of man: man as subject was the measure and centre of all that is. And Francis Bacon was a stimulus particularly to the practical consequences of this view, for he taught that man is capable of realizing his autonomous position with the aid of technology and science. Technology can be used to control nature and to create the kind of culture that surpasses all restrictions of time and space and that obeys the hand of man. In the modern age, man is caught in the clutches of the infinite, committing himself to the limitless possibilities of science and technology, particularly after the age of Enlightenment.

Earlier I stated that positivists and Marxists consider Christianity an obstacle to the growth and progress of the technological-scientific culture. Roszak, by contrast, says that this culture originated in the Judeo-Christian tradition and was particularly advanced by later Calvinism. Roszak attributes the present cultural problems to Christianity because the history of Christianity produced thinkers like Descartes and Bacon and also the pretensions of the Enlightenment.

Such contradictory views indicate a problem that deserves our attention. In my opinion, both views are wrong. Undeniably, the Judeo-Christian and especially the reformational view of man's unique position regarding nature was a decisive factor in the development of the exact natural sciences. But more has to be said about the manner, condition, extent, or purpose of this unique position of man. For example, did the authority man received to exercise dominion in created reality imply unrestricted sovereignty? People draw this conclusion too readily when the biblical mandate found in the beginning of the book of Genesis is taken out of context. It is a mandate whose hallmark is service. The Bible clearly points out that man must resist the human temptation to misuse his given mandate to manage and to have dominion in creation.[21]

It is the philosophy rooted in the thought of Descartes that has focused its attention from the outset on the exact natural sciences. The idea of human autonomy posited in Descartes's philosophy stimulated the development of the natural sciences; this development, in turn, appeared to confirm the idea of autonomy. Since the Enlightenment, this idea has also been made effective by the application of scientific knowledge and by the projection of scientific characteristics into culture. Guardini has shown that when basic religious meaning is replaced by pretended autonomy, the resulting void will be filled by violent, practical, manipulative science. The biblical mandate was continually perverted during the course of the historical development of Western society and finally amounted to technocratic exploitation. A passion for perfection and completion has caused Western culture to secularize the biblical mandate: man takes his destiny into his own hands, while Christian eschatology is reduced to a dogma of horizontal progress in history.

When Christians, at least those Christians who hold a dualistic view of life, associate themselves with this dominant persuasion, Roszak is correct in accusing them of complicity in the realization of our science-infused culture. However, positivists and Marxists are often correct as well in charging Christians with opposing science and technology. They reach this conclusion because they frequently observe Christians who oppose the development of science and technology as a result of their failure to integrate a responsible attitude towards it with their dependence on a transcendent reality. Such Christians neglect the fact that the Christian faith embraces values pertaining to created reality.

Indeed, faith thus understood rejects the idea of human autonomy, but not responsibility for science and technology. With Roszak we reject man's excessive

---

[21] Cf. Genesis 11 about God's judgement on the tower of Babel. Cf. also Matthew 4:8–11, a description of the resistance offered by Jesus Christ to the temptation by Satan to take possession of the whole earth, a temptation that is met with radical rejection.

pretensions for science and technology, but we cannot therefore support the proposal of the counterculture, since they do not give up the idea of autonomy but merely gives it a different content. Roszak does not see man as autonomous regarding nature. Rather, man and nature belong together in an indissoluble bond with its own point of reference. Hence Roszak's pleas for transcendence should be understood as a restoration of that bond. Despite appearances, the transcendence Roszak advocates is immanent; it remains within this world and its multifaceted reality of man and nature. According to the spokesmen of the counterculture, the limitless mystique of transcendental experience establishes a harmony between man and nature and a harmony of man with himself. In that experience, escape from an alienated and threatened existence in the technological-scientific culture is thought possible.

An Alternative

Roszak rejects the prevailing trend of our culture. So do I. But unlike him, I would like to defend an alternative approach in which a transcendental orientation includes responsibility for technological-scientific development.

For a better understanding of this approach, we must take a brief look at the spiritual history of the Western world. The most fundamental basis inspiring man's scientific and technological activities has been the idea of autonomy. Within the tradition of Western philosophical thought, philosophy has been shaped and developed to support and confirm the pretended autonomy of human thought. Philosophy was thus accorded the function of a religion or a pseudo-revelation. Subsequently, philosophy, in the line of rationalism, which generally orientates itself to the exact natural sciences and modern technology, was reduced in scope until it was confined within the natural sciences and their methodology. In this context, we can readily grasp C.F. von Wiezsackers's remark that scientism as a faith in science has increasingly assumed the role of religion in Western culture.

Thus philosophy is used as a stepping-stone to combine the idea of autonomy with science and its methodology. People then begin to find assurance and confidence in a tacit, religious devotion to the scientific method, used in gaining mastery of practical affairs, particularly the affairs of modern technology. Philosophy is even equated with the most characteristic feature of science, namely, abstraction. Accordingly, the scope of philosophy is increasingly reduced and constricted in its continuous adjustment to the nature of scientific knowledge.

As man humbly submits himself to the dictates of this abstract scientific thought, he will very likely absolutize it. He forgets the reducing character of science and suggests instead that science produces complete and concrete knowledge of the whole of reality. Whenever people hold this view of abstraction and wherever in triumphant anticipation of progress, they apply scientific knowledge to project this abstraction into reality itself, the result will be a constricted reality, that is, a reality of what is technologically possible. In fact, it leads to a fragmented distortion of reality and a dislocation of culture.

It becomes clear, therefore, that an uncritical appeal to science as a panacea for the problems facing our culture today will indeed bring temporary relief only. In the long run, in view of the inherently limited structure of science, this reliance on science can only confront us with new and even more distressing problems. In fact, the solution it offers provides us with a more involved and more obscure complex of old and new problems.

View of Science

Rationalistic philosophers never concerned themselves with the problem that there is a still more fundamental way of knowing that precedes a conceptually qualified, scientific knowledge. It apparently did not occur to them that there might be a mutual relationship between these ways of knowing. Historically and structurally, the kind of knowledge used by the special sciences is restricted to definite boundaries and is based on a knowledge that is fundamental, uncompartmentalized, and concrete. This pre-scientific knowledge, which consists of both actual and factual knowledge of changes effected by human action, is in its turn guided and kept on course by an original and irreducible ultimate trust. This trust-knowledge constitutes man's point of reference for all his philosophical and scientific endeavours, regardless of whether he realizes it. But in autonomous theoretical reflection, inspired since the beginning of the modern age by a religious faith in philosophical and scientific thought, this pre-scientific trust-knowledge has been thoroughly perverted and falsified. Meanwhile scientific knowledge and its practical application have been accorded superior status.

Reformational philosophy, however, rejects the autonomy of man and therefore the autonomy of philosophical and special scientific thought as well. All reformational thought begins with the acknowledgement that nature, man, and culture are not anything in themselves. On the contrary, it recognizes that man, who, after all, did not create himself, needs a revelation in order to find out who he is, for what purpose he exists, and what is the meaning of the history that encompasses him and all things. Guided by divine revelation, man is in a position to become aware of the origin of everything, of the cause of his distress, and of the two-fold meaning of nature, of human life, of work, of culture, and of history. At the same time revelation enables man to learn about the way of redemption from his distress.

This leads to a knowledge whose content is open to understanding by faith; it is the most profound knowledge, which our thinking cannot fathom. It is a central, deep, and comprehensive knowledge that concerns the very heart of man and that, for man's own sake, should have decisive influence on scientific thought and on technological endeavours. It is a knowledge of acknowledgement, of confession, springing from the heart itself. This knowledge is dynamic rather than static; it is a gift as well as a mandate. It is unfolded continually as long as there is an obedient and receptive understanding of divine revelation. That is to say, when man renounces every pretension to autonomy, he finds opening up before him the very way to life, wisdom, and insight. The content of this insight is made explicit in faith-knowledge. It is neither philosophy nor science, for these are both characterized by abstraction. Faith-knowledge is radical and integral, for it concerns the central, fundamental choice of human life in the very heart of human existence. It shows man which direction he is to take in history, and it motivates him to follow it through.

What does all this mean for science? This question is critical in view of the theme of this essay. It certainly implies a view of science critical of the views widely held. It is also significant for the development of technology.

First of all, the nature of scientific theories needs to be recognized. The characteristic feature of these theories is not that they are objective or value-free, as is so often claimed even today. Rather, these theories have a basic religious foundation, although personal and social conditions may also play a role. For that reason, scientific theories should *not* function as if they were absolutely independent, nor should they be permitted to straight-jacket our knowledge and our action.

A theory starts out from a hypothesis. A hypothesis, which is an expression of human creativity containing man's faith-determined view of reality, aspires to the status of a scientific theory. This status can be granted only if the hypothesis has not been falsified either by observation or by experiment, particularly in the special sciences. Therefore, in view of faith-presuppositions and a hypothesis that is not necessarily the only one possible, it is clear that scientific theories are always *conditioned* and *partial.* They are also relative, for they relate to the knowledge of a particular aspect of reality, such as the physical aspect.

Scientific knowledge, therefore, is achieved by means of a method of analysis and abstraction based on a certain hypothesis. This means, however, that scientific knowledge is also abstract and *restricted.* That, of course, is not all, for as conditioned, relative, abstract, and restricted as scientific knowledge may be, it can also grow and even change by varying hypotheses, by a refinement of methods, and by increasing specialization. There is no end to this process of acquiring scientific knowledge.

Concerning the relationship between pre-scientific and scientific knowledge, we can say that the latter must continually be re-embedded in full, direct, concrete human knowledge and experience that has fundamental trust-knowledge as its core. Abstract, scientific knowledge must continually be integrated into and corrected by pre-scientific knowledge, so that the restrictions inherent in scientific knowledge are lifted. In this way, man's integral, pre-theoretical knowledge is gradually enriched and his measure of responsibility is increased.

Reality consists of many more aspects than the one abstracted by a particular science. This must always be kept in mind when the knowledge of one aspect is integrated into pre-scientific knowledge. From a normative standpoint, pre-scientific knowledge requires many-sided enrichment by the integration of numerous forms of scientific knowledge, but this is virtually impossible because of tremendous scientific specialization. It is also important to be open to a multi-disciplinary approach, provided that the sum of various forms of abstract knowledge is never equated with complete knowledge. To achieve this, we need a community of people who can work with one another on the basis of a common view of science. But in this day, with its diffuse religious convictions, this may be extremely difficult to realize.[22]

---

[22] Schuurman, op. cit., section 4.6.6.

I shall venture a conclusion at this point. Our fundamental trust-knowledge is the heart of our full and concrete human experience with its factual and applicable knowledge. This pre-scientific experience is enriched by the absorption of new scientific knowledge. It is not only our practical and factual knowledge that profits from this enrichment: our faith-knowledge benefits from it as well. For scientific knowledge is well-suited to the disclosure of more dimensions of the revelation that creation is meant to be. Accordingly, we may say that science brings out truth, but on condition that in all scientific endeavours the Truth is both the starting point and the constant point of return. Thus considered, scientific activity can be carried out with faith as its starting point and a strengthened faith as its result. Then our scientific endeavours will serve the cause of wisdom and of an increasingly comprehensive insight; it will buttress and enlarge our responsibility.

Philosophy of Technology[23]

The most important question that ought to occupy us at this time concerns the consequences of this reformational view of science for the development of technology and for a possible redirection of its present development. It will be clear by now that I would be the last person to renounce the scientific basis of modern technology. But I do oppose *the faith in autonomy,* which has combined with science and technology, particularly in Western culture. Under its influence, the conduct of science has become the high road to the entire field of knowledge and action, whereas science should be neither more nor less than a helpful supply route.

To gain a better understanding of the auxiliary function of science relative to technology, it is important to note both the motive that ought to drive technological development and its meaning. From God's Word revelation I take the proper motive to be the cultivation and preservation of creation. To limit it to preservation alone would imply a choice for nature and against culture, a renunciation of technology and a submission to natural fate. To limit the motive of technological development to cultivation alone would imply a presumptuousness that fails to consider what is and what is not essential or prudent. God's Word reveals our culture as one that bears the marks of its own extinction within

---

[23] Ibid., ch. 4.

it, threatening both man and nature; hence this Word highlights the cultural destitution we experience.

Within the harmonious calling to cultivate and to preserve, man, the image-bearer of God, is to serve in love. In cultivating and preserving creation he confirms his love towards his Creator and Redeemer and at the same time lovingly represents all of creation. That means, among other things, that man is responsible for unfolding the meaning of creation and that he must resist every attempt to disturb, disintegrate, and destroy this meaning.

Whenever man submits himself to the guidance of this meaningful motive, he will be in a position – *coram Deo* – to accept his task in technology willingly and responsibly. He will pay special attention to the meaning of technology and will attempt to deepen it. For the meaning of technology is rich and manifold; in fact, it is inexhaustible.

Although no one can supply the full meaning of technology, we can state it in part. Technology will be able to alleviate the fate forced on man "by nature." It will offer greater opportunities for living: reducing the physical burdens and strains inherent in labour, diminishing the drudgery of routine duties, averting natural catastrophes, conquering diseases, supplying homes and food, augmenting social security, expanding possibilities for communications, increasing information and responsibility, advancing material welfare in harmony with spiritual well-being, and helping unfold the abundance of individual qualities in people. Moreover, in science and in its own field, technology will develop new possibilities for promoting a variegated disclosure of culture. Technology will also make possible labour that is meaningful as well as productive; it will provide room for work that is marked by creativity, service, love, and care. It will also provide room for leisure and reflection.

This picture of technology, however, is not how it actually functions today. Inspired by wrong motives, modern technology has been made into a threat to nature, culture, and man; whereas the right motive would lead technology to contribute to the unfolding of nature and to the enrichment and deepening of culture and human life.

To seek both cultivation and preservation, with due respect paid to the meaning of technology, is to pour new and profound content into the extremely high moral purpose of scientists and engineers. For it means that they should no longer arbitrarily follow their own will. Instead, they should eagerly seek to be of service in the unfolding, deepening, and enrichment of meaning. They should not strive to do all things possible, but they should be able to do all things necessary. The purposes, values, and norms for technology should be made explicit in an ethics of technology developed on this basis.

I realize that these observations, too, represent a position contrary to prevailing practice. Very frequently engineers permit themselves to be lured into weighing the advantages. That is what they call ethics. Today such an ethic is impossible, for the scales continue to tip in favour of the disadvantages. It cannot serve to map out the direction technological development ought to take. The actual course and process of development has been accepted as the norm in terms of which a technological ethic weighs the pros and cons. Similarly, no new direction can be expected from an ethic of technology developed after the characteristics of science have first been projected into technology, an act that was inspired by man's determination to gain control and by his concept of technology as applied science.

Ethics based on continued abstraction reduces and restricts ethics itself. It testifies to an urge to continue in a direction that inevitably leads to a dislocation of the meaningful coherence of reality. Although we expect to restore this coherence later without at any point having chosen a new direction, our hope will inevitably remain unfulfilled because of the point of departure. Thus all we can do is fight the symptoms. The cause of the problem will not be eradicated as long as its root is left intact. It is not science and technology that should set up and determine the ethics, for the latter will not be left untouched by the current problems surrounding science and technology. Rather, ethics should *precede* science and technology, so that it may decisively influence their development. The direction technological development will take ought to be determined by our response to ethical issues rather than to technological ones.

A philosophy of technology also requires that an analysis of the relationship between science and technology be given a central position. Such an analysis should examine the basic sciences technology may draw on, the meaning of science for technology, and the precise nature of the technological-scientific method. Ultimately, this philosophical analysis of modern technology must locate the decisive points of contact present in the making of technical designs. That is to say, the engineer's creaturely originality and inventiveness, enriched by a full scientific knowledge, should be assessed as to their place in the designing process.

This done, we need to point out meaningful directions that do justice to the proper motive and meaning of technology. We must also warn against the dangers of wrong motives and wrong choices. As we foster insight into the meaning of technology, we should also seek insight into the dangers and possible nonsense of technology. In this way, the engineer's responsibility for the development of technology is appropriately increased.

Moreover, when the prospective engineer realizes who he is, namely, a human being marked by short-sightedness, shortcomings, and a tendency to underestimate the unfavourable side-effects of his work, he will not be tempted to dominate technological development presumptuously, nor aspire to unlimited achievements. Instead, he will practice wisdom, level-headedness, carefulness, prudence, patience, modesty, and scrupulousness. He will also be prepared to expose his work continually to critique and scepticism. He should desire to interact with his peers in order to find and accept communal responsibility. From that vantage point alone will it be possible to give due attention to the development of technology within the marginal conditions of a historically developed cultural situation and to the necessity for continuity of both the environment and the supply of energy and raw materials.

By emphasizing the responsibility of the engineers, we will be able to slow down current technological development with its ironclad logic and gigantic dimensions and dangers. We will be able to penetrate this development and rework it into a multifaceted, richly varied, and enduring development. There is a tremendous gap between the small-scale technology proposed by the counterculture and the large-scale technology of present culture. It is a gap that needs to be closed in a surprising way by the engineer who seeks to develop the true meaning of technology.

Interaction Between Technology and Culture

In the long run, philosophical reflection on technology should result in an analysis and normative assessment of the societal consequences of technological development. It should also lead us to examine the influence of technology on other cultural sectors. By creating a technostructure, technology forms the basis for other sectors and should contribute to their disclosure instead of to their constriction. Constriction takes place when both the technological-scientific method and the technological results are absolutized instead of serving to disclose various cultural sectors. The meaning of the technostructure, however, is that it serves as a basis for the unfolding and realization of individual and communal responsibilities in areas such as family nurture, education, housing, health services, labour conditions, politics, economics, aid to developing countries, and so forth.

On the other hand, we must also investigate how technology is influenced by culture, particularly from the socioeconomic sector. We should explore the purposes, values, and norms of culture so that we can critically compare them to the normative principles that enhance the meaning of a full cultural life.

The mutual interaction and conditioning of technology and culture constitute a major part of a philosophy of culture. With it we can enrich the engineer's scientific and technological sense of responsibility with a social dimension. We must recognize that other, non-technological points of view ought also to be included in the development of technology.

Finally, keeping in mind the "jolts of our age," in which technology plays a dominant role, I would like to plead for an expansion of the curriculum to enable the student to reflect philosophically on the technological-scientific culture. A thorough study of the origin, motives, and meaning of technology is required, especially for the training of engineers. Without a responsible philosophy of technology, the engineer is likely to remain unaware of his multifaceted responsibility as a bearer of culture. This is even more important now that the technological-scientific culture is faced by nearly insurmountable problems.

In short, a philosophy of technology is essential. It will critically examine the development and problems of technology and the relationship between technology and culture. A philosophical ethics, elaborated on the basis of the proper ethos, must clarify and disclose the motives, goals, values, and norms that guide technological science. In this fashion, we will be able to unfold the originality, creativity, communal spirit, and, above all, the responsibility of the prospective engineer and of all those who are involved in the development of technology.

Select Bibliography

Adorno, Th. W. *Negative Dialektik.* Frankfurt, 1966.

Becher, P. *Mensch und Technik im Denken Friedrich Dessauers, Martin Heideggers und Romano Guardinis.* Frankfurt, 1974.

Beck, H. *Philosophie der Technik—Perspektiven zu Technik—Menschheit—Zukunft.* Trier, 1969.

Bloch, E. *Das Prinzip Hoffnung.* Frankfurt, 1959.

Brinkmann, Donald. “Aufstieg oder Niedergang unserer Kultur?” *Universitas* 2 (1947), 1291–1296, 1435–1440.

----. “Geistige Grundlagen der modernen Technik.“ *Universitas* 8 (1953), 289–294.

Dooyeweerd, Herman. *Reconstruction and Reformation.* Unpublished translation of *Vernieuwing en Bezinning* (Zutphen, 1959), available from the Association for the Advancement of Christian Scholarship, 229 College Street, Toronto, Ontario.

Ellul, Jacques. *The Technological Society.* London, 1965.

Gehlen, Arnold. *Die Seele im technischen Zeitalter.* Hamburg, 1957.

Guinness, Os. *The Dust of Death.* Inter-Varsity Press, 1973.

Habermas, Jürgen. *Technik und Wissenschaft als “Ideologie.”* Frankfurt, 1968.

Hetman, F. *Society and the Assessment of Technology.* OECD, 1973.

Heiss, R. *Utopie und Revolution.* Munich, 1973.

Hommes, J. *Der technische Eros—Das Wesen der materialistischen Geschichtsauffassung.* Freiburg, 1955.

Hooykaas, R. *Religion and the Rise of Modern Science.* Edinburgh and London,1972.

Howe, G. *Gott und die Technik.* Hamburg, 1971.

Hunig, A. *Das Schaffen des Ingenieurs—Beitrage zu einer Philosophie der Technik.* Köln, 1974.

Jungk, R. and J. Galtung, eds. *Mankind 2000.* Oslo and London, 1969.

Kahn, Herman and A.J. Wiener. *The Year 2000.* New York, 1967.

Klaus, Georg. *Kybernetik: Eine neue Universalphilosophie der Gesellschaft?* Berlin, 1973.

Lenk, H., ed. *Technokratie als Ideologie.* Stuttgart, 1973.

----. *Philosophie im technologischen Zeitalter.* Stuttgart, 1971.

Mekkes, J.P.A. *Teken en motief der creatuur.* Amsterdam, 1965.
Meyer, H.J. *Die Technisierung der Welt—Herkunft, Wesen und Gefahren.* Tübingen, 1961.
Müller-Schwefe, H.R. *Technik und Glaube.* Göttingen, 1971.
Mumford, Lewis. *The Myth of the Machine.* New York, 1967.
----. *The Transformations of Man.* New York, 1956.
Musgrove, F. *Ecstasy and Holiness—Counter Culture and the Open Society.* London, 1974.
Popma, K.J. *Evangelie en geschiedenis.* Amsterdam, 1972. English translation: *Gospel and History.* Trans. Harry van Dyke. Aalten: WordBridge, 2021.
Rapp, F., ed. *Contributions to a Philosophy of Technology.* Dordrecht, 1974.
Reich, C. *The Greening of America.* New York, 1970.
Riessen, Hendrik van. De *Maatschappij der Toekomst.* Franeker, 1952; 5th ed. 1974. English translation: *The Society of the Future.* Philadelphia: The Presbyterian and Reformed Publishing Company, 1957.
----. *Wijsbegeerte.* Kampen, 1970.
Rohrmoser, G. *Das Elend der Kritischen Theorie.* Freiburg, 1970.
Roszak, Theodore. *The Making of a Counter Culture.* New York, 1969.
----. *Where the Wastelands Ends.* New York, 1972.
----. "The Monster and the Titan: Science, Knowledge and Gnosis." *Daedalus.* 1974, 17–32.
Sachsse, H. *Technik und Verantwortung — Probleme der Ethik im technischen Zeitalter.* Freiburg, 1972.
Schelsky, H. *Der Mensch in der wissenschaftlichen Zivilisation.* Köln, 1961.
Schilder, Klaas. *De Openbaring van Johannes en het sociale leven.* Amsterdam, 1924; 4th ed. 1972.
Schumacher, E. F. *Small is Beautiful.* London, 1973.
Schuurman, Egbert. *Techniek en Toekomst.* Assen, 1972.
----. *Technology and the Future: A Philosophical Challenge,* trans. by Herbert D. Morton (Toronto: Wedge Publishing Foundation) 1980.
Senghaas, D. and C. Koch, eds. *Texte zur Technokratiediskussion.* Frankfurt, 1970.
Toffler, Alvin. *Future Shock.* New York, 1970.
Waskow, A.I. "Creating the Future in the Present." *The Futurist* 2 (1968), 75 ff.
Weizsäcker, C.F. von. *Tragweite der Wissenschaft.* Stuttgart, 1964.

White, L. "The Historical Roots of our Ecologic Crisis," in *Philosophy and Technology*. C. Mitchen and R. Mackey, eds. New York, 1972.

Zuidema, S.U. *De revolutionaire maatschappijkritiek van Herbert Marcuse.* Amsterdam, 1970.

THE SCIENTIZATION OF MODERN CULTURE

(1979)

Introduction

In this paper I would like to present a short philosophical analysis of the spirit of our time and the dominant position of science in our culture and its future. Science is a growing power in our practical, daily life. The characteristics of science are increasingly becoming the characteristics of modern culture. Sometimes we can observe that the characteristics of science are overshadowed, especially by economics and politics; but economics and politics, too, usually employ the scientific method nowadays, and they, too, therefore often serve to strengthen scientific power and its characteristics in our culture.

Generally speaking, we can say that it is in particular practical rationalism which has fostered the cultural power of science. This power gives every appearance of being marked by objectivity and autonomy. This appearance is enhanced by modern technology, and by modern organization, which makes use of modern technology. All philosophical reflection on culture has to deal with this seemingly objective cultural power of science. Every philosophy of culture has to reckon with it. Meanwhile, it is clear that it is very difficult to overcome this power. The case of the neo-Marxists, especially, shows clearly that this difficulty is very real. The neo-Marxists try to overcome the cultural power of science and technology, but in the very act they become its victims. The real cultural influence issuing from the neo-Marxist philosophy is certainly small, and even negligible. The neo-Marxists are well aware of the problems accompanying the current development of culture. It is a development dominated by science and technology, and it is oppressive for man and nature. Therefore they choose for "freedom," and they oppose this power. They work for a revolution. But their proposed revolution turns out to require science and technology in its own turn. When they recognize this situation, many neo-Marxists return to the philosophy we view as orthodox Marxism. In orthodox Marxism, science and technology are viewed as productive tools that form the basis of cultural development.

Today one can even speak of the scientization of culture, as an on-going process caused by an absolutizing of science and scientific method as the universal basis for all cultural activities. This absolutizing of science is an outworking of the religion of science: *scientism*. Scientism is the origin of the problems of scientized culture. To describe those problems will be my first task in this paper.

My second task will be to examine the main features of modern systems philosophy or systems analysis as they relate to these problems. In systems philosophy one finds the problems of the current cultural situation presented quite well. Systems thinkers try to solve these problems by means of a universal systems science and with the help of a worldwide political organization or world state. I will argue that systems theory can solve the problems of scientization in the short term, but that in the long run it can only aggravate them. In my critique and evaluation of systems philosophy and the spirit of the present age – my third task in this paper – I will try to make this clear.

In analyzing the problems and dangers besetting scientized culture, I shall restrict myself to the influence of science on technology and organization. I shall not deal with scientized economic practice, for example.

Furthermore, to the extent that I deal with systems philosophy, I shall refer only to the publications of Ervin Laszlo, the "ideologist of the Club of Rome." The main contours of his thought and the motive force behind it are germane to the subject of this paper.

Finally, I should say before proceeding that the spiritual background of the scientization of modern culture involves the deification of science and its method. By science I mean the natural or exact sciences. But it seems to me very important to recognize that the scientization of modern culture does not find its *origin* in the absolutization of the natural sciences. This absolutization became possible only after the prior scientization of the Christian faith in Western culture by absolutizing theology.

Older and newer forms alike of modern theology provide examples of such scientization of the Christian faith. More and more, concrete personal faith in God in the light of the Holy Scriptures was replaced by theology. The over-estimation of theological science in Christian circles lowered people's resistance to the imperialism of scientific thought in general as it grew beyond its boundaries to assume a well-nigh all-important place in Western life and culture. The scientization of faith destroys, in fact, the basis for resistance against all other forms of scientization. In short, scientism with its narrow, ironclad logic first appeared in theological thought. Then it went on to overshadow all other areas of human action, thought, and life. If we are to analyze the spirit of our time, we therefore need to understand the connection between man's pretended autonomy and his scientific thought. For the struggle of Christians in the future will be in the first place a struggle against scientism. In this regard, it also seems that the proper relation between Christian faith and theology will have to be restored.

This matter belongs to my third point, but it is perhaps the most important one. Certainly we can neglect it only to our own harm.

1. The Problems of a Scientized Culture

The problems and threats facing our culture are known to all from many current publications. Many of these problems involve the relation between man and technology. Man in his labor is increasingly enslaved by the machine. He is no longer the master of technology; often he is reduced to a component of the machine. If he refuses to be part of technology, he is faced by the likelihood of unemployment. In the meantime, the dynamic, large-scale and seemingly unlimited development of technology is itself threatened by the finiteness of natural resources, both of matter and of energy. The publications of the Club of Rome have made us aware in recent years of our potentially catastrophic situation in this regard. Moreover, the frontiers of modern technology lie in dangerous territory. I shall only mention the problems of nuclear energy and the computerization of many cultural fields. To present a full view of these and other developments and their possible consequences in this paper is, of course, not possible.

An important related problem area involves the relation between man and nature. The small-scale technology of earlier days was still characterized by a gap between man and nature. Today, however, aided by modern technology, man continually interferes with nature on a grand scale. As a consequence, the given coherence of nature is disturbed and upset or even shattered. Nature is not allowed the chance to repair the damage it sustains. Furthermore, the adverse side-effects of modern industrial technology, modern transportation, and an almost fully industrialized agriculture reinforce each other, making the dangers of a collapse of the environment ever more real. We draw from nature more than it is able to produce, and we leave more waste products behind us than nature is able to break down – not to mention those that cannot be broken down. The possibility of the destruction of the environment is becoming very credible indeed. Modern technology's threat to man, therefore, can be looked at from two angles. On the one hand, it consists of the structure of a gigantic, massive technology and its advancing frontiers. On the other, it consists of a disruption of the coherence in nature and the severing of the intrinsic bonds between man and nature.

All in all, it is understandable that many should experience the explosive and ambitious development of technology as an apparently autonomous one. Furthermore, given the socio-economic context within which technology is developed, this sense of having to do with an autonomous development is reinforced. Fundamental research – usually secret – and the development of technology take place inside mammoth concerns and within the so-called military-industrial complex. Should this research and development ever be taken over totally by the state, we would experience technological development as the lowering of a limitless, fatal doom. Specialization and massification are also typical of the present situation. The resultant technology often fails to meet actual needs and desires. A discrepancy arises between real, actual needs and the systematically determined goals dictated by a systems-engineering founded on growth ideals. Unscrupulous technology, together with a resultant dehumanization, continues to grow in scale and complexity.

I have worked all these problems out at some length in my doctoral dissertation[24] and in "Reflections on the Technological Society."[25] I am convinced that the problems concerning man and technology, man and nature, and the socio-economic coherence can be better understood when we look at the predominant influence exerted on *technology* by *science*, particularly natural science. The characteristics of natural science project themselves into technological activities, both into the technological production process and into the results of this process. This is a result of the scientific basis of modern technology, and of the technological-scientific method of designing. These characteristics are operative wherever technological products are introduced. They leave a decisive mark on the various cultural sectors in which this method and its technological results – for instance the computer – are dominant. The characteristics of science become characteristics of culture. Culture is "scientized."

Before I go on to explain, let us consider an analogy. Scientific knowledge is equivalent to a map of a landscape. The map is not the landscape itself; it only represents it. By making use of the map, a person is better able to orientate

---

[24] *Technology and the Future: A Philosophical Challenge.*

[25] See pp. 21ff.

himself in the landscape. One may not identify the map and the landscape. If one were to do so, he would have to live in the map, not in the landscape! Then the landscape would be reduced to the narrow lines of the map. The landscape would be dislocated and disrupted. The person in question would be restricted as well. He would no longer be a part of the landscape but a part of the map. He would certainly have his troubles! This example can help to make clear to us what happens in a scientized culture. The characteristics of science are projected into culture itself.

The characteristics of (natural) scientific knowledge are as follows: It is *universally valid;* it is *abstract* and *detached* (it is, that is to say, a reduced knowledge of reality because it is knowledge of the laws for only one aspect of reality); it is *enduring* and *durable;* and it possesses *logical coherence.* The relation between the first three characteristics of scientific knowledge and the dominant basic features of reality – namely, the *uniqueness, full coherence* (that is, interconnectedness and interrelatedness) and *mutability* of everything real – is one of tension. The *universal,* the *abstract* and the *durable* as characteristics of scientific knowledge become characteristics of technology, too, whenever science and the technological-scientific method are made to display a purely logical coherence in their bond with technology. Unprecedented compulsion is consequently brought to bear on man by technology and its development. In such a situation there is no appreciation of man's historical vocation: man should be active, shaping culture creatively with the aid of technology. Human activity in technology is regulated, and in consequence man is robbed of his responsibility and his freedom as well as of his inventiveness. The upshot is the rise of a one-sided, monotonously levelled technological culture in which people are graded to be equal and levelled to be average. That is to say, they can be plugged into various work-situations and consumption patterns interchangeably. They become statistically manageable and manipulable. There is an effacement of personality. Because modern technology has assumed large-scale, universal proportions in its designing activities and because its scientific basis and method have been absolutized, it has come to be a part of all the world's cultures. It molds these cultures into the uniform and monotonously levelled-out pattern of a technological-scientific world culture. The universal character of scientific knowledge comes to be reflected in the global nature of the technological-scientific world culture.

Even so, it should be emphasized that the technological-scientific culture itself is chronically in danger of becoming an internally torn, fractured and segmented culture, since the scientific abstractions projected into it reflect differing scientific approaches. On account of the dominant influence of scientific applications, no integration of culture as a whole is accomplished; that is, unless we want to identify such integration with the purely technological-scientific approach, which would mean accepting a tremendous impoverishment and reduction of cultural life. Such reduction is to be observed nowadays in the standardization of nearly everything, in the compartmentalizing of human life, in the isolated existence of the various sectors of human life, in the great gap between living, working and recreation and in the disruption alluded to above of the coherence within nature and of the intimate bond between man and nature. The future, moreover, comes to be fixed beforehand by the same process of reduction. The future is stifled and restricted; a rich variety, complexity and stability have no place in it. The substantial bias and dynamics of the scientized culture make that culture an extremely unstable monoculture with tyrannical tendencies. This is illustrated by the current world energy crisis and the imminent danger of a global socio-economic depression. But the coercive dynamics and totalitarian character of the scientized world culture make it very difficult to break this pattern.

On balance, we are led to conclude that the problems and dangers of a scientized culture can no longer be described as incidental or acute. Rather, they are *structural* and *chronic*. These problems are an expression of a growing cultural crisis and global malaise.

The background of this development is to be found in modern man's having linked the old idea of human autonomy to modern science and its methodology. In our time man finds assurance and confidence in a religious, generally tacit devotion to the scientific method as it is used in gaining mastery of practical affairs. Modern man, because of his faith in it, tends to submit himself to abstract scientific thought. He forgets the reductive character of science and succumbs to the illusion that science produces full and concrete knowledge of reality as a whole. Whenever people hold this view of abstraction in triumphant anticipation of progress, and whenever they apply scientific knowledge technologically in order to project abstraction into reality itself, the end result is a constricted, distorted reality and a dislocated culture. It is clear, then, that this view, with its uncritical appeal to science as a panacea for the problems facing our culture today, can bring only temporary relief. In the long run, given the structure of science and its inherent limits, this approach can only confront us with new and even more distressing problems. The solution offered affords us only a more complex and murkier constellation of old and new problems.

2. Systems Science and Systems Philosophy

The publications of the Club of Rome have been very much in the public eye in recent years. This Club has made clear to many people what the problems and dangers are that are threatening modern culture today. Given the analysis I have presented above, it is important to examine the substance of the suggested therapy. The matter is all the more urgent if the Club expects solutions from science and technology itself. In other words, will the recommendations of the Club provide impetus for a further scientization of culture? In the first report of the Club of Rome we were told that there would have to be a fundamental change in the direction of culture, that there would have to be a Copernican Revolution in the spiritual sphere. Man's cultural and social values and norms would have to be changed. Otherwise, we would not be able to achieve a state of equilibrium in world culture.

But how is this new situation to be realized? The Club has described that state of equilibrium in terms of world models. Now it seems to me that the general tendency of the Club to this day has been to expect people to adapt their values to the best of all possible future-models. The scientific model is to be allowed to prescribe human values, and these values are to be implemented by means of political power. The scientistic character of the worldview that has generated the problems is not examined or revised; the only change the Club has been able to envision is that the direction of cultural development should be guided increasingly by a new, global politics. Peccei, the president of the Club, has said only that this change to a global approach will require a new philosophy of life, a *new humanism* on the basis of which people will be willing to follow the recommendations of the Club.

It is my cautious conclusion that the Club of Rome envisions a new future based on a new, scientific approach to the problems and dangers besetting a growing world culture. To realize the new future, modern man is thought to need a new philosophy, a new religion, a new ethos, and a new politics.

Ervin Laszlo is perhaps preeminent among those who have attempted to fill the order of the Club of Rome. One can gain insight into the major contours of his thought from several of his publications. Two important articles are: "Why I Should Believe in Science?" (*Phil. and Phen. Research*, 1973–1974, p. 477–478); and "Goals for Global Society: A Positive Approach to the Predicament of Mankind" (*Analysen und Prognosen*, 1975, p. 15–19). His most important books are: *The Systems View of the World* (N.Y., 1972); *Introduction to Systems Philosophy; Toward a New Paradigm of Contemporary Thought* (N.Y. 1973); and *A Strategy for the Future* (N.Y., 1974).

According to Laszlo, the new way of thinking he has adopted – that is, systems theory and systems philosophy – is destined to be the only science and philosophy of the future in the future. He has set out to unify science on the basis of a natural systems view of the world. Moreover, he wants to restore the relation between philosophy and science. He is opposed to the philosophies of the positivists and of the irrationalists. According to him, the practical upshot of science today is reductionism, and it is his intention to provide a new, holistic approach. He prefers a science of the entity, of the whole, and not a science that is an approach to only a part or aspect of reality.

So he begins with the largest whole (the world system) and descends from there to the smaller systems. In addition, he wants to bring about a close relation between systems science and politics.

It is systems science, Laszlo argues, which makes possible a simultaneous consideration of all aspects of humanity's evolution, from individual values and attitudes to ecological and environmental conditions. Systems theory counters narrow nationalism and it focuses on the long term. Laszlo states:

> We may be on the threshold of a world civilization of justice and peace through cooperation; of global unity through regional and national diversity. In this perspective the rapid quantitative growth of the past hundred years is a necessary if chaotic cumulative growth phase on the way toward achieving a healthy global society. As the world level is reached in population size, communication, mobility, energy and resource flows, and eventually in political process, growth could change from a quantitative and cumulative process to a differentiated and partly qualitative one. In the maturity of mankind, progress could be measured by

justice and stability of the intricate webs of global balances and the quality, rather than the material correlates, of human life ("Goals ...." p. 16–17).

To understand Laszlo's ideas we have to know something of the substance of a view called General Systems Theory. I can only suggest its outlines here.

General Systems Theory is an approach that originates in modern technology. Thus Laszlo often refers to Norbert Wiener, the "father of cybernetics." He also refers to Von Bertalanffy, who developed systems theory in the field of biology. But I would emphasize that Bertalanffy regards organisms as cybernetic machines. It is important to note that General Systems Theory is a technological way of thinking. Central to it are the fundamental principles of cybernetics: feedback, information, communication, self-stabilization, self-organization, and so on. Moreover, in General Systems Theory the whole of the history of man and society is considered to be a cybernetic structural self-evolution.

What, then, is a system? one might first ask. A system can be defined as a complex of interacting elements, or as the sum of the parts and the interrelations between the parts of an entity. These interrelations are characterized by feedback or circular causality. So the entity is a cybernetic system. The relations between a system and its environment are expressed as input and output of matter, energy and information. That means that the relation between a system and its environment is likewise a cybernetic relation.

In the Preface to Laszlo's *Introduction to Systems Philosophy* we read that systems philosophy is "founded on the natural scientific world picture" and "centered on man and society." Systems philosophy is proposed as a basis for the unification of science and the consolidation of scientific control. It is to bridge the antithesis between the sciences and the humanities, between technology and history. Systems philosophy addresses the roots of our present cultural crisis, from the dangers of modern technology to the disequilibrium of the ecosystem to the countless psychological, social, economic and political problems of the present situation.

The pretensions of systems philosophy are clearly great. It will usher philosophy back into the mainstream of present-day science so that it can deal with the many urgent problems confronting science and society. General Systems Theory cannot be equated with a special science. It tries to approach the whole of reality. Reality as a whole is held to be a hierarchy of systems. The smallest system is an atom; the largest one is the world system. Systems philosophy is thus a form of *integrated pluralism* which "proclaims both the diversity and the unity of the world."

Every system is characterized by four "invariances." For all systems the form of the invariances is the same, while the content is different. What are these invariances? The first invariance is that all systems can be analyzed in terms of cybernetic principles. A system viewed as a whole has no mystery any more; all can be explained by the sum of the parts and the cybernetic interrelations between the parts. The second invariance is the self-stabilization of systems in their environments by means of negative feedback. The third invariance is the self-organization of systems in their environments by means of positive feedback. And the fourth invariance is that every system is a system in a hierarchy of all systems. These four invariances guarantee the isomorphism between the systems. An important fact to notice is that the four invariances can always be expressed in the categories of mathematical information theory. In consequence, each system can be simulated by computer.

Laszlo distinguishes various systems. There are natural and cognitive systems. The natural systems are exemplified by physical, biological and social systems. The evidence for the natural systems relies on the evidence for the empirical existence of those systems. The evidence for cognitive systems rests on introspection. The relation between the two kinds of systems is that of correlation. So reality can be approached from two perspectives. The conclusion is that each natural system has a consciousness. Laszlo mentions the consciousness of an atom only in passing; he deals at length with the consciousness of man and society. As I have already said, the largest natural system is the world system; its cognitive complement is the world consciousness.

At the top of the hierarchy of all terrestrial systems is the "currently forming global sociocultural ecosystem." Every other system has to adapt itself to this hierarchy. The world system ought to be in optimum condition, but at present it is not. The optimum world system will be obtained when every subsystem has adapted itself to the goal of the world system. That goal is survival. That goal has to form the framework for the ethos of the new age and for a new world religion. The content of the new ethos ought to be reverence for all natural systems. But what, then, is the content of the new world religion to be? Laszlo calls the world religion of the future "universal humanism." It is composed of all the basic values of man. But how can such a humanism arise as a universal world religion? According to Laszlo, people have to know and to accept that this religion is universal and that every present-day religion is merely a particular differentiation of the world religion. Everyone has to accept it, and will accept it, as he comes to accept the rules of cybernetics as basic values for the survival of humanity. After accepting the world religion, everyone can differentiate it in his own way – on the condition that the universal world religion be able to restrict that individual differentiation.

Laszlo states and concludes the matter as follows:

> Our present commitments need revision and updating. They were functional in the more limited contexts of national and regional societies but may prove to be counter-functional in the global context. Existing ideologies unite by division – something we can no longer afford. They unite Christians by separation from heathens; socialists in contrast to capitalists, the initiates in distinction to outsiders, the developed world's "haves" from the poor world's "have-nots." They may continue to play a functional role on the level of specifying regional value systems and commitments. But they need to be supplemented by a global ethic and a commitment that unites without division. We must find the ideals of achievement which are functionally equivalent to local and regional myths, religions, and ideologies of healthy society but function on the *global* level. The ideals which could bring together needs, services and resources, this time for all the people on this earth, through a commitment to the survival and development of mankind as a whole ("Goals ...", p. 18).

According to General Systems Theory, both the boundaries between the various disciplines and the boundaries between geographic nations must disappear. That means that the General Systems Theory and systems philosophy are to have important consequences for politics, too. General Systems Theory provides a motivation for world politics as the highest form of politics, and for a world political order. The purpose of General Systems Theory is to provide us with normative reference points for envisioning a functional and humanistic world order.

Many people might be led to conclude that Laszlo's proposal would be one of a world government. That is not the case. According to him we need a *provisional* world government, until we have reached the optimum steady-state of the world system. The provisional world government must be a universal functional organization with just sufficient power and authority to regulate critical decisions concerning global security, economy, ecology and population growth. In the meantime, we may expect consciousness, world ethics, and world religion to arise, bringing with them an awareness of the interdependence of man and nature, of the unity within diversity of all mankind and of the common long-term fate of our societies in the coming global culture. The constraints on this evolution are objective. Once man has accepted them all, political coercion on a world scale will disappear. The optimum state of the world system will be one of cybernetic self-stabilization and self-organization. Once this strategy for the future is effected, the greatest number of persons will attain the highest level of satisfaction, thanks to an equitable distribution of those resources which constitute the necessary, if not sufficient, condition for the fulfilment of the hierarchy of human needs. Laszlo says at the end of his book, *A Strategy for the Future:* "This truly is the route to paradise on earth" (199).

3. Criticism and Evaluation

I now come to my third and final task in this paper: to present a critique of systems philosophy and an evaluation of it. I would draw attention to those points especially which relate to my theme: the scientization of culture as an expression of the spirit of our time.

It should be clear that systems philosophy is not a very subtle or nuanced philosophy. In fact, as philosophy it is superficial. It is easy to criticize it on immanent and transcendental grounds. But I will not do that. Although systems philosophy is superficial, it is nevertheless very consistent. Moreover, it has gained tremendous influence on modern cultural development. The reason for this is that it gives a very important, almost exclusive place to objective cultural possibilities and to cultural power. It is dominated by the passion to control the future fully, and it seems to many people that we need such control in the face of the menacing dangers and catastrophes of our day.

We can characterize systems *philosophy*, when it is absolutized, as technicistic thought. It is a form of *technologism*, or perhaps we can better say: *cyberneticism*. It interprets all natural systems as cybernetic machines. It calls the mind, too, a feedback mechanism. I must conclude, therefore, that systems philosophy interprets all natural systems as artificial systems. According to Laszlo, systems philosophy is not a reductionistic philosophy since it pays attention not to functions or aspects but to the whole. The question is, however, can the whole, the entity, be fully reflected in the system? The system is scientifically qualified; however, the living entity may have another qualification. I must therefore conclude that systems *science*, no less than systems philosophy, is reductionistic. Cyberneticism as an expression of practical neo-rationalism contributes to the scientization of reality. Cyberneticism is, finally, a refined form of scientism, and therefore it is also more dangerous. This form of scientism is not based on linear causality, as mechanistic thought is; rather, it is based on circular causality.

The reductionist character of systems philosophy is clearly reflected in Laszlo's explanation of evolution. According to him, evolution is a cybernetic process. He attributes to the past the characteristics of cybernetic principles. As he reduces the past, so too does he reduce the future. The future will be a good future if everything is ruled by cybernetics. We have already seen that Laszlo accommodates even ethics and religion to the cybernetic world system. The consequences implied by this cyberneticism for the future are indeed considerable.

The future implied by systems philosophy is, in short, a cybernetic future in which all cultural pursuits and societal relations are progressively levelled down to become components of the one gigantic, all-embracing totalitarian system. In that system man is no more than a cybernetic subsystem. Human freedom is a co-determining and inter-determined function in the hierarchy of natural systems. Man is gradually reduced to a standardized component of a cybernetic machine; he becomes interchangeable and replaceable. This all-devouring collectivism will culminate in a technocratic world state which will be in complete control of the future. Its iron control will turn our world into one great prisonhouse for free and responsible people. The future is de-futurized: there will be no surprises, nothing unexpected anymore. The future is determined by the self-stabilizing and self-organizing cybernetic process of history. According to Laszlo, at the end of this development the world state will disappear automatically, so that there will no longer be any compulsion. Everybody will be free, as a cybernetic subsystem of the cybernetic world system. In reality, however, I am compelled to say, an elite of experts will always be required to direct the development of culture. Culture can never be self-directed. People always have to make and be responsible for cultural decisions. They can indeed get on the wrong track. They do so whenever they seek to establish the norms and values of life themselves, in autonomy.

It is clear that systems philosophy is founded on the autonomy of man; in this philosophy no one tries to reckon with the transcendent. The world of the future will therefore be a godless world. Moreover, the opposite of all that is promised will be realized.

I would remark in closing that although I have not been able to discuss Marxism in this paper, there are many differences between the Marxist conception of using cybernetics to establish the communist world culture on the one hand and the General Systems Theory conception on the other. Nevertheless, I would stress that the similarities between these two ways of thinking are clear too. In practice there could be a growing convergence. It is necessary to note that the systems philosophy now in vogue in the so-called free Western countries is unable to offer any resistance to the ideas of Marxism at the level of fundamental principles. The question therefore arises: will systems science and philosophy be swallowed up by Marxism, or vice versa?

In either case we can only conclude that the scientization of culture will be on the increase in the future, on a world scale. Man's engineering intellect and skill may well appear to be on the way to establishing the Kingdom of Man. It is good to know that this apostate kingdom ideal can still be entertained only because Christ's Kingdom is coming; and it, too, is coming with haste.

SYSTEMS THEORY, MODERN TECHNOLOGY, AND CULTURAL CHANGE (1979)

Introduction

The title of this paper indicates its emphasis clearly. I do not deal here with the subjects of systems analysis and systems theory as such. I simply attempt, rather, to draw attention to the cultural consequences of systems theory when it is harnessed to modern technology – especially information technology or computer technology. Alternate subtitles I might have used are "From Systems Analysis via Information Systems to Computer Systems" and "How Systems Theory Fulfills Itself in Cybernetics as the Science and Practice of Control." I suggest an approach to the problem of analyzing the origin, the development and the future possibilities of systems theory in connection with computer technology – a complex called *cybernetics*.

This paper is a sequel to another, "The Scientization of Modern Culture,"[26] an early version of which was published in *Philosophia Reformata* in 1978. In the present paper I show that, broadly speaking, the *scientization* of modern culture has a complement: the *technologization* of modern culture. Systems theory has strengthened the main tendency of our culture, which is the drift towards scientization and technologization. Many philosophers have even called our modern culture a technocracy. Eventually, with the help of systems theory and computer technology, the technocracy could very well become a computerocracy. This main tendency of modern culture confronts us with a lot of problems. First, what are we to think of the present cultural situation? Next, what conditions must be met in order to set modern culture in a more responsible direction? And finally, how are we to deal with the difficulties of actually changing its course?

I am concerned here to examine the dominant influence which systems theory now has on the *shaping of our culture*. I shall not deal with systems theory as a science as such. Nor shall I deal with the influence of cybernetics on such particular fields as sociology, economics and politics. My concern is with cybernetics and the *spiritual* background and dynamics of the development of modern culture as a whole.

---

[26] See pp. 99ff.

A number of prominent writers and philosophers have been inspired by systems analysis and cybernetics and have written about them and related fields. Norbert Wiener, Karl Steinbuch, and the Marxist Georg Klaus take their orientation from cybernetics as a part of modern technology. Ludwig von Bertalanffy, the biologist and philosopher of biology; David Easton, the philosopher of politics; Kenneth Boulding, a philosopher of economics; and Russel Ackoff and Ervin Laszlo, the sociologists, have all come, through their different scientific backgrounds, to take a keen interest in the new science of systems theory and in the new technologies of information. Can we perhaps conclude from this fact that there is a new and peculiar development going on in science and technology?

Origins of Systems Theory and Cybernetics

Systems theory and cybernetics originated on the ground of *modern technology.* Norbert Wiener, of course, is the "father of cybernetics." Before Wiener, Ludwig von Bertalanffy had already called special attention to the organization of a living organism as a whole (1938). But he was only able to develop his theory of systems (1968) after the pioneering work of Norbert Wiener in the field of cybernetics, or control and communication in the animal and the machine (1948).

Since Wiener, one no longer interprets systems as mechanisms or "clockworks." No, the concept of information has made it possible to focus on a system as a whole. Before Wiener, the whole of a system was viewed as being the sum of its parts. Since Wiener, the whole of a system is generally regarded as being more than the sum of its parts. It has been argued that with this development we have passed beyond the period of mechanistic thinking and that we have returned to the organistic thought of Aristotle and the Middle Ages. But that is a fundamental misunderstanding. Aristotelian and medieval organistic thought derived its organistic quality from biology. Modern so-called organistic thought is inspired by and oriented to technology. The new interpretation of a given entity as a whole is a *technological* interpretation. A given entity is seen as a constructed, artificial, technical whole. An entity is seen as being governed by the cybernetic principles of information, communication, feedback, equifinality, systemic steady state, and so on.

Now when some people disagree and argue that it is not, after all, possible to conceive all given entities as cybernetics systems, they usually have in mind the existence of an unknown, X, within the particular system being interpreted in terms of cybernetics. This X may be *life, growth, play, creativity, freedom* and so forth. The danger of such a limited critique, however, is that this X can conveniently be forgotten as soon as cybernetics is implemented to *control* the system in question. That is to say that the inherent limitations of the cybernetic interpretation may initially be recognized to some extent but that reduction takes place later on anyway. Here, I believe, lies the root of many of the difficulties one encounters in current philosophical discussions of the relation between man and the computer. Instead of speaking of a new organicism, it is better, I believe, to speak of *Cyberneticism.*

Cybernetics designates not only the science of describing or explaining a system; it designates also the technology (or better: technique) of controlling a system. Cybernetics is science and technique in one. One encounters this conception of cybernetics quite early on, in the practicalist neo-rationalistic background of Norbert Wiener. Wiener's own philosophy was inspired by the philosophy of pragmatism. More recently, progress in technology and especially in its new branch, "information technology" or "cybernetics," has inspired many others, too, to use the new technological-scientific method – the method of cybernetics – as *the* approach to all sectors of modern culture and its future. The result has been a burgeoning of systems analysis and systems designing. Cybernetics has been absolutized and hence has become imperialistic. Cybernetics has become, in fact, a new philosophy.

A New Philosophy

During the sixties and seventies a number of studies have appeared devoted to the subject of cybernetics and philosophy. *Kybernetik, Brücke zwischen den Wissenschaften* (Frankfurt am Main), edited by Helmar Frank, was first published in 1962. The Marxist Georg Klaus had already written his *Kybernetik in philosophischer Sicht* (Berlin), as early as 1961. In 1973 he wrote another book, entitled *Kybernetik, eine neue Universalphilosophie der Gesellschaft?* (Berlin). Another noteworthy work is *Philosophie und Kybernetik* (Munich, 1970); it is a joint study made by the professor of information theory at Karlsruhe, Karl Steinbuch, and the philosopher Simon Moser. Besides these studies several introductions to systems theory have also appeared. In 1956 the economist Kenneth E. Boulding published "General Systems Theory – The Skeleton of Science," in *General*

Systems: Yearbook of the Society for the Advancement of General Systems Theory, Volume I. In 1973 the sociologist Ervin Laszlo published his main work, *Introduction to Systems Philosophy: Toward a New Paradigm of Contemporary Thought* (New York). A critical survey of systems theory within the social sciences by R. Premo, J. Ritsert, E. Strache, *Systemtheoretische Ansätze in der Soziologie* (Rowolt), appeared in 1973. And Russell L. Ackoff published his *Redesigning the Future: A Systems Approach to Societal Problems* (New York) in 1974. The tendency in all these studies is not to promote science as it is but to develop a new scientific approach – an approach which results, via information theory, in a new technology: computer technology. Ackoff in one of his studies says: "I believe, and will try to show, that our society is now in the early stages of a change of age that results from a radical change in our point of view, our way of thinking, and the kind of technology they are producing. We are going through an intellectual revolution that is as fundamental as that which occurred in the Renaissance. The Renaissance ushered in the Machine Age, which produced the Industrial Revolution. The currently emerging intellectual revolution is bringing with it a new era that can be called the *Systems Age* which is producing the *Postindustrial Revolution.* I believe these changes give rise to most of the crises we face and simultaneously offer whatever hope there is for dealing with them effectively." Since 1950 teleology, the study of goal-seeking and purposeful behavior, has been brought into science, and it has begun to dominate our conception of the world. The cultural effect issuing from the new conception is the automation and planning and control of the future through cybernetics. It is, that is to say, no less than the redesigning of the future. Through systems theory, systems design, and systems control the disadvantages of the industrial age – focussing on a part, reductionism, analytic mode of thought, mechanism, determinism and dehumanization – are being transformed into the advantages of the post-industrial age: holism, expansionism, synthetic mode of thought, teleology, information and self-control.

We can tentatively conclude therefore that these modern thinkers have a strong tendency to take their orientation no longer from physics but rather from cybernetics viewed as a sector of modern technology.

Physical parts are exchanged for *technical,* which is to say for artificial, *wholes.* As a result of this development, the computer has been introduced in recent years into several areas of modern society. With the help of the computer, one tries nowadays to control and steer the fields of economics, politics and management.

The bridge between systems analysis and systems control is built by means of the concept of information. With the help of mathematical information theory, it is possible to simulate a system by computer. It is even possible to simulate the consequences that specific decisions will have within a system so that the optimum decision can be made. Hence there is a harvest of publications which can rightly be seen as fruits of systems theory and information theory. A few of the most important works are: M. Greenberger, ed., *Computers and the World of the Future* (Cambridge, M.I.T. Press, 1962); Zenon W. Pylyshyn, ed., *Perspectives on the Computer Revolution* (New York, 1970); and Joseph Weizenbaum, *Computer Power and Human Reason* (San Francisco, 1976). Yet another is R. B. Coats and A. Parker, *Computer Models in the Social Sciences* (London, 1978). All these studies make it very clear that the social sciences have become technologically oriented.

The interest in systems design and systems control by means of computers has grown especially strong since the activities of the Club of Rome, beginning with 1972. The possibilities of cybernetics have become apparent to many social scientists. Ervin Laszlo is a case in point.

Ervin Laszlo's Thought

What I have said so far can be nicely illustrated at this point by a short abstract I have made of Ervin Laszlo's thought. Readers of my antecedent paper, "The Scientization of Modern Culture," will already have encountered this abstract there in somewhat expanded form, and they are therefore invited to pass over this section and continue reading at *Criticism.* Ervin Laszlo is perhaps preeminent among those who have attempted to fill the order of the Club of Rome. One can gain insight into the major contours of his thought from several of his publications. Two important articles are: "Why I Should Believe in Science?" (*Phil. and Phen. Research*, 1973–1974, p. 477–488); and "Goals for Global Society: A Positive Approach to the Predicament of Mankind" (*Analysen und Prognosen*, 1975, pp. 15–19). His most important books are: *The Systems View of the World* (N.Y., 1972); *Introduction to Systems Philosophy: Toward a New Paradigm of Contemporary Thought* (N.Y., 1973); and *A Strategy for the Future* (N.Y., 1974).

According to Laszlo, the new way of thinking he has adopted – that is, systems theory and systems philosophy – is destined to be the only science and philosophy of the future in the future. He has set out to unify science on the basis of a natural systems view of the world. Moreover, he wants to restore the relation between philosophy and science. He is opposed to the philosophies of the positivists and of the irrationalists. According to him, the practical upshot of science today is reductionism, and it is his intention to provide a new, holistic approach. He prefers a science of the entity, of the whole, and not a science that is an approach to only a part or aspect of reality. So he begins with the largest whole (the world system) and descends from there to the smaller systems. In addition, he wants to bring about a close relation between systems science and politics.

To understand Laszlo's ideas we have to know something of the substance of General Systems Theory. I can only suggest its outlines here.

What, then, is a system? one might first ask. A system can be defined as a complex of interacting elements, or as the sum of the parts and the interrelations between the parts of an entity. These interrelations are characterized by feedback or circular causality. So the entity is a cybernetic system. The relations between a system and its environment are expressed as input and output of matter, energy and information. That means that the relation between a system and its environment is likewise a cybernetic relation.

In the Preface to Laszlo's *Introduction to Systems Philosophy* we read that systems philosophy is "founded on the natural scientific world picture" and "centered on man and society." Systems philosophy is proposed as a basis for the unification of science and of scientific control. It is to bridge the antithesis between the sciences and the humanities, between technology and history. Systems philosophy addresses the roots of our present cultural crisis, from the dangers of modern technology to the disequilibrium of the ecosystem to the countless psychological, social, economic and political problems of the present situation.

The pretensions of systems philosophy are clearly great. It will usher philosophy back into the mainstream of present-day science so that it can deal with the many urgent problems confronting science and society. General Systems Theory cannot be equated with a special science. It tries to approach the whole of reality. Reality as a whole is held to be a hierarchy of systems. The smallest system is an atom; the largest one is the world system. Systems philosophy is thus a form of *integrated pluralism* which "proclaims both the diversity and the unity of the world."

Every system is characterized by four "invariances." For all systems the form of the invariances is the same, while the content is different. What are these invariances? The first invariance is that all systems can be analysed in terms of cybernetic principles. A system viewed as a whole has no mystery anymore; all can be explained by the sum of the parts and the cybernetic interrelations between the parts. The second invariance is the self-stabilization of systems in their environments by means of negative feedback. The third invariance is the self-organization of systems in their environments by means of positive feedback. And the fourth invariance is that every system is a system in a hierarchy of all systems. These four invariances guarantee the isomorphism between the systems. An important fact to notice is that the four invariances can always be expressed in the categories of mathematical information theory. In consequence, each system can be simulated by computer. It is in information theory that we find the bridge between *systems theory* and *computer technology.* In other words, the systems approach provides, with the help of information theory, the basis for applying computers in several sectors of *culture.* In this way, society becomes, finally, a computerized society.

At the top of the hierarchy of all terrestrial systems, according to Laszlo, is the "currently forming global sociocultural ecosystem." Every other system has to adapt itself to this hierarchy. The world system ought to be in optimum condition, but at present it is not. The optimum world system will be obtained when every subsystem has adapted itself to the goal of the world system. That goal is survival. That goal has to form the framework for the ethos of the new age and for a new world religion. The content of the new ethos ought to be reverence for all natural systems. But what, then, is the content of the new world religion to be? Laszlo calls the world religion of the future "universal humanism." According to General Systems Theory both the boundaries between the various disciplines and the boundaries between geographic nations must disappear. That means that General Systems Theory and systems philosophy are to have important consequences for politics and, via politics, for social and economic practice. General Systems Theory provides a motivation for world politics as the highest form of politics, and for a world political order to determine the goals for a global society.

Criticism

I now come to the task of presenting my critique of this development from systems analysis via systems design to systems control. Systems philosophy is neither subtle nor profound. On the contrary, it is superficial. It is easy to criticize it on immanent and transcendental grounds. Importantly, however, it has gained a tremendous influence on modern cultural development, largely because of the almost exclusive place it gives to *objective cultural* power. It is a philosophy dominated by the passion to control the future fully. For many people, of course, it is precisely this kind of control that we need in the face of the menacing dangers and catastrophes of our day. I want to focus my critique, therefore, on the consequences of this cultural influence.

Anyone informed by Christian philosophy will easily recognize the new philosophy as a new expression of the old philosophy of human autonomy. We are dealing with a revolution in scientism, not with a conversion of it. The revolution now going on in scientism means a strengthening and enlargement of it, because the influence of scientism is now being extended beyond the theoretical fields into actual form-giving and the direction of culture. In short, the spiritual background of the new philosophy is the old religious worship of the ratio, which in this case has assumed the form of a magic devotion to the imperialism of the *instrumental* reason as it reaches its crowning achievement, its completion and its fulfilment, in the computer. Because of his *faith* in abstract scientific thought – in abstract cybernetic thought – modern man tends to humbly submit himself to that thought. Remarkably, he forgets that science has a reductive character. He assumes that the new science of cybernetics produces full and concrete knowledge of the whole of reality! In fact, of course, it does no such thing. Its leading principle much sooner distorts reality and dislocates culture. Cyberneticism can only encumber us with new problems more distressing than the ones we already have.

Cyberneticism reinforces the scientization and technologization of culture. The characteristics of cybernetics become *cultural c*haracteristics.

Logical coherence, universality, abstractness and durability as characteristics of cybernetics are in tension with the uniqueness, full coherence and mutability of concrete reality, and with the responsible creativity of man. For instance, when systems goals and computer ethics prescribe the way in which cultural problems are to be solved, there can be no appreciation of man's *historical* situation. Man is a historical being. As such, he should be active. He should shape creatively. But in cyberneticism human cultural activity is regulated. In cyberneticism man is robbed of his responsibility, his freedom, and his inventiveness. Of course, within the boundaries of the system man is "free." Yet if he does not accept those boundaries, he will be oppressed: he will be compelled into obedience, compelled to be "free."

Laszlo and Ackoff stress that cybernetics, or systems philosophy, is not reductionistic in character since it draws attention not to a function or an aspect but to the whole, even to the top of the hierarchy of all systems, the world system. In reality, however, despite their protests, we do have to do here with a new kind of reductionism. I believe that what we must learn to do is to distinguish between a given entity as a whole, on the one hand, and the (theoretical/scientific) "system" as an interpretation or artificial construction of such a given entity on the other. The simulation of the artificial system with the help of information theory in a computer produces a further transformation, which is in effect just another reduction of the given original system viewed as a whole. Given the introduction into culture of an absolutized cybernetics, we can therefore rightly speak of a new scientization and technologization of culture.

Cybernetics, furthermore, is characterized by largeness of scale. This largeness of scale finds expression especially in the proposals for a world politics. But for all its grandness, the global conception remains reductionist. In the cyberneticized culture we accordingly encounter not only a reduction of man and human communities but of political systems as well. These are to be controlled by computers. From this point of view, we can say that technocracy reaches its fulfilment and consummation in cyberculture, or computerocracy. The future of that culture is a cybernetically controlled future. Because it is not based on the principle of linear causality but of circular causality, the cybernetic future is more flexible but at the same time more stable than the technocratic future. In the long run, however, the disadvantages are greater, in proportion to the largeness of scale and the dynamics and range of the effects of the decisions produced by the cybernetic system.

That means that the future, too, is reduced. It is a cybernetic future in which the goals to be realized are the goals that will have been chosen. But who will have chosen those goals? Someone has said that everyone will have chosen those goals. Study reveals, however, that in actual fact it is the experts of cybernetics who will have done so. An elite of experts will be required to direct the development of the cybernetic culture. This elite will in all likelihood be a *hidden* elite, and it is as a hidden elite that it will plan the direction of the culture. It is the elite who will have to make the decisions. Yet this elite will most certainly be going wrong when it chooses an absolutized cybernetics as its guide for controlling the future, whether according to its own ideas or those of economic and political authorities acceptable to it. Observe that when a situation is very complex, the goal of the cybernetic process may be hidden. In that case nobody will know the 'wherefore' of various decisions; all attention will be devoted to solving the complex problem. People will be manipulated all the more, simply for the sake of solving the particular problem that has captured the fancy of the elite. Some people believe that democracy will prevent such evils, but they forget that democracy is incompatible with cybernetics. Democracy has a very slow feedback, whereas cybernetics needs a dynamic, that is, automated feedback.

In politics this approach now confronts us with a centralized, nearly autonomous cybernetic world-culture, or worldwide computerocracy. It is clear that there will be a growing regimentation of life in such a culture. In management we are confronted by the same tendency. Abbe Mowshowitz says in his study. *The Conquest of Will: Information Processing in Human Affairs* (Amsterdam, 1976): "Centralized management is facilitated by the achievement of technology in transportation, communication and information-processing." George Grant has provided a very instructive contribution to the discussion of the cultural influence of the computer in his article: "The Computer Does Not Impose on Us the Ways It Should Be Used." He makes it clear that this pronouncement is a lie and that it hides the problems of the Computer Age and man's responsibility within that Age. All existing communication and information processes in computers are scientifically and technically controllable processes. This means that they are reduced processes. We take a universalized information system and reduce reality to make it fit that system. The existing flow of information is moulded in a universal way. *New* information systems can be fashioned. The characteristics of all such artificial systems oppose the characteristics of reality, as I have already shown. The result is that the individual user or receiver of information is made a universal, standardized user or receiver of selected or formed universal, uniform information. So the computer does, indeed, impose on us the ways it should be used. Man is forced to obey the structure of the computer.

The increasing use of the computer has reinforced many observers in their view that people are no longer in control of the direction of culture but that they are instead controlled by a seemingly autonomous development. The computer works faster and more accurately than people, has nearly inexhaustible endurance, and works automatically. In this sense it is, indeed, *relatively* autonomous. It gives the appearance, furthermore, of becoming entirely independent of people. For the fact is that while people always set the conditions under which the computer must function, the functioning machine runs quite independently of its makers and operators. People simply cannot maintain supervision of the functioning of the computer. The results, therefore, do have, albeit within certain limits, the character of a surprise.

In addition, a computer's user is not necessarily its programmer – a fact which again increases the gap between people and the computer. The gap is widened even further as one user replaces another. At the last, a computer's user may have not a glimmer of the criteria by which it may be functioning. The program has become incomprehensible to him. He *must* yield to it, and trust it. In the future this problem will grow in scope and seriousness, as machines that learn and machines that reproduce themselves are brought into use and as the computerocracy perfects itself.

Perspective and Challenge

I shall now attempt to evaluate this program for the development of our culture by means of systems design and control. I shall do that by discussing three points:

1. First, I want to call attention to the reformational view of science, which provides a basis for assessing the real meaning of systems analysis. I want to say something about *the idea of the inner reformation of science.*

2. Secondly, I want to describe the reformational view of the relation between theory and practice, that is, between abstract theory and full, concrete practice.

3. Finally, I want to point out the conclusion to be drawn from the first and second points with reference to the increasing use of the computer and the problems resulting from that.

1. The reformational view of science recognizes that, historically and structurally, the kind of knowledge used in a science is restricted to definite boundaries and that it is itself based upon another kind of knowledge, one that is fundamental, integral or uncompartmentalized, and concrete. This other, pre-scientific knowledge, which consists of both actual and factual knowledge of changes effected by human action, is in its turn guided and kept on course by an original and irreducible ultimate trust. This "*trust-knowledge*" is a knowledge in love that constitutes man's basic point of reference for all his philosophical and scientific endeavours, regardless of whether he is aware of the fact or not. But in autonomous theoretical reflection (a kind of reflection which since the beginning of the modern age has been inspired by a misguided religious faith in philosophical and scientific thought) this pre-scientific trust-knowledge has been thoroughly perverted and falsified. Meanwhile, scientific knowledge and its practical applications have been accorded superior status. Reformational philosophy, however, rejects the autonomy of man and therefore the autonomy of philosophical and special scientific thought as well. Divine revelation leads to a knowledge whose content is open to understanding by faith. It is the most profound knowledge, a knowledge which our thinking cannot fathom. It is knowledge springing from the *heart.*

Now what does all this mean for science? It means we need to de-absolutize science and the scientific method as they are now being used as a basis for cultural activities. It is possible, given the openness of man's heart to the full revelation, to relativize science and the scientific method. While science is indeed a very fascinating activity of man, scientific knowledge has to be seen as conditioned, partial, restricted, abstract and tentative knowledge. On the basis of the reformational point of view one can accept the openness inherent in science, and a pluralistic methodology. He can also appreciate cybernetics as a new, valuable scientific approach.

Concerning the relation between pre-scientific and scientific knowledge, it should be said that the latter must continually be re-embedded in the full, direct, concrete human knowledge and experience that has fundamental trust-knowledge at its core. Scientific knowledge, including knowledge of systems theory, must continually be integrated into and corrected by pre-scientific knowledge, so that the restrictions inherent in scientific knowledge are lifted. In this way, man's integral, pre-theoretical knowledge is gradually enriched and his measure of responsibility and stewardship is increased. Seen in this way, scientific activity can be carried out with faith as its starting point and a strengthened faith as its result. Then our scientific endeavours will serve the cause of wisdom and lead to an increasingly comprehensive insight; they will buttress and enlarge our responsibility and better enable us to orientate ourselves in the direction of the future.

2. My second point deals with the relation between abstract theory and concrete praxis. Standing in the tradition of Western thought, the philosophers of systems theory try to build the bridge between systems theory and praxis from the side of theory. This means that autonomous thought is the beginning and end of all practical activities. The result, therefore, is that systems theory leads to new scientization and new technologization of culture and to many problems, old and new. Every time people are confronted with these problems, they try to overcome them with a new form of thinking. How difficult it is to overcome them in such a way is demonstrated by the various trials and tribulations of modern thought. Systems thinkers try to make the bridge with the help of politics. That gives them an advantage and a lead over many other thinkers. But in the long run, the problems they encounter will only be the larger, as we have already seen.

From a Christian philosophical point of view, the bridge between theory and praxis ought to be built by man from out of the deep sense of responsibility that has its root in faith. No science is acceptable as a guiding principle for chartering the future. Philosophy and science may not become the instruments of human power to conquer the future according to some model of man's own making for the future. As we have seen, systems thinking sometimes tries to create a new ethic and a new religion. That is the world in reverse. Man is called to disclose the creation in faithful responsibility, under the guidance of the *given,* revealed normative principles that are concentrated in the commandment of love. Science can have a useful function in man's giving form to culture. In several fields of our modern culture we can also use systems theory; but we should only do so in such a way that our insight into complex matters is strengthened, so that we can take ethical, responsible decisions. The central point in such decisions is the meaning of the creation: the Kingdom of God, the meaning of which is love, peace, and salvation.

3. I now come to my third and last point: how is it possible to make responsible decisions in a culture in which the prevailing spirit is secular? In answer to that question I want to stress again that the problems and even threatening dangers and catastrophes of our culture arouse reactionary spirits, which try to give, and sometimes do give, good ideas and alternatives, while the spiritual background remains just as secular. Christians can learn from these ideas and alternatives, but they should also undertake activities to redirect and to deepen them. At the same time, we must not underestimate the problems that confront the Christian. These problems are not merely cultural; they pertain to the religious and Biblical background. In short, we have to be concerned about the Kingdom of God. We believe that this Kingdom is absent because of the Fall into sin, and that it is present in Jesus Christ. At the same time, however, we have to say that this Kingdom is not yet fully present and that it will not be fully present until the second coming of Christ. Given especially the sins of secularization, all this means that we cannot be optimistic about our mortal efforts to redirect the future. The tendencies of the age are increasingly apocalyptic. Nevertheless, I will close with a few suggestions that it might be helpful to keep in mind.

There has been mention in reformational circles of the "idea of an inner reformation of the sciences." We should learn to speak in the same way about what have erroneously been called "the applied sciences." These include the field of computer technology. We have seen that the development of computer technology gives the appearance of being almost totally autonomous. The use of the computer has led to such tremendous accumulation of information relevant to decision-making that no one making a decision nowadays can manage to grasp it all. Moreover, there is a strong tendency to centralize. As that happens, a central information system becomes autonomous in its selection of relevant information, in its interpretation of it, and in its sway over the decisions grounded in it. Little attention has been given to this development because the threat it presents is not quantifiable. It is not in the first place material but spiritual. We must therefore acquaint ourselves with the views of the people who are using computer technology. We must analyze especially their views of man, of communities, of society, politics, and the future – views which, though hidden, are present in all that they do and say. The objective of our work in this field should be the reformation of thought and practice.

I believe we shall also have to face the question of when and where to use computers. Sometimes we have to use them, for instance in space travel. We also often have to use them where the norm of efficiency is the most important one. That can be the case when there is work to be done under very dangerous conditions, as in the construction of a subway system. But in most cases we are simply free to make use of the computer, whether it is necessary or not. In such cases we need to make the incomprehensible programs comprehensible, for the user has to bear his responsibility for the results. He is accountable to those people who are affected by the consequences of his decisions. In short, this means that we have to integrate computer systems into the faithful responsibility and stewardship of the computer's user. It is precisely that responsibility that forestalls the restriction of the creation and promotes much more the disclosure of creation.

In conclusion, information technology need not bring centralization as an inevitable consequence or autonomous development. Computer technology, if kept within its normative limits, can make a powerful contribution to enhancing the Christian's normative stewardship of his cultural task. It can promote a decentralized, pluriform, stable, wholesome and enriched culture. That means there is a lot of work to be done! It can be done responsibly only on the basis of a conversion and reformation of the secular spiritual background of the development of Western science, technology, and culture.

TECHNOLOGY IN CHRISTIAN-PHILOSOPHICAL PERSPECTIVE (1980)

Preface

Many people are involved in technology. Not all of them are Christians, and of those who are, only a few are able to express philosophical views on technology and the relevant sciences. Ultimately only a very few will actually do so.

This perspective is really necessary to restore confidence and to relegate fear to where it belongs.

Mankind came to a fall right in the beginning. However, through his subsequent arduous efforts man has always succeeded in wresting sustenance from the earth. Science and technology are encouraging rays of light along the path of man's endeavours and should be received with grateful recognition from the Almighty – they provide, with the rest of created reality, so much pleasure to practitioners and users alike.

Alas, like the history of Israel, man, true to his nature, has not only abused creation but soon boasted of his acquired self-sufficiency, his autonomy. Nations grasped this human aid and developed it into a power that soon came to be regarded as a demonic power. If only the people behind this power could still be perceived, it would be a consolation. But now the whole of creation "groaneth and travaileth" because for many the only possible faith now is in science and technology, while the same things have come to constitute an over-riding fear in the minds of many others. Thus, as long as man sees no further than this earth-bound human activity, it is a realm destined for destruction.

Technology in itself is not a demonic power, for it is but a tool in the hands of mankind. Many probably believe in its efficacy because it provides such tangible results. But even though it is so easily controlled (compared for example with other ways in which mankind implements power) it alone, or even technology with the support of politics, economics, etc. cannot determine the destiny of man. God mercifully did not entrust that to man himself.

It serves no purpose to call technology a power (in either the good or the bad sense). What man misses out on is the lesson to be learnt from history. This is repeated with regular monotony. The lesson is that many things on earth are provided by the Creator for the benefit of man. However, no sooner does man discover such a benefit than he applies it towards his own destruction. If this sounds illogical, it must be remembered how easily man misleads himself into believing the contrary. His autonomy allows him to do so.

Prof. Schuurman clearly arrives at the basic problem and points to the solution which has always existed. Man's obsessions debar him from this great discovery, and he needs a special revelation from outside himself.

Prof. W. J. Taute
Head of the Department of Electrotechnology
Potchefstroom University for Christian Higher Education

1. Introduction

The development of modern technology is fraught with many problems. There are tensions between the power of technology and the freedom of man. There are economic problems. There are the problems of modern technology itself, such as those related to nuclear energy and the environment. And all the dimensions of the problems are growing, so that the development of nuclear energy, for instance, presents problems which we have never encountered before: the dangers have assumed new proportions in space and time. Man's technological innovations tend nowadays to be at once irreversible and negative, and destructive to the natural environment. Man values nature only as an area and object for human action. He is literally consuming the very foundations upon which his life is based. He is using natural material resources and energy resources as if they were limitless. In reality, however, these resources are not his income but his property, which is to say, again, that they are limited. With an enormous, ever-growing dynamic man is approaching the limits – and that means he is heading for a catastrophe. The picture is very gloomy if we but think of the threatening possibilities (or impossibilities!) of nuclear weapons.

In the meantime, people all know that the solution to the problems is no longer to be found in more and better technology. What we need before all else is a fundamental discussion about the spiritual background and roots of the modern development of our culture; we need a discussion not on the level of science and technology itself but on the level of philosophy, ethics and religion.

The discussions on those levels are already going on. Jacques Ellul, especially, and the circles around him have already done a lot of work. Although we can learn much from their contributions, the direction of their discussion is not, in my opinion, wholly satisfying. They are asserting that the problems of modern technology are so weighty that they are too heavy for people to bear. As Christians they think that modern technology is an autonomous, demonic power. In *The Technological Society* – but also in his books of later date: *Ethics of Freedom* and *The New Demons* – Ellul says that man is not the master of technology but rather its slave and its victim. Man is the victim of a universal, artificial, monistic, self-directing power. Technology means doom; in any case, it cannot be a blessing. Now it seems clear to me that thinking about the future of technology within this view, which I think does not meet the requirements of a biblical approach, leaves no room for a perspective of deliverance.

I do not agree with Ellul. I acknowledge, however, that it is *difficult* for people to control modern technology. The *idea* that we have to do with an autonomous development arises, I believe, precisely from that difficulty. Take computer technology, for example. The computer works autonomously, and as such it gives the impression or appearance of *absolute* autonomy, whereas in reality it is only *relatively* autonomous. I maintain that technology is not an autonomous force. The fact that the development and direction of technology proper are guided and set by norms given to it from the outside and the fact that it is precisely humans who determine the character of its development make it clear that technology is not an autonomous force. Nevertheless, the problems are considerable. To understand these problems better and to gain a possibility for a new, liberating perspective, we need in the first place to take a Christian, biblical view of history. I shall therefore begin by suggesting the basic outlines of such a view.

We shall then be in a position to consider together the predominant motives in science and technology. This consideration will enable us to understand the problems much better. I shall then turn our attention to a Christian-philosophical perspective on modern technology and conclude with a reflection upon the meaning of technology and upon the blessing and the curse of modern technology.

2. Some Thoughts on A Biblical View of History

Man himself cannot give a total overview of history. Many people, however, especially philosophers, seem to think they can. They try to take history in their grasp and to show their power over history. Most modern thinkers propose to use science and technology to subserve history, to make the future fully subject to man's ideas. We can find this inspiration and motivation on a grand scale in Marxism and in modern "system" philosophy. But the outcome of such a tension is inevitably the reverse of what is promised. Man becomes not the master of history, but its slave. The reason for this is that despite his assumed autonomy, man cannot really secure his own mastery. Mortal man cannot lord it over history. He himself is fully historical. His pretence of giving meaning to history must therefore result finally in historicism, relativism and even nihilism. He cannot uphold his autonomous pretension, and he is confronted instead by the autonomous development of history. He thinks to find a way out, a perspective, on his own; but the result is that he knows nothing for certain any more. And in the meantime– as in our time – science and technology manifest themselves as apparently autonomous cultural powers.

The cause for this development has to be sought in the secularization of the Christian idea of history. This secularization begins with man himself, with his idea of autonomy, and it leads to man's being confronted with seemingly autonomous powers. This already indicates my disagreement with Ellul. He begins with the autonomous powers, and he neglects their root: the idea of the autonomy of man. In relation to the questions of history and of human freedom and responsibility, he is therefore struggling, in my opinion, with an incorrectly formulated problem.

According to God's revelation, man has to believe and to know that he cannot speak the first and the last word on history.

He is not the giver of history's meaning, he is not autonomous, not self-sufficient, and not sovereign.

To know the meaning of history, we have to acknowledge that the light must come from outside history. For in that light man knows by faith what the meaning of history is. And it is the Revelation of God that places man in that light. Neither human reason nor human technology can give such light. In Christian perspective the cornerstone of history is not man, but the revealed Word of God in Jesus Christ. Christ is the light of the world. If man pretends to give light in history by his reason or technology, in reality all becomes darker. That is clear from the experience which neo-Marxist thinkers like Adorno and Horkheimer have had with their *Dialectics of the Enlightenment.*

The important question, however, is: What is the *content* of the biblical light for an understanding of history, and of human technology within history? The Bible as God's Revelation gives us a foundation.

When I speak about basics I do not mean that Christians can answer all questions. They can give a biblical perspective but cannot solve all puzzles and problems. The deepest meaning of history is a divine mystery (see Deuteronomy 29:25 and Revelation 10:4). This mystery stresses on the one hand that man is not the master of history; but on the other hand it stresses that he *is* responsible for the development of history, and of culture within that history. Men are responsible for their doings, for they are servants of God and not their own masters. Let us now look first at four basic biblical givens.

2.1 The Cultural Mandate

One basic biblical given is the cultural mandate for man. It is found in Genesis 1:28. Man has received the calling to exert dominion over, to build, the creation, and to keep it (Gen. 2:15). Having been created in the image of God, man must work in the Kingdom of God so that everything in creation will unfold and find its proper place. It is via man's cultural task in history that the fulfilment of all possibilities in the creation is to be achieved.

2.2. The Fall

A second biblical given is the fall of man from communion with his Creator. Man forsook his original task. He himself wanted to be God the creator. After the Fall man himself, although it remained his task to develop culture, could no longer fulfil the cultural mandate. History was therefore no longer wholesome but disastrous. The way of life was changed to a way of death. What was meant to be a disclosure of creation became a distortion. The wholeness of history was broken, and nature was cursed and became a threatening environment for man; man became mortal. The sin of Adam and Eve incurred a lessening of the earth's spontaneous abundance (Genesis 3); the earth refused to give her strength to the sinner Cain (Genesis 4). Sin always involves a loss of 'earth' in some sense or another: alienation from God and alienation from the creation go hand in hand.

Since the time of the Fall, history has no longer been regarded as the unfolding of creation through the fulfilment of man's cultural task. On the contrary, history has since that time been running aground. Of this the Flood, the building of a Babel culture, and the biblical history of Israel are clear manifestations. Nor can man himself restore history. Rather, he is the cause of its many dislocations and destructions. Skills and techniques of all kinds may be admirable, but the tyrannical or greedy use of human power over nature is a failure deriving from human sin, not from God's intention in creation.

2.3 Redemption

A third biblical given is the promise of Genesis 3:15 as fulfilled in Jesus Christ. God himself provides redemption. Jesus Christ is the second Adam. He has done what the first Adam was supposed to do and in addition has provided for the reconciliation of all things. In Him the redemption and the fulfilment of creation are assured. He has all power over history to bring about the Kingdom of God (see Hebrews 2:14b, Galatians 4:4, 1 Corinthians 15:20–28). The destructive power of Satan is broken; that power is still manifest, of course, but destruction is no longer the final word. In Jesus Christ there is a definite, new perspective for history. In Him the Kingdom of God has come and is coming. The full disclosure of creation and the redemption of the creature have been provided for by Him.

Although this perspective of the Kingdom of God now dominates history, we nevertheless continue to see and to experience dislocations, destruction, and death. Yet behind this development, as we know in faith, is the working, saving power of Jesus Christ. Through Him history is placed under the sign of a total re-creation, the full revelation of the Kingdom of God. Christ gathers together all things which are in heaven and which are on earth (see Ephesians 1:10, Colossians 1:15–23).

The Gospel of Jesus Christ opens up history's dead-end roads, its ways of death. The last enemy that shall be destroyed is death. There is, and there will be, eternal life in the Kingdom of God. We are on the way from Creation and the Fall to re-creation. We are on the way from the garden via the wilderness (and sometimes monstrous cultural achievements) to a new and holy city. At the last, creation will be completed in a new heaven and a new earth. Then alienation, not only between man and God and man and man, but also between man and animal, between man and nature, between man and technology, will come to an end (see Isaiah 11 and Revelation 21).

This perspective of history places people in expectation and gives them hope. In this perspective man can once again carry out his cultural mandate, in obedience to Jesus Christ and as a follower of Him and under the guidance of his Spirit. He is on the way of the Kingdom of God. Signs of that Kingdom are already manifest, if but darkly, here and now.

2.4 Disobedience and Secularization

The first basic given I mentioned, the cultural mandate, is expanded in the third, that is, in redemption. The second, the line of the Fall, is expanded in the fourth, which is disobedience and secularization. For it is clear that not every one lives in the motive power of the Kingdom of God. Many people do not seek the Kingdom of God; in fact, even the deeds of many Christians are not in line with the spirit of the Kingdom of God. But for all that, people cannot escape the motive power of the Kingdom of God. It is true that instead of seeking it, they seek themselves. It is true that they take the third basic given (the Kingdom of God) and secularize it again and again in seeking the kingdom of man. Yet even when people do not accept, or do not accept any longer, the way of escape and salvation in Jesus Christ, they still cannot separate themselves from the predominant force of history; no, they parasitize this force. In Western culture especially people want to go their own way. And as they attempt to do so, the fourth feature intensifies (especially through the possibilities of modern technology) the characteristics of the second feature, in an expansion of history's chaotic, destructive, and demonic powers. In other words, modern secularization is a particularly destructive manifestation or expression of the Fall.

We would make a mistake, however – and it seems to me that Ellul makes this mistake – if we were to conclude that the fourth line is the decisive line of history. In reality we have to do here only with a perversion of history's meaning, a perversion which, because of the power of modern technology and its destructive effects, seems quite overwhelming in our time. The Kingdom of Man, however, even in its most monstrous manifestations, is but a perverse imitation of the Kingdom of God. And it is a constant consolation to know that man on his own and by himself cannot make the meaning of creation, the Kingdom of God, impossible. On the contrary, the Kingdom of God that is already on the way means that at any time people can be converted and led to seek that Kingdom again – even in a technological society.

It is very difficult to perceive the interrelation of redemption and secularization, the third and the fourth features of history I have mentioned. While all is related to the Kingdom of God, all is not related to it in a positive way. We can even say that the fourth line will be judged by the third line: redemption will triumph over disobedience; salvation will triumph over secularization. In this interrelation there is also manifest the divine mystery of history.

3. The Spiritual Roots of Modern Western Culture

The historical development of Western culture since the Renaissance throws special light on the problem of the interrelation of secularization and redemption; we will be able to understand that interrelation better if we look at that history, and especially at its spiritual roots.

Generally speaking, in the Reformation people accepted their divine calling to develop creation. For a long time the influence of the Reformation was strong, especially on the daily praxis. But from its inception, the Reformation was confronted with the influence of the Renaissance and humanism, especially among philosophers and scientists. Since the Enlightenment of the eighteenth century, the influence of humanism has predominated. It is here that the secularization of Western culture begins. Man himself becomes the centre of all reality. Christian eschatology is increasingly secularized and made over into the expectation of a technological salvation, of a technological futurology. More and more, man is convinced that he can make a new world, a paradise on earth. The promise of the Gospel will be realized by man himself, given especially the powerful aid and the possibilities of science and modern technology. That is the secularized faith. Moreover, issuing from Western culture, this secularized faith in progress through science and technology, in combination with politics and economics, has flooded the whole world. Orthodox Marxism and American pragmatism are – notwithstanding differences in the politico-economic systems – the most important expressions of this faith in progress. In this faith man thinks to find the way to life, but the actual situation of the Western welfare state with its large-scale menace of destruction and all its problems makes it clear that this, too, is the way that leads to death. Never before were the cultural problems so huge and threatening as they are in our time; and never before has the influence of the secularized expectation been so deep-seated. Man expects salvation through technology, but its opposite, destruction through technology, obtrudes. Technological development was expected to be wholesome, but it has turned out to be pernicious. Modern technological society is experienced as a monstrous, demonic society. The direction in which people seek the solution to their problems is, once again, a direction characterized by science and technology. The spiritual roots of that direction do not change.

Many Christians – following Jacques Ellul – have concluded that the actual cultural situation is fully demonic and that they should therefore seek to transcend culture in order to rediscover their freedom. Their opinion is that there can be no Christian perspective for modern, secular culture.

I agree, but only to the extent of saying that it *seems* as though there is no way out, that there is no perspective for culture. For even if man were to accept his responsibility, it would presently be difficult to change the massive, dynamic structures of modern development. Their influence can only be a destructive one for a long, long time to come even if people were to be converted this very day to seeking the Kingdom of God in science and technology.

Nevertheless, these sad states of affairs will not always continue; we have to judge the present state of affairs in culture in the light of the given, biblical view of history. People cannot undo the power of Jesus Christ; they cannot undo the restored meaning of the creation. They may well deny this power and this meaning. And this denial may lead to dislocations and distress in our technological culture. Nevertheless, perhaps we can even say that the menacing problems of the secularized society provide us with signs of the second coming of Jesus Christ, with signs of the full re-creation. "We know that the whole creation groaneth and travaileth in pain together." "For the earnest expectation of the creature waiteth for the manifestations of the sons of God" (Romans 8:22, 19).

At this point, you may wonder what the immediate relevance of the Christian view of history is for technological development today. You may ask: Is it not necessary to know what people can do in a responsible way in modern technology itself? Before dealing with this question, we need one more word about the structure and history of technology and about the dominant motives in modern technology.

4. The Structure and History of Technology

Briefly then, we may say that we are speaking about technology whenever people use tools to give form to nature for human purposes. These purposes should not be based on the pretension of human autonomy, of human power, but on the divine calling to develop created reality to the honour and love of God and to the love and well-being of our neighbour. In technology, too, man is called to be a responsible steward.

Man's present tools have become very refined and have taken over many of man's functions, which have themselves multiplied. In the gigantic and dynamic technological development of modern times, the position of the tool has become one of increasing independence. Following a tool technology (think of the hammer) and an energy technology (think of the steam-engine) we are now in the stage of information technology (think of the computer). More often than not, technology today implies an enhancement of the possibilities and is characterized by automation.

The latest stage of technology is attended by many problems. We have a pollution problem as a result of the industrial application of technology. We face an unemployment problem. Furthermore, man often seems to be the slave rather than the master of technology; he seems to be a subservient part of the machine itself. Technological society is fragmented and abstracted: its spiritual background is the predominant or absolutized influence of modern science in technology, and the characteristics of science – *abstractness, universality,* and *logical rigour durability* – have therefore become the characteristics of modern technological society. The complex of interrelated techniques, products of technology, and systems of technology has become a seemingly omnivorous, independent force which shapes and moulds human life in technological society. Because modern technology seems to obey its own laws, the impression arises that it is now very difficult, if not impossible, to control it.

In this context many anti-technologists and many Christians think that we have to escape from technology. The anti-technologists say that we have to return to the past. Their position is the unrealistic way of romanticism and counter-culture that does not see the continuity of history. Many Christians, on the other hand, think that their work in the technological culture should only be to witness to the coming Kingdom of Jesus Christ. But in my opinion this valid and essential witness may not be made to compete with the work to be done in technology itself. Otherwise, the Christian will find himself having to work on two levels at once. On the ground floor he will be working in technology itself, but on the second floor (and then only on the second) he will be witnessing. Along the lines of this approach, there is no integration between Christian faith and technology, and any opportunity to re-direct and to re-form modern culture and modern technology in interrelationship with science, politics and economics is precluded from the outset.

Yet Christians do not see how it should be possible to achieve this integration, this redirection and reformation. I am often asked, "Is technology not in itself bad? Does the Bible not teach that technology is an evil human power?" The building of the tower of Babel is a commonly used example. In Genesis 11 God says: "... and this they began to do and now nothing will be restrained from then, which they have imagined to do." As we know, God put a stop to that aspiration for limitless power and mastery. The fact that technology first developed in the line of Cain easily leads to the conclusion that technology is in itself sinful and bad. However, we should not overlook other very important revelations in the Bible as well. First of all, it is easily forgotten that technology in the Bible, as in the building of Noah's ark, is a sign of salvation. Besides, the Bible teaches that God himself gives wisdom and insight to man for technology. Exodus 35:3 says that Bezaleel and Aholiab could not ply their technical trade without the Spirit of God to inspire them.

Technology is not bad in itself. The question whether we shall have to deal with the blessing of technology or with its curse will always depend on man's motivation in technology.

So the first big question confronting us today is: Which motive is in fact guiding modern man in the development of technology? If the motive can be identified and if the cause of the wrong direction of the development of modern technology and of so many problems and threats can thereby be made clear, then the question arises concerning the identification of the motive which ought to re-direct modern technology and, furthermore, of what the real consequences of that motive for technology and its development would be if it were allowed to exert an unrestricted influence.

5. Motives for the Development of Technology

The chief motive discernible today in Western culture is man's determination to gain control. It is the will to power in association with the motive of applied science. Religiously, the motive is faith, belief in science; and behind that belief there is faith in man and in human possibilities regarding science and technology, a faith in progress. The predominant character of science can be seen most clearly in its application via modern technology. In other words, in man's present technological endeavours he attributes absolute power to science and to the technologico-scientific method. Technological development accordingly mirrors scientific knowledge. This leads to an extremely scientific technology (particularly on the level of design) and, as a result, to the diminution if not elimination of human creativity and responsibility in modern technology. Creativity seeks expression in invention, and responsibility implies possibilities for re-direction and reformation. So modern technology's absolutizing scientific base restricts man's liberty in shaping technology. The bent of scientific knowledge towards continuity (since it is knowledge of a fixed and determinate subject-matter) is projected, as it were, onto technology, which in turn becomes fixed and determinate as well.

Rationalist dominance in technology has led to a dynamic technological development of immense proportions, but that development has at the same time displayed constricting tendencies. New inventions are made with greater difficulty, and old technologies are less rapidly changed, because with logical relentlessness rationalism puts its stamp on modern technology and renders it a stifling and rigid development.

The tremendous restriction entailed in this development is not generally recognized because people are obsessed and fascinated by the accomplishments of technology. For although rationalism is a major motive, it is not the only one. On the contrary, it is intertwined with several other motives which are sometimes quite proper but which more often serve to buttress the dominant trend.

First of all, there is the motive of technology for technology's sake. Let us call it the imperative of technological perfection. Many engineers are so stimulated by this motive that they hold that whatever can be made and perfected must be made and perfected. Things must always be made bigger and bigger. This motive leads to unchecked and aimless technological power, which engineers pretend to control and master but which in fact victimizes them. They are under the spell of their own works. That means that the results are the opposite of what they intend. Technology gains absolute dominance over man. Even nature and culture are threatened by this absolutized technology.

Another motive for technology is to serve economic powers. These powers dominate the development of technology, and in the process, the profit-principle is absolutized. The goal is economic and material growth, with its concomitant emphasis on the acquisition and consumption of material goods. That is to say that the content of the Western idea of progress has indeed been much too narrowly conceived, with the result that the pattern of life in Western society exhibits severe distortions. There is wild growth in the scale of enterprises and technologies, and there is an accompanying devaluation of human work. Furthermore, technical and economic growth also imply a technology developing without any regard for the norms which ought to apply in the management of the environment. Under the influence of this economism, technology ceases to answer to its essential purpose. It no longer serves human purposes represented in existing needs but creates artificial needs instead. The result is a one-sidedly directed technological development, an overcropping or rapacious force that leaves waste, pollution and destruction in its wake and which, in addition, yields technological products that are redundant, and superfluous, and even priced too low; for not all damage to man, environment and nature is economically priced. In a certain sense we can say that under the guidance of this motive we are neither technical nor economical enough. For we typically find a tremendous overdevelopment of the technical and economical aspects of social institutions – the gigantic scale for instance – but at the same time we find a corresponding underdevelopment of other aspects which, although they are present, have been ignored. Among these other aspects are human freedom in, and responsibility for, work; a great variety in economic and technical scale; more stability; the aesthetic relation between technology and nature; social justice; fair relations with the third world and so on. All those features which might have consequences detracting from economic and material well-being are ignored. In the beginning it was possible to neglect subsidiary effects related to such aspects; but at the moment we cannot neglect them anymore, because they give so many problems.

The three motives I have just referred to dominate particularly those who are actively engaged in managing the development and social direction of technology. The motives driving those who are not active in technology itself may vary; for example, some may think that technology is neutral or autonomous. Yet as it happens, they, too frequently close their eyes to the dangers of the technological development. Popular hopes frequently assign to technology a messianic role in the resolution of human problems. Blinded by an insatiable quest for welfare, people are eager and willing to adapt to the prevailing process of development in order to derive private benefit from it. So people outside economic and technical management are also responsible – in fact we are all responsible – for the huge problems facing the technological society of today.

Having observed this distressing situation in modern culture, we come now to the question concerning the basis for these motives. Is it possible that these large-scale problems and threats issuing from modern large-scale technology have arisen because the motives I have mentioned are based on man's large-scale pretensions? In other words, are the motives not always directed to and concentrated on man himself, and does that not perhaps imply the pretension of man to be the centre of reality?

To answer these questions we must look briefly again at the spiritual history of the Western world. What is the most fundamental basis inspiring man's scientific and technological activities? In fact, the inspiration originates in the idea of autonomy, which means simply that man is his own measure. The prevailing *philosophy* in Western culture originates from this idea of autonomy, and it strives to confirm this idea of autonomy. In this way philosophy serves a religious function, following the Enlightenment, philosophy oriented itself to *science,* especially to the natural sciences, and to their methodology. In this way scientism, as a faith in science, as absolutizing *instrumental* reason, increasingly assumed the role of religion in Western culture. Thus philosophy is the stepping-stone whereby the idea of autonomy has been linked up with science and its methodology. Philosophy is used to provide assurance and confidence – in a religious, generally tacit devotion to the scientific method as it is used in gaining mastery over practical affairs, particularly those of modern technology and its technologico-scientific method, but also in relation to economic and political power.

The nodal point, however, is science. Science, as I mentioned earlier, is characterized by abstractness. Now, given the religious function of science and scientific method as I have just described it, this abstractness – and the reduction of reality it implies – is lost from view. This abstractness is forgotten, and it is suggested instead that science produces concrete knowledge of all reality. In fact, however, the exclusive application everywhere of science and of reason as an instrument – what we might call scientization – reduces reality to a potentially exclusively technological accomplishment. And this construction distorts reality and dislocates culture. An uncritical appeal to science for the solution of cultural problems can therefore only mean that we will be confronted in the future by even bigger problems. Technology-as-curse, with its monstrous and demonic qualities, can only deepen if the motive underlying the use of technology is not changed. At the same time, this pernicious development of technology itself reinforces the tendency towards a technocratic political dictatorship of planetary dimensions. In such a political situation there will be no place any more for human freedom and creativity. In that case Ellul is right: man becomes a prisoner in a universal concentration camp.

6. The Old Motive as a New One

If the direction of culture is to be one of deliverance and unfolding, the basic guiding motive of man will have to be changed. The new motive that is required is really a very old one. It is the motive in which man is not the centre of reality, in which man is not autonomous. It is the motive which gives expression to the fact that man has been created in the image of God. This motive gives the address of the responsibility of man. This motive expresses itself in the love of God and the love of one's neighbour. The consequence of this motive for politics is the abolition of the motive of the will to power and the institution of the motive to do justice and to bring righteousness; for economics the main principle then is no longer the absolutized profit-principle but responsible stewardship; for science the motive ought to be not that knowledge is power, but rather that knowledge can serve wisdom. We have to see clearly that science – and the same can be said of technology – can be a good servant. We have to see clearly that it has been under the influence and guidance of wrong motives that science and technology have been bad masters.

In relation to technology I have to say that we need not deny the scientific basis of modern technology. I am opposed only to the faith in autonomy that has been associated with science and technology. The conduct of science has been singled out by that apostate faith to appear as the highroad in the entire field of knowledge and action, while in fact it ought to be considered as neither more nor less than one of the pathways to be followed in the service of practical knowledge and action.

We get a better understanding of the proper service function of science in technology if we take note of the original motive that ought to inspire man in technology.

The biblical motivating force in human history can be regarded as the task of dressing and building, keeping and preserving creation. To limit the motive to "preservation" alone would imply a choice for nature and against culture and that would be a choice for natural distress, a choice for fate. To confine this motive to "building" alone would imply a presumptuousness on the part of man in which he would neglect to consider and weigh what is wise and essential and what is not. It would be to choose for cultural upheaval. Both nature and man would be threatened by a nearly autonomous and destructive technological power, and the future would face down a dead-end road.

Within the twofold harmonious calling, however, to build and to preserve, to dress and to keep, to progress and to conserve, man, the image-bearer of God, is called to a twofold service of love. In building and at the same time preserving creation, he both confirms his love towards his Creator and Redeemer and at the same time lovingly represents all creation. That means, among other things, that man is responsible for the unfolding of the meaning of creation in dressing and keeping it, and that he at the same time must resist every attempt to disturb, disintegrate and destroy this meaning including those attempts which lead to the tremendous problems of the technologico-scientific culture of our day. Guided by the right motive, man in his cultural activity can be a blessing for nature (1 Kings 4:33, 34) and at the same time enter an unobstructed road to the future.

We have to stress again that this perspective is not a very easy one to be achieved. First of all, there will always be many who will maintain that science and modern technology make religious faith and commitment irrelevant. Although the present cultural situation reveals again that science and technology make religious faith and religious commitment more necessary than ever, many people are still blindly following the old idol of progress, of the faith in science and human power. But nothing less than the transcendent God who lives and rules and who is active in our world can free us from being imprisoned by a closed, self-contained universe. At the same time, however, we have to say that it is high time that people acknowledged that God is at the centre of the reality of our lives and culture, and that He gives people hope. The horizon of that hope is the horizon of the Kingdom of God as the fulfilment of the disclosure and redemption of the creation. On the other hand – and again this basic given sometimes makes it very difficult to understand what is going on – the Kingdom of God reminds us of a transcendent dimension, for the Kingdom is both here and yet to come; it is very much a reality of this world, and it is at the same time something which comes to us from beyond. For that reason we have to say of people who acknowledge this given that they are in the world but not of it, that they are at home in the Kingdom of God and at the same time pilgrims in this world seeking that Kingdom.

7. The Meaning of Technology

To seek both "cultivation" and "preservation" in technology is to pour new and profound content into the extremely high moral purpose of scientists and engineers. For it means that they should no longer arbitrarily follow their own will. Instead, they should eagerly seek to be of service in the unfolding, deepening and enrichment of the meaning of technology within the meaning of creation. Technology, since it is but a part of man's activity, can be properly evaluated only if it is set in the context of the total reality and good of man and not judged as a self-sufficient whole exclusively in terms of its own inner laws and dynamics (See: W. Norris Clarke, "Technology and Man – A Christian Vision" in: *Science and Religion,* New York, 1968, p. 284). So while engineers should not strive to do all things possible, they should be able to do all things necessary. The purpose, values and norms for technology should be made explicit in an ethics of technology developed on the basis of this new ethos.

I realize that these observations represent a position contrary to prevailing attitudes and practice. Frequently engineers permit themselves to be lured into considering the advantages of a project. That is what they then call ethics. Today such an approach is no longer possible, for the scales continue to tip on the side of the disadvantages. Can it be of any use to map out the direction technological development ought to take when the actual course and process of development has been accepted as the norm, and when a technological ethic weighs the pros and cons exclusively in terms of that norm? Similarly, no new direction can be expected from an ethic of technology developed only after the characteristics of science have been projected into technology when that projection has been inspired by man's determination to gain control and by his concept of technology as applied science and by his concept of reason as instrumental reason.

An ethics based on continued abstraction is a reduced and restricted ethics. Such an ethics testifies to an urge to continue in a direction that leads inevitably to a dislocation of the meaningful coherence of reality. Although man expects to restore this coherence later without at any point having chosen a new direction, his hope will inevitably remain unfulfilled, because of the mistaken point of departure. In that case all man will ever be able to do is fight the symptoms.

The cause of the problems will not be eradicated as long as its root is left intact. It is not science and technology that should determine ethics, but ethics that should precede science and technology, in order to decisively influence their development. In other words, the direction technological development is to take ought to be determined by our response to ethical issues rather than to technological ones.

Moreover, if the prospective engineer will only realize who he is – namely, a human being marked by short-sightedness, shortcomings, and a tendency to underestimate the unfavourable side-effects of his work – he will not be tempted to dominate technological development presumptuously, nor will he aspire to unlimited, megalomaniacal achievements. Instead, he will practice wisdom, level-headedness, care, prudence, patience, modesty and scrupulousness. He will also be prepared to subject his work continually to critique and scepticism. He will desire to interact with his peers in order to define and accept communal responsibility. Only from that vantage point will it be possible for him to give due attention both to the development of technology within the marginal conditions of a historically developed cultural situation and at the same time to the necessity for continuity of both the environment and the supply of energy and raw materials.

By emphasizing the responsibility of the engineers in such a way, we will be able to slow down current technological development with its ironclad logic and gigantic dimensions and dangers, and we will be able to give attention to those areas of our cultural life which are now suffering from underdevelopment. On the basis of a new system of values – such as rest for nature, rich variety rather than dull uniformity, balance between centralization and decentralization, social justice, harmony between man and technology and nature – we will be able to promote not the "quantity" of life (for instance the enlargement of the standard of consumption); rather we will be able to give more attention to the "quality" of life. We will therefore be humanly able to penetrate technological development and re-work it into a multi-faceted, richly varied, enduring, stable development of technology with less risk in time and space. The gap between the large-scale technology of our present culture and the small-scale technology of earlier times needs to be closed in a creative way, by the engineer, under the guidance of the meaningful motive I have suggested and in keeping with all the non-arbitrary normative principles which people ought to obey. Such a reformed technology would create the conditions for a stable political society in which man could live in freedom and responsibility.

If the engineer will do his work *coram Deo,* "before the face of God," he will be able to disclose the meaning and the blessing of technology. Although no one could possibly supply a statement of the full richness and diversity of the meaning of technology, we can nevertheless express it in part. Technology is able to alleviate man's fate as forced on him "by nature." Technology can offer greater opportunities for living. Technology can reduce the physical burdens and strains inherent in labour. It can diminish the drudgery of routine duties, release the working animal, avert natural catastrophes, conquer disease, supply homes and food, augment social security, expand possibilities for communication, increase information, enhance responsibility, advance material welfare in harmony with spiritual wellbeing, and help unfold the abundant qualities of individuals and nations. Moreover, technology makes room for more leisure time, and it promotes the development of new possibilities in the fields of science and of technology itself. Thus it clears the way to a varied development of culture which is at the same time in harmony with nature. Technology also helps make possible work that is more meaningful in character. In addition to facilitating productive labour, technology creates room for work that consists in the creative giving of aid and service, of love and care. It also provides time for rest and reflection. I should like to conclude by stating that technology seems to me to be a pilgrimage of obedience, a mandated way to greater insight into the meaning of creation as the Kingdom of God. We are called upon to honour the Lord in technology and in every sphere of life. Is it not time then, that the *proclamation of a Kingdom* approach to technology be made an integral component of our evangelical witness to today's modern, technological society?

INFORMATION SOCIETY: CULTURAL IMPOVERISHMENT OR ENRICHMENT? (1984)

We are well upon 1984, the year of George Orwell.

In the years leading up to 1984 it has become realistic to speak of the all-pervasive character of information and communication technologies. Although "Big Brother is watching you" is not yet a reality, the existence of these technologies means that the possibility that such a situation could develop is greater than ever. The political dictatorship behind the Iron Curtain has clearly been strengthened by the adoption of the latest technological possibilities, and is therefore more firmly in the saddle than ever. The Orwellian Society is now distinctly within the realm of possibility.

Perhaps you will already have concluded that I am a melancholy pessimist. At a time when many are suffering from technophobia, would it not be preferable to spotlight the positive aspects of the latest technical possibilities? I share heartily in that wish. The problem is simply one of discovering how to insure that our culture will be enriched and not impoverished as an information culture.

Orwell can teach us something pertinent to this problem. The society he described becomes possible, he says, when people cease to give a *spiritual* answer to the question of the meaning of existence and instead assume tacitly that the answer is to be found in a continuous development of science and technology (accompanied by a strongly enhanced specialization of experts) in a purely *materialistic* approach. Technology in that case becomes a modern form of *magic*.

The ancient legend of King Midas illustrates the point. The gods granted his request that all he touched be turned to gold. What an example he is of greedy, economico-technical, materialistic man! After a few days King Midas implored the gods to deliver him from this blessing, since literally everything he touched – his food, his wife, and so forth – turned into a lump of gold. All that gold could not satisfy King Midas's profound personal yearning for love, and he was threatened by physical and spiritual starvation.

The prescient Spanish philosopher of culture José Ortega y Gassett sensed long ago that our culture with its gigantic technical possibilities was on the road to becoming a spiritually suffocating society. He observed that many specialists promote this development in a barbarous manner, and that the masses surrender themselves to it in materialistic faith. "Therefore these years in which we live, although they are the most intensely technological that human history has known, are at the same time the most empty."[27]

The question I wish to address is how we can prevent that kind of disastrous development and promote a meaningful development in the information society. Or, to take to heart the lesson of King Midas, it is that of how we can prevent life's suffocation by an uncontrolled development of information technology while ensuring that we do justice to the proper value of this technology throughout the whole of culture.

To answer these questions I shall examine the present trend of development in science and technology. Critical attention to this development makes clear that we face entirely new ethical problems. If no solution to these problems is forthcoming, society will be technicized and we will be confronted by increasing social dislocation. If, however, we can find ways of effectuating responsibility in technology and society both individually and communally, then culture will be enriched rather than impoverished by the newest techniques.

The Technological Culture

Modern information technology did not appear out of the blue. Rather, it is the continuation of the whole development of modern technology. Just as modern technology has its basis in the modern natural sciences, so computer technology has its basis in the natural sciences and in systems theory.

The influence of science on computer technology is often overlooked in discussions of this technology. The inevitable result is a certain superficiality. In that case computer technology is discussed in terms of traditional craft technologies – as if it were a means to an end, for example. Then it is forgotten that the computer is not an instrument or a tool but a component incorporated into systems. And these systems are more important than instruments. Yet computer technology is often thought of in instrumental terms. Computer specialists in

---

[27] See José Ortega y Gasset, "Man the Technician," in *History as a System and other Essays toward a Philosophy of History*. Trans. Helene Weyl. New York: Norton, 1962, pp. 87–161, p. 151.

particular often tend to say that they run the computers and will use them as they wish.[28]

That such statements are misleading can be made clear by substituting the word "automobile" for the word "computer." In that case it would be said, "We can use the automobile as we wish." However, such a statement completely overlooks the problems that have accompanied heavy use of the automobile, such as the problems of modern traffic control, traffic jams, the great number of accident victims, urban congestion, environmental pollution, and excessive damage to the landscape. Also overlooked are the many economic, social, and political problems that have accompanied increasing use of the automobile. We can expect a similar effect from the great number of applications of computer technology which will undoubtedly be developed in the future. The development of micro-electronics – chips – has enhanced the likelihood of such an eventuality. Before computers can be applied in existing information and communication processes, these processes must be made subject to scientific control; only then can they be technically and efficiently managed.

Making the existing information and communication processes subject to scientific control involves enhancing these processes according to the norm of efficiency and likewise reducing them to universal, uniform, and homogeneous processes. In other words, the reality that exists must be made to fit the scientific-technical structure of the computer, and that is a structure which has been universalized. New information and communication processes are also characterized by that uniform and homogeneous quality. Of course, the consequences will be less conspicuous and burdensome in the case of small scale than of large scale systems.

In short, the use of computers wherever feasible entails many consequences – cultural consequences that I believe will prove more significant and that will overtake us more rapidly than those accompanying increased use of the automobile. Although it is impossible to predict these consequences precisely, it is clear what the trend will be. The characteristics of scientifically controllable systems run counter to the fundamental characteristics of our everyday experience

---

[28] See George Grant, "The Computer Does Not Impose on Us the Ways It Should Be Used," pp. 117–31 in *Beyond Industrial Growth*, ed. by Abraham Rotstein. Toronto Press, 1976.

and reality. The characteristics of systems theory and information technology are universality, abstractness, impersonalness, and logical coherence; together these form an efficient technical network. These characteristics run counter to the unique, the concrete, the subjective, the full coherence of reality, and to our creative responsibility.[29]

Thus if computer systems are employed wherever it is feasible to employ them, we will be increasingly victimized as we are forced to adapt ourselves to fit these systems. Resistance to this development could be marked by numerous social and political conflicts. Many today are unconcerned about the unnormed use of the computer. People are succumbing to the expectation that they will be better off when the development is finished and the harvest is in. The fact that this dangerous development is sometimes irreversible makes the situation all the more perilous.

New Ethical Problems

To gain a somewhat better view and command of the new ethical problems of modern technology, and of information technology in particular, it will be helpful to compare the main differences between traditional and modern technology. It will then also become clear that we can no longer speak of modern technology in terms of the categories of traditional craft technology.

In traditional technology – think of a blacksmith, for example – everything is characterized by interpersonal relations. A traditional technology is comprehensible, its effects are short-term, and the negative consequences are few and predictable. Traditional technology is static. It does not affect the entire culture but is merely a sector of it.

All this is quite different in modern technology. Modern technology leaves its mark on the entire culture, to which it has brought an enormous dynamic. It touches the entire world. Its consequences affect both outer space and the distant future. The development of modern technology is increasingly incomprehensible. Moreover, it seems to have become unarrestable. Thus some now speak of an autonomous, self-governing development to which people contribute but

[29] See Schuurman, "Concern About Responsibility in Applying Science" (trans. Herbert Donald Morton) in *Research in Philosophy and Technology* 5 (1982): 77–86.

which they no longer control. Many obviously feel they are no longer the lords and masters but the slaves of technology.

Our perilous human position is aggravated by the many unfavorable side effects that impact upon us before we can predict them. The devastation of the environment, the energy crisis, the many victims of traffic accidents, cultural protests against technology – all are examples of such side effects. Furthermore, the scope of these problems is still increasing since the various negative impacts tend to reinforce each other, creating. a snowball effect.[30] This is especially conspicuous in the field of computer technology. When this technology is used wherever feasible, labor problems increase and a chilling routinization settles in upon society. Every individual seems to become a number – and an erasable one at that. Alienation and estrangement become the order of the day. Furthermore, large organizations and centralized control contribute to a growing incomprehensibility and anonymity of systems. Viewed from this standpoint, information technology leads to a growing, impersonal technocracy.

Further, the American philosopher Victor Ferkiss[31] has correctly pointed out that the old tension between power and freedom in our culture will intensify in the computer era. On the one hand impersonal technocracy will be able to develop in the direction of a computerocracy. On the other hand, however, the freedom ideal will also adopt the new technology and thereby introduce the possibility of a computer anarchy. Computer criminality is a sign of that side of the development. In other words, the tension enveloping science and technology in our culture will increase rather than diminish given the latest information and communication technologies. Social conflicts of no little scope may be the result. The almost unavoidable irreversibility of this entire process of modern technology comes into sharper focus at times when the problems and perils assume catastrophic proportions, for it is precisely at such times that we are made most keenly aware of our lack of experience in dealing with such developments. We have still not drawn any lessons from the past, though in view of the latest developments it would seem highly desirable to do so. From an ethical

---

[30] See Hans Jonas, "Technology and Responsibility: Reflections on the New Tasks of Ethics," *Social Research* 40, no. 1, pp. 31–54.

[31] See Victor C. Ferkiss, *Technological Man: The Myth and the Reality* (New York: Braziller, 3rd ed., 1970), pp. 157, 201, 205.

standpoint, too, we do not seem able to keep up with events. Inexperience, ineptness, and ignorance go hand in hand, impeding our search for solutions to new problems.

When we add up the differences between the older craft technologies and modern scientific technology, the question arises whether it is possible to be really responsible in modern technology. Certainly it is clear that the weight of personal and communal responsibility has grown much greater, both internally to science and technology and externally, that is, in society. Yet it needs to be understood that besides growing weightier, responsibilities have become more difficult to bear and to discharge.

One reason for this is the fact that the scientist or engineer has become more and more of a specialist. The work of specialists may have implications for an ever widening terrain at the very moment the nature of that work precludes their having a comprehensive view of that terrain.

Specialists know "more and more about less and less." They can no longer comprehend the whole. Hence they must place their confidence in associates who work with them in equally specialistic ways. Yet the specialization at issue undermines and saps the very community which, as a whole, has to shoulder the increased responsibility. Thus that is clearly a negative point.

For the rest, even where adequate insight exists to make possible the bearing of personal and communal responsibility, it appears there are not infrequently scientific, technical, economic, or political factors which impede its effectuation. To break through the existing dynamic, preeminently massive trend in present-day culture is supremely difficult. Given the preceding observations, I am prepared to join Hans Jonas in speaking of an *ethical vacuum*.[32]

The problems involved in a responsible development of information and communication technologies are great. The ethical vacuum looms all the larger when it is recalled that our technological society no longer possesses a common view of man, history, culture, and the future. The more the current development requires a communal sense of norms, the less possible it is to find it. The background to this state of affairs is the fact that technical integration is being matched by spiritual-cultural disintegration.[33]

---

[32] See note 4.

[33] See Schuurman, *Reflections on the Technological Society*, pp. 41ff. (pp. 70ff. above).

Scientization and Technicization

Meanwhile, the development of information technology is continuing apace. When scientists and engineers derive their norms for practice from the possibilities of science and technology as such, we see a process of scientization and technicization. Not least among those who foster this approach are philosophers who consider their ideas to have an affinity to computer technology, or who actually move in computer circles. I have in mind the father of cybernetics, Norbert Wiener[34] and just as importantly, the publications of Karl Steinbuch,[35] who is attached to the Technical University in Karlsruhe as a professor of informatica. Thinkers from the orthodox Marxist camp, too, such as the East German Georg Klaus,[36] hold an equally optimistic view of culture, calling for unimpeded use of computer technology wherever feasible. In a certain sense they have no brake or means of restraint on the way to strengthening the information society. In fact, they are of the opinion that the problems and perils of present-day science and technology as science and technology of the first degree can be solved by a science and technology of the second degree, namely, by systems theory and information technology. The consequence of this attitude is the derivation of the norms for practice from the systems themselves. This leads to the strengthening of these systems.[37]

Obviously, philosophical resistance to this development is vitiated if the relation between man and the computer is understood wrongly. All too often philosophers start out speaking anthropomorphically of the computer – it thinks, it learns, etc, – only to end up espousing a computeromorphic view of man.[38] In thinking about the computer, they abstract from humans at the very outset, whereupon the computer threatens to become an autonomous power. As a

---

[34] See "Norbert Wiener: The Father of Cybernetics," pp. 177–211 in Schuurman, *Technology and the Future.*

[35] See "Karl Steinbuch: Cybernetics and Futurology," pp. 213–59 in Schuurman, *Technology and the Future.*

[36] See "Georg Klaus: Marxism and Cybernetics," pp. 260–313 in Schuurman, *Technology and the Future.*

[37] Ellul, *The Technological System* (New York: Continuum, 1980), p. 117.

[38] See F. H. George, *Philosophical Foundations of Cybernetics* (Kent: Abacus Press, 1979), p. 5 and passim.

result people are "thingified" or thought of as mechanical objects. This *technicization* is at the basis of our culture's becoming artificial, routine, chill, uniform, levelled, abstract and totalitarian as a technological culture. This technicization results in the exclusion of the particular, of the unique, and of a rich cultural variety. The tendency towards technicization is apparent today in modern urban planning, domestic architecture, health care and social work, in numerous production processes, in economics, politics, and so forth.

Perhaps the prevalence of abstract, specialistic, scientific thought is nowhere more apparent than in the disappearance of love from modern society and the increase in estrangement and loneliness – in a word, in dehumanization. The love that builds communities cannot flourish in fragmented, universal structures. That is why so many in the technocratic state are heard to complain that no one cares for them, that no one loves them. In the technicized culture the essential bonds of community are severed and exchanged for merely artificial ones. It is for this reason that love cools, that sympathy and compassion vanish from the technocracy, that loneliness and estrangement increase, and that people cry out in protest after protest for love and pity.

In short, if the technicization process is continued and enhanced by means of information technology, then the consequences will be comparable to those of the last phase of the Industrial Revolution, which was not only destructive of nature and the environment but also disruptive of society. In this case, however, the most important consequences will be psychic and social.

New Initiative

The question is not of rejecting the latest information technology. It is precisely one rather of doing information technology *justice* by relativizing instead of absolutizing its *meaning*, in order to generate a meaningful countermovement to the technicization process. A first requirement is that we should resist surrendering ourselves to science and technology and delivering ourselves naively to computer systems. We should assume instead a position of *critical distance*. Having done that, we shall have created room for responsible reflection about the meaning of this technology and its significance for culture. *We should not make all that we are able to make but only just what we need!*

Now, that means we should be sensitive to the threatening dangers of human loneliness, the abuse of power, the aggrandizement of political dictatorships, growing unemployment, and social dislocation.[39] Insight into all these processes requires not only specialized knowledge but also and especially *wisdom* regarded as comprehensive insight.

To attain wisdom we must acknowledge and accept both the individual and communal responsibilities of scientists and engineers, and also their professional and social responsibilities. In general we need not be unduly concerned about the acceptance of the professional or technical responsibility. The educational system and the professional organizations pay a great deal of attention to that. With respect to *social* responsibility, however, there is reason for serious concern.

Education and the professional organizations pay too little attention to that.

Modern technology is not an abstraction but a many-faceted cultural phenomenon involving enormous social consequences. Hence it imperative to render the respective roles of technology and society more transparent. It is also essential, furthermore, to build bridges between technologists and society. That will restore society's confidence in the engineering world. By the same token it is mandatory, as Joseph Weizenbaum of the Massachusetts Institute of Technology has so lucidly stated, that the computer specialist not become addicted to his own theories. As an addict, he would apply his theories without restraint, thereby exacerbating the technicization process.

There are two ways of meeting the need to accept social responsibility. In the first place, scientific and technical work must be responsive to society. Education must meet a clear requirement here.

Specialists shall have to be *generalists* as well, sensitive to the coherence of their respective professional fields with society. *Technological potentiality may no longer be permitted to technicize the social situation. The social situation should relativize technology.* The materialistic attitude must yield to a profound

---

[39] See (1) J. Reese et al., *Gefahren der informationstechnologischen Entwicklung.* Frankfurt: Campus Verlag, 1979; (2) U. Kalbhen et al., *Gesellschaftliche Auswirkungen der Informationstechnologie.* Frankfurt: Campus Verlag, 1980; and (3) S. Nora and A. Mine, *The Computerization of Society.* Boston: MIT Press, 1978.

spiritual conviction in which the material – and technology as well – is relativized. In this way justice can be done to the meaning of technology.

The social responsibility accepted by engineers fosters responsible technology. A first step in the right direction is the methodology of "technology assessment," which studies the economic, ecological, cultural, and social aspects of new technologies. It would be better to go even further and add philosophy and ethics to the educational requirements of scientists and engineers. This is all the more urgent now that the computer has begun to play an important role in education, with the possible side effect of narrowing the student's intellectual horizons. The danger is great that from the student's point of view everything the equipment cannot control or that cannot be programmed will simply cease to exist. Furthermore, an exclusively technical education is patently inadequate given the many and powerful interactions between technology and society.

In the second place, society shall have to assume a most critical prudent stance with respect to the latest information technologies. Wherever computers are to be introduced an inquiry should be conducted of the view of science and technology and the view of society involved. The result of such an inquiry should be decisive for approval or disapproval of the introduction of the latest technology. Measures must certainly be taken in all cases to guarantee that those use the computer are also made *constantly* accountable. In all cases users must be required to make clear *why*, to *what end*, and *how* the computer is to be employed. In a certain sense, this is to speak of an obligatory *social* responsibility, a social testing (or certification, if you will), the proper framework for which must be shaped via politics. Moreover, "incomprehensible" computer programs must be made and kept "comprehensible," so that users are not tempted .to transfer their own responsibilities to others.[40] These personal and social responsibilities are very extensive indeed, since information technologies are fraught with consequences for human beings as individuals and as communities.

When all the conditions I have described are met, it can be seen that computers will not be employed to solve economic, social, or political problems as such. Rather, such problems must first be reduced to *technical* problems, whereupon the *solutions* advanced will likewise be nothing more than technical

---

[40] See Joseph Weizenbaum, *Computer Power and Human Reason* (San Francisco: Freeman, 1976), p. 228.

solutions. Where insight concerning this key point is lacking, the process of technicization will go forward with all conceivable intensity. However, where insight concerning this key point exists, computer technology will acquire a place not of supremacy but of service. The relativity, the abstractness, and the provisionality of the computer's solutions will then be obvious. We shall interpret these solutions cautiously and wisely. Furthermore, we must never forget that even salutary use of the computer can give way to misuse. It is for this reason that, given the growing trend towards information storage, many people are increasingly concerned about protecting their personal freedom and privacy.

Enrichment Or Impoverishment of Culture?

We have returned to the question posed at the beginning: will information technology impoverish or enrich culture? The answer to this question must depend on our view of technology in culture as a whole. If we proceed from the standpoint that all problems can be solved through technology, then the technicization process will impoverish culture and cast it into crisis. Another direction is possible too, of course, namely, that of the enrichment of culture. In that case, however, information technology will have to satisfy ethical criteria.

Bringing norms to bear upon technology will not dislocate or disturb but rather disclose the real meaning of technology. It is our responsibility to choose the path of the enrichment of culture. To make the point at issue perfectly clear, I shall mention a number of *ethical criteria*, or principles, the simultaneous realization of which discloses to us more of the meaning of technology.

In the first place, it is necessary to satisfy the *cultural* norm of differentiation and integration, of continuity and discontinuity, of large-scale forming and small-scale forming, of uniformity and pluriformity. These various components must not be regarded as contradictions. In forming culture it is necessary to do justice to both elements in each case, in order to prevent one-sided and dangerous developments. Naturally, if justice is done only to centralization while the principle of decentralization is disregarded, the result will be an unstable, vulnerable culture or an oppressive technocratic rigidity. If, however, justice is done to both these elements, the result will be the promotion of a stable and richly varied cultural development. This point needs to be taken into consideration – if I may be permitted a topical Dutch example – in planning a national payments circuit for the Netherlands. The development in that direction is irreversible, and it comes at the cost of variety – a variety which while it does satisfy the requirements of pluriformity, nevertheless very possibly fails to satisfy the requirements of efficiency.

Personal and social responsibility likewise demand that we do justice to the *social norm* of communication. It is not superfluous to seek consideration for this norm in an age of communication technologies, so that the danger of manipulation can be recognized from the outset and the fact acknowledged that all should be involved in various ways, shouldering their various responsibilities, in the formation of culture.

Furthermore – to mention yet some additional ethical criteria – justice needs to be done to the norms of economy (stewardship), aesthetics (harmony), jurisprudence (justice), and ethics (concern, love).

All these norms taken together assure that justice is likewise done to information technology itself. In that case information technology comes to its proper meaning and makes its proper contribution to the meaning of technology in general. This perspective is diametrically opposed to the technicization process. Pursuit of the cultural perspective of *service in responsibility* affords every conceivable opportunity for technical creativity and inventiveness while at the same time providing insight into the social and cultural limits of technical activity. Given the perspective of service in responsibility, it is possible to break through the dilemma that juxtaposes something desirable – increased possibilities for human creativity, provided especially by information technologies – against something undesirable – the suffocation of human creativity that results from ever more complex, more massive technologies. As the amount of *technical* work we need to do is diminished by information technology, we should take every opportunity to engage in other kinds of work, such as social, health, and artistic work. Reflective and intellective work will also flourish. At bottom, the insight at issue is based on the *meaning of technology*.

The meaning of technology is vastly rich and deep. Certainly it can be said to include the following: Technology can ease the difficult circumstances in which people live "by nature." Technology can lighten burdens and conquer diseases. It can afford an enlargement of life's opportunities, i.e., it can relieve the burdens and difficulties of physical labor. Technology can free people from deadly routine, avert natural catastrophes, provide food and shelter; work can be made mental, and social security and prosperity can be advanced. Real information and communication can be expanded. In all these matters, of course, human responsibility is increased. Material prosperity and spiritual well-being should advance together. Through technology, the cornucopia of gifts and qualities of individuals and nations should be enhanced and enjoyed. Technology creates room for a rich cultural development, for numerous and many-faceted cultural activities, and much more.

In short, every over-estimation of science and technology – together with the ultimately pernicious threat such over-estimation inevitably entails – should be rejected in favor of the responsible pursuit of a cultural perspective offering the prospect of a meaningful development of the most modern information technologies. Thus, far from being forbidden, *computer power should be advanced in the perspective of service.*

CHRISTIANS IN BABEL (1985)

Introduction

This book calls attention to our culture's advances in science and technology. I plan to focus on the problems and threats which accompany this development, problems engendered by phenomena such as nuclear energy, nuclear weapons, computer technology, unemployment, biotechnology, genetic manipulation, environmental pollution, the depletion of natural resources and energy sources, the legalization of abortion, and the demand for world government.

Analysis of these developments and their spiritual background reveals that we no longer live in a Christian culture. There is a startling correlation between the development of science and technology on the one hand and the process of de-Christianization on the other.

To understand this correlation, we must evaluate both the development of science and technology and the spiritual forces behind this development in the light of biblical prophecy concerning the end of time. The closer we come to the day of Christ's return, the more we are able, through *faith,* to see the approaching Kingdom of God. The light of biblical prophecy shows how the development of culture will bring enormous scientific and technical accomplishments which will, ironically, bring about the decline and doom of culture.

There was a time when people spoke of our Christian culture. That term is now meaningless.

From a biblical point of view our culture can probably be identified as *Babylonian.* Here man worships various gods as he builds whatever his science and technology enable him to build.

I intend to assess this Babylonian culture and call attention to the position of Christians as *exiles* in it. The exile theme is eminently biblical. As exiles, we can have no grandiose pretentions, and yet we need never lose heart. We shall have to learn more about our *responsibility* within our current culture. Exiles yearn to return to the living God. We must reject the false deities of our society and continue to fight the good fight.

1. Currents in Futurology

To provide a sound analysis of our current culture we will need to know more about the field of spiritual forces behind the development of science and technology.

We will start by investigating two major currents in humanistic futurology. When we compare both currents to Christian expectations of the future, we will see the problems resulting from science and technology in proper perspective.

Evolutionary Futurology

Within modern futurology science has a religious function. Man entrusts his life to science and its powers. Scientific technology has become a deep source of inspiration for people anxious about the future. Contemporary man assumes an evolution process, and then applies his knowledge to eugenics to try to provide future generations with better "inner equipment." Concomitantly, modern scientific technology is seen as a means to improve man's "external equipment." In the past, evolution was thought to operate automatically, without man's intervention; today, man has begun to believe that he can direct his own and society's development.

Controlling the future. The process of rapid and unceasing change Western culture has known since the Second World War has been frightening. We find ourselves in a headlong rush which, if not controlled by man, threatens a dark and dangerous future.

Futurologists believe that this rapidly closing future can be rescued only if we learn to control it through the judicious use of science. Their efforts to control the future depend on technology and its scientific method. Applying the scientific methods of control to non-technological activities such as economy and politics will presumably bring them under control. Thus they see technology as the engine of cultural progress and our increasing scientific knowledge as the fuel that fires the engine. They have pinned their highest hopes on the computer in relation to the fields of systems design and cybernetics. The computer, they say, is the most potent tool to help us research, guide and control the future. Whatever problems and threats now exist – even those spawned by technology itself – could be solved by even more recent discoveries in science and technology.

Scientific method in the hands of the modern engineer has proved to be very successful in controlling matter. If this successful progress is to continue and expand, futurologists argue, the scientific method will also have to be applied to areas that are not specifically technological. This means that man, his society, and the future will have to be manipulated by the scientific method.

Modern planners think in this evolutionistic and technocratic way. Therefore, they advocate a kind of technological imperialism within the supremacy of science.

Motive. We know that our culture is on the wrong track when we analyze the dominant motive guiding the technocrats. That motive is completely secularized. "We will have to be as certain about tomorrow as we were about the past," said one technocrat. The old humanist ideals of "knowledge is power" and "to know is to foresee" are highly operative. Man strives to redeem and maintain himself. Through his knowledge and capabilities, his demands and desires, he makes himself *the measure of all things.* He is determined to regain Paradise lost through his own strength. The futurologist Olaf Helmer promises that if the technological-scientific control method is applied consistently to man and his society, not only will suffering disappear altogether and wars become a thing of the past, but man will be able to luxuriate in unparalleled material prosperity.

The Christian must ask himself whether he should concern himself in a scientific way with the future. The answer is yes. Our society has become so complicated in its dynamics that it is virtually impossible to say anything about the future without resorting to scientific analysis. But the Christian should not allow scientific knowledge to become the norm. Scientific knowledge should serve to enhance, not eradicate, the Christian's responsibility.

When man's responsibility, not science, is made central, we may still expect new, positive opportunities, especially in the sense of unexpected solutions to difficult problems. However, many secularized planners seek to obtain solutions in an arrogant way, using only the tools of technocratic imperialism. They find then, to their dismay, that the solutions cannot be reached. *To succeed in our struggles concerning the future we need to reject every pretension that science and technology can help us save ourselves and secure the future.*

If we fail to recognize the norms for the future, fail to see man as the created image of God, and fail to put away secular motives, evolutionary futurology will give us the opposite of whatever we seek. Man will be imprisoned by ever-expanding forces which will ultimately be united and mobilized in one single global power. That is, the future sketched by the evolutionary futurologists will be a technological society in which everyone and everything will be reduced to component parts of one large, comprehensive, totalitarian system. Within that system, a human being will be a cog, fully interchangeable and replaceable. And a technological world state will be the personification of the future's power.

Such a future will have no future. It will bring oppression and insurmountable problems, some of which have already been signaled by the Club of Rome. However, most people, united in their common materialist concerns, seem content to follow the course outlined by the technocrats. Those who resist this trend with a passion are the revolutionaries.

Revolutionary Futurology

The *revolutionary* futurologists object strenuously to the static, rigid future perceived by the technocratic elite. Their guess is that such a future would serve only the self-interests of that elite; present injustice, suffering, evil and oppression, will be reinforced rather than eliminated. Moreover, they fear that nuclear war, increasing environmental pollution, and the ongoing depletion of minerals and energy resources – all ethically unacceptable to them – will be logical, viable options for the technocrats. The technocrats take only limited measures against these problems, merely postponing them. At a later date, they will surface again in a much more acute form. According to revolutionary futurologists, this happens because technocrats subordinate history to the progress of science and technology. Their mistake means that current economic cultural forces, together with all their evils, will only be reinforced in quantum leaps.

Revolutionary versus evolution. Revolutionaries agitate for revolution. To them, revolution, with its protest, conflict and action, is the eternal combustion engine of history. The fuel for this engine is utopian fantasy – which oddly enough uses science for its fulfilment. Revolutionary futurologists, most of whom are neo-Marxists, sense that the fulfilment of history and the future will be frustrated by the technical-scientific approach. They fear that man will become a prisoner of production forces; as his labour is reduced to purely productive labour, its cost to the profit system must be offset by escalating consumption. Revolutionaries therefore feel that man will become a beast of burden in the technocratic structure of society. They jump to the defense of people who are not even aware of their predicament, partly because of the mind-numbing forces exerted on them by the production-consumption syndrome and partly because of their own unlimited lust for consumption. Revolutionaries insist that the materialism which is basic to all evolutionary futurology cannot redeem man; instead, it will bring him to his doom.

From this vantage point, it seems that revolutionary futurologists and evolutionary futurologists (planners) are polar opposites. Revolutionaries refuse to rationalize the current situation and stridently oppose the main tendencies of current culture. Instead of presenting history as an evolutionary process, they emphasize the discontinuity of history, insisting that unique, innovative, and individual acts are the keys to escaping man's impending doom, the keys to liberty. Their ideal is not a planned, determined future, using the tools of science and technology; instead, they adopt a wait-and-see attitude concerning the future. The past, they argue, cannot simply be prolonged into the future by extension, and existing political and economic forces which use science and technology as their tools may not be allowed to continue and consolidate their strength indefinitely. These forces must be dismantled and replaced by more imaginative approaches which again using science and technology can lead mankind into a truly utopian future. Their utopia will give to man a kaleidoscope of options that should ensure a free, enriched future.

Such a utopia, however, cannot remain wishful thinking. Even while the current structure of society is being dismantled, the new utopia should be taking shape. Revolution is the beginning of a developmental process in which man can once again become himself, no longer alienated from himself, as he is in a burgeoning technocracy.

Method. The method of the revolutionary futurologist is therefore based on a radical negation of the premises underlying past and present.

Revolutionaries resist existing social forces – "the establishment"– and hope to bring them down in open conflict. Only when the establishment has been brought down will there be new potential for the future. Only then will the rule of independence, liberty and freedom have a chance. However, not even they are certain that their goal will ever be achieved. They are not particularly optimistic about the future, partly because conditions in Marxist countries show that revolution offers no guarantee of freedom. There is no guarantee that the revolution will not be channelled into a more comprehensive and technocratic dictatorship, as happened in the Soviet Union. Current revolutionaries are therefore much more radical in their perception of revolution than old-guard Marxists. Orthodox Marxists felt that a single revolution would guarantee redemption. They were wrong. Current revolutionaries think that revolution must be an ongoing, perpetual process. Thus, only when a society is in a state of perpetual revolution can the progress of science and technology be guided into avenues of freedom and peace for mankind.

To prevent suppression, exploitation and disaster, science and technology should not be placed in the hands of the establishment. The power of science should instead be placed in the service of revolutionary creativity and creative revolution in order for man to fulfil himself. The road of perpetual revolution is the road to salvation. Karl Marx's cardinal error was to propose a single, culminative revolution. However, when that single, terminal revolution came to an end, oppressive forces again emerged in the form of an elite which proceeded to turn man's newfound freedom into slavery. By contrast, when revolutionary destruction is perpetual, freedom – in the sense of creativity – will feed on the perpetual disorder.

I noted that current revolutionaries are far from optimistic. For one thing, the establishment seizes every opportunity to curb revolutionary forces and thus prevent fulfilment of the revolutionary, utopian ideal. Moreover, the masses see no need for a revolution, let alone a *perpetual* revolution. The masses, revolutionaries generally feel, are quite content to sleep the sleep of the dead. For this reason revolutionaries appoint themselves as the protagonists of the struggle, thereby creating a new revolutionary elite. But it was precisely their aim to do away with all forms of elitism!

The Spiritual-Historical Background

For technocrats, history is absorbed into the continuity of scientific-technical progress. But, revolutionaries counter, such history yields a society of power expansion, suppression, lack of freedom and mammoth cultural catastrophes.

But do the revolutionaries really offer an "open" future? Despite their claims, they are not so sure.

And in fact, they merely *assume* that destruction of the old will produce a better "new." Moreover their faith that revolution will somehow ennoble man's ambitions cannot conceal their fear that revolution could well lead to arbitrary excesses. This makes the future of revolutionaries an unknown factor. Beneath their stridency, despair lurks.

Conflict between the two. A cursory reading of the two positions would suggest that they were complete opposites. And, indeed, a constant, almost antithetical battle rages between them. That conflict intensifies as the problems and catastrophes faced by Western culture intensify. To truly understand the conflict, we have to locate its origin and course in history.

Why is it that one part of Western culture has been so arrogant in its use of science and technology to support the dominant economic and political edifice, while another part, the revolutionary segment, has been so hesitant and guarded and, with respect to current technological-scientific culture, downright despairing?

Reformation and Renaissance. The development of science and modern technology became possible in Western culture when early modern European man gained new understandings of history, nature, the development of culture, and human freedom. Since the end of the Middle Ages, people have come to see this world as one that must be developed.

Yet there has never been any unanimity concerning how to go about this development. From the beginning, there were two spiritual movements that contrasted sharply. For example, they answered the question of origin differently. The Reformation defined man as called to freedom, yet responsible to God. Thus, his freedom meant service and stewardship. By contrast, the Renaissance saw man as autonomous and self-sufficient, a creature who could discover and support himself. The fundamental disagreement between the Reformation and the Renaissance was not immediately clear, and so each side borrowed from the other. However, the Renaissance quickly secularized certain Christian ideas. This was especially true in philosophy and knowledge. Because philosophy and knowledge had little impact on practical day to day life, the influence of secularization began slowly. As the influence of philosophy and science on culture intensified, so did the influence of secularization. At that point the idea of human autonomy, always a central thesis in humanism, began to assume leadership throughout Western culture.

The god of the philosophers. Timidly at first, but more aggressively later, philosophy posited human freedom as a right, not a gift. Western philosophers and scientists came to see man as a creature who deserved his own freedom. They attempted to found this freedom in man's autonomy, and later they tried to give scope to this autonomy in the fields of science and technology.

The first renowned philosopher of the new age, René Descartes, placed rational man at the center of the universe. Descartes' anthropocentric philosophy gradually permeated Western culture; eventually, people came to worship man's rational endeavors, especially in the fields of science and technology. This religion of man also surfaces in terms like self-interest, self-determination, self-realization and self-sufficiency.

At first the philosophers continued to talk about God, but this new god was clearly not the living God of Holy Scripture. The god of the philosophers was a god created after man's own image. For a while this theoretic god still functioned as a necessary cause, but as man became more rationalistic and anthropocentric, the deity gradually disappeared from the scenario, leaving a thoroughly humanistic philosophy and science. Knowledge and science became the tools with which man would clear a path into the future. That path began to open when modern technology began to develop. Gradually the idea took hold that both man and the world could be brought to fulfilment through the use of science and technology. And *Christian eschatology steadily retreated in the face of technological expectations of redemption.*

When the combined efforts of science and technology began to yield material fruit in the nineteenth century, the secularized idea of progress made deep inroads among the masses. Many Christians also embraced the idea of progress, and then found themselves with an ambivalent lifestyle. Thus the Christian element of Western culture was attacked by pretentions of human autonomy and undercut, even from the inside by secularized expectations of the future.

Apparent antithesis. But this does not explain why evolutionary and revolutionary futurologists nearly always oppose each other. For that we must once again return to the spiritual history of the West. Both social evolutionists and social revolutionaries, proceed from the premise that man is self-sufficient, free, and autonomous, a creature who can stand on his own two feet without the help of a living God. This "free man" now pretends to be lord and master both of himself and of the world around him and, finally, of the future.

Social evolutionists, planners, and technocrats base their certainty and trust on science. They tie their idea of man's freedom to the idolization of his reason.

Therefore they also deify the results of their scientific enterprise. The natural sciences are of ultimate importance to them, because the reality in which they live conforms precisely to the laws of mathematics, physics and mechanics. Everything is absorbed into the chain of cause and effect, even, in the final analysis, autonomous man. This, of course, destroys human freedom, though man also fights the destruction of his freedom.

The Enlightenment. The question of man's freedom and its destruction constitutes the inner tension of the philosophy of the West. The deterministic science postulated by free reason imperils man's freedom. Sometimes the struggle in philosophical-scientific thinking between science and freedom is subdued; at other times it becomes quite fierce. Since the days of the Enlightenment that conflict has grown beyond the merely philosophical and theoretical to the cultural sphere. After all, it is in the spirit of the Enlightenment not only to know reality logically but also to order it logically. The intent may be to create, through scientific reason, a society that serves the self-realization of human freedom. However, then science becomes supreme and autonomous, and the results of science determine society's history and shape its future. Evolutionary futurologists (planners) have fallen into this trap. Under their influence, man is seriously threatened. The threat intensifies as cultural development becomes more dynamic, more complex and more obscure to the common man.

The degree to which modern man becomes ensnared in the net of scientific-technical culture determines the intensity of his struggle to liberate his freedom from the technocrats. His resistance to evolutionary futurology takes shape in his cultural activity. Through perpetual revolution, he seeks to cast off the chains of the forces of science and the yoke of massive technology. Today it is primarily the neo-Marxist revolutionaries who act as spokesmen for resistance against the ever-burgeoning technocracy. Standing squarely on the principle of the autonomy of man, they turn against an establishment that seeks to usurp and control both history and the future. This is why they plead for imagination, creativity and for the subjective historical freedom of man. They intend to redeem man's autonomous freedom through the revolutionary destruction of the existing order.

Thus, the battle between evolutionary and revolutionary futurologists is at heart a battle between two types of humanists who agree on the fundamental claim that self-sufficient man needs no God. Yet that fundamental agreement is what divides Western culture against itself.

The next section will show how this struggle between types of humanists will come to dominate the future of the West. Any proposed future in which man and his kingdom are central will then have to be evaluated in the light of Christian eschatology, which is the light of the future of Jesus Christ and the coming of God's Kingdom.

Views on the Future: Futurology versus Eschatology

So far, I have discussed evolutionary and revolutionary futurologies. They give two extreme views of the future which are diametrically opposed to each other. I have tried to describe their forces fundamental to their operation and the tendencies that characterize these two futurologies.

What do these two futurologies mean for the future of Western culture? There are problems and dangers in both positions. We will evaluate them in terms of the Christian expectation of the future, otherwise known as eschatology. As we shall see, humanistic futurology both notices, and fails to understand "the signs" mentioned in Christian eschatology.

Evolutionary futurology defines the future in scientific terms and seeks to control history and development by the power of technology. It seems to believe that its own image of the future takes into account all the diverse possibilities for the future. Evolutionary futurologists lack the modesty to anticipate completely unknown factors. They try to integrate various contingency plans for the future into one comprehensive, universal plan. The execution of such a universal plan, however, means the increasing restriction of personal freedoms, since the execution requires a collectivization and concentration of power.

In the twentieth century, the drive to rule the world through technology has had disastrous consequences. The humanistic ideal of permanent peace, presumably attainable through science and technology, has been blown up more than once. The ideal of universal material prosperity and progress has been realized only in the Western world, and at the expense of the rest of humanity and a battered environment. We face an alarming shortage of natural resources and energy. The idea of progress, with its belief in limitless production and consumption, has been checkmated by the limitations of creation and the finiteness of its resources. And although modern man strives to make both himself and his world more "human," he only manages to alienate himself from his fellow man and from his world.

Whenever technocrats are confronted with such problems, they go back to science and technology to try to solve them. They develop new strategies in which they retain their power to violate human freedom. Various world congresses and forums held to deal with critical problems only reenact this pattern.

This concentration and escalation of power, and the catastrophes it brings – as, for example, irreversible environmental pollution and the perils of nuclear energy – are opposed by revolutionaries, who always come to the defense of oppressed humanity. They insist on absolute freedom in an historical or cultural sense. But when they strive to give cultural form to this freedom, they, too, must utilize the powers of science, technology and modern organization. Invariably, a revolutionary elite arises.

For this reason, revolutionaries will either cross over into the camp of existing powers or will themselves mobilize a force that seems more totalitarian and dictatorial than the power they are fighting. Thus they betray their own revolutionary ideas, and their revolution devours its young. The cultural tension between technological-scientific control of the future and man's attempt to escape this control keeps escalating and intensifying. To remain consistently revolutionary, the revolutionaries have to become more and more radical, resorting ultimately to violence. In the end, there will be chaos.

Technocrats control the reality of this cultural tension, and therefore enjoy an advantage over revolutionaries. They do not put their faith in man as a purely historical being, but rather, in man as a rational, scientific being, and in the technology itself – in, for example, systems analysis, cybernetics and the computer. Additionally, the masses have no alternative but to put their trust in the technocrats, and they choose to do so because their faith in the blessings of science and technology has not yet been severely shaken.

Man without God

Science and technology are often blamed for the mess we are in, but I assert that this is a devastating misunderstanding. The origin of the problem is man. Man has come to see himself as the Alpha and Omega. Beyond the here and now, modern man reasons, there is nothing. The meaning of history and of life is confined to observable reality. Since access to a living God has been closed, mankind has put hope in what science and technology can make of the future. The spirit of Western man has been touched by the vision of perfection and fulfillment. After all, he has learned of Paradise. But when he rejects God's love and His revelation, Western man no longer uses his God-given freedom in his capacity as God's steward, but instead usurps it as he tries to create a utopia which reflects his own beliefs. When man divorces his power, freedom and calling from obedience to Jesus Christ, he does not relapse into paganism but instead becomes post-Christian man. With no religious trust in God, he bases his new, secular religion on science and technology, two areas whose development, ironically, were made possible because of the Christian faith.

Lawlessness and Meaninglessness

When man rejects revelation and history and exploits modern technology and its potential to unprecedented proportions, technology assumes sinister and demonic traits. When he is in control of the earth and all its treasures, modern secular man becomes tyrannical, exhausting both himself and his environment.

Moreover, as secularization intensifies and Christian resistance to the process decreases or disappears, modern secular man will become lawless and nihilistic – a figure aptly described as the man of the end of time.

Nowadays, the idea of unlimited progress is obviously in conflict with the limited potential within creation. Thus, material progress and its values have become matters of discussion. Secular man's hope for an earthly paradise seems to be arbitrary. Instead of cherishing an optimistic view of the future, people have become profoundly pessimistic. But the "answers" provided by these pessimists are also far from uniform. The house of secular humanity is badly divided against itself.

An arbitrary, technocratic dictatorship stands diametrically opposed to a revolutionary idea of freedom which believes that every technocratic success must necessarily be destroyed. There is a nihilism fundamental to the mechanical order of the technocrats, and it stands opposed to the nihilism of revolutionary chaos. Technocrats will mobilize all their forces to create a technically streamlined society, in the hope of preventing catastrophe; meanwhile, in the name of freedom, revolutionaries seem bent on the self-destruction of culture.

Culture Heading for the End

Unless man reverses himself radically and turns to God, the conflict and nihilism of Western culture will probably intensify while it travels down the road to cultural decay.

A culture without God always carries the seeds of decay, as so ably demonstrated by Klaas Schilder in his book *De openbaring van Johannes en het sociale leven* [The Revelation of St. John and Social Life] (1924). The forces of decay will ultimately bring its total ruin. Nevertheless, since God will not abandon even such a culture, man will come to experience the meaninglessness of a culture without Him.

The course plotted by humanists whether they be evolutionary or revolutionary, will come to a dead end. The death already evident in our culture is a sign of the expectation confessed in Christian eschatology: the return of Jesus Christ for the purpose of establishing God's Kingdom. Christian eschatology is diametrically opposed to humanist futurology. Eschatology does away with the dilemma of pessimism and optimism because, taught by God's Word, it confesses, that God Himself, through Jesus Christ, is the Lord who constantly controls history. His decree includes both the believer and the unbeliever, and it unfolds towards the coming of His Kingdom. In his Christian faith, hope and expectation, the Christian is called to serve that Kingdom, a service that includes his scientific work and technical achievements.

2. Man's Culture Reduced to a Scientific and Technological Model

Because of the dominant influence science has on culture and because of the stamp modern technology puts on our culture, that culture is based, more and more, on a scientific and technological model. Once we have defined the scientific and technological model, we will have a better understanding of the background to our culture. We will do so with six points.

1. Science keeps becoming more independent, autonomous, and self-sufficient. People are asked to accept science as the only truth, and thus to accept with religious certainty the conclusions science draws.

2. Science is also being called on to be the main instrument of human control over the world. Man increases his power over reality by exploiting scientific

means, particularly in the development of industrial technology. Many believe that modern technology is nothing more than applied science, and so culture according to a scientific model becomes culture according to a technological model. Expressed differently, scientific, rational control over nature and man's society leads to technological control of reality.

3. This concentration on control comes from man's religious yearning to realize and liberate himself. The universal longing for liberation is directly related to the potential of science and technology. Science and technology take a messianic role. Through science and technology, man hopes to be liberated from misery and suffering and to find material happiness.

4. Until the modern industrial revolution, the idea of progress motivated only men of science. But once the material prosperity of the industrial revolution had become available to the masses, the masses accepted progress as an article of faith. Philosophies such as positivism, Marxism and pragmatism all contributed to the belief that modern technology would function as a liberating force.

5. Economic and political forces have done much toward building the scientific model and the technological model. Only through these forces could large scale development have taken place. For this reason, the hidden economic and political forces should be criticized – which neo-Marxists are happy to do. However, neo-Marxist criticism does not penetrate to the root of the matter; it may lead to a change of players, but the new players will simply continue to build the scientific and technological model.

6. Because the religious dynamic of the world is apostate, culture will become more and more secular. The deification of science and technology goes hand in hand with resistance to the Christian faith. Transcendent reality has become a myth, a projection. Rational, technological man himself, it is believed, will eventually achieve a utopia which he himself has designed and will control and subject. In this scientific-technological world, man will be lord and master, independent and sovereign. This world is a thoroughly *godless* world. All its problems are to be solved through democracy, which can channel science and technology into truly redemptive paths.

Problems and Perils

In this way of thinking, reality is robbed of its meaning. Reality is no longer creation, a richly differentiated, deeply alive entity borne up by the Word of God. Instead, man "creates" a world impelled by a technological dynamic, and then tries to accept it as the real world, though it is devoid of meaning. Modern man equates the technological world with total reality. Of course, created reality does not allow for such a reduction. All the aspects of created reality cohere in a meaningful unity. But if man denies this God-centered coherence, man's development of reality bring about his own doom. Doom may come slowly and cumulatively, but it will come.

Establishing an independent technological world is impossible. The growth of technological development is limited to the potential in created reality. Energy sources and mineral deposits are limited. Environmental problems, such as the pollution of seas and oceans and the contamination of soil, water, and air, show how current technology dangerously exploits the environment. Technology also betrays serious internal tensions around such issues as nuclear energy and biotechnology. Increasing reliance on the computer has already caused a great deal of unemployment, social dislocation, loneliness and alienation. The specific and unique functions of every person, the individual and creative responsibility of man functioning within the context of a full-orbed world of experience, are being systematically eliminated from the technological model of reality. A culture defined as a scientific-technological unity becomes torn apart internally. Externally, it is the reflection of a cold, uniform, impersonal and homogenous abstraction.

Creating an independent scientific-technological world and letting it dominate and destroy the full-orbed world of experience brings about the problems and perils I have mentioned. The problems show that reality is one, created by God and maintained by God. The problems also show that scientific knowledge is always impelled and permeated by a pre-theoretical or supra-theoretical knowledge. The uniqueness of the Christian view is that pre-theoretical or supra-theoretical knowledge is based on faith founded in God's revelation. This enables the Christian to be both critical and appreciative of science and technology. Seen from a Christian point of view, science and technology can be meaningful only if they remain limited areas of the totality of human experience and do not become models for how all of the other aspects of life are to be organized – to their hurt.

"World of Experience" and "Scientific-Technological World"

What do we mean by "our world of experience"? That world is the world in which we live, hope, suffer and struggle; it is the world in which we see things simply, in which we feel and love; it is also the world of faith and trust; in fact, faith and trust are the pivotal point of that world. This world of experience is original and primary; it cannot be fully comprehended; it is complex, concrete, full, richly varied and profoundly inscrutable. Every human activity and its meaning-science and technology and their meaning, belong to this world. Our knowledge of this original, primary world of experience comes from being inextricably related to it and involved with it. It is an intuitive knowledge which both precedes and transcends all scientific knowledge.

The second "world" is the world of philosophy, science and the application of science. Thus it is also the world of scientific and technological control. To construct a scientific and technological model for the whole of reality, as many people do, is to subordinate the first world, the world of primary, intuitive knowledge, to the second. Then the scientific-technological world begins to dominate the everyday world of experience.

Technocrats have the illusion that science provides the only, true, complete and concrete knowledge of reality – a belief that arose during the time of the Enlightenment. However, when the abstract, sharply reduced world of science becomes the primary world, the genuine primary world of total experience is reduced to a scientific abstraction. That reduction will eventually end in destruction. The tendencies towards scientific and technological models of reality can be seen in modern urban growth, industrial policy, housing, health care, social work, the economy, politics, and defence. Fortunately, because the real world of experience refuses to disappear, technology will never be completely successful. Nevertheless, as power is concentrated within a technocratic society we become aware of the disappearance of love, which cannot possibly flourish within the cold, uniform structures of such a society. After all, love orients itself primarily to the specific and unique. The degree of social welfare provided in a technological state cannot alter that complaint that "Nobody really cares about me." Within a technocratic culture, essential bonds of human communion are severed and replaced by artificial ones which cripple love, destroy compassion and empathy, and increase alienation and loneliness. People who suffer the agony of such a cold, impersonal world make public a host of protests and claims.

What lies behind man's drive to develop science and technology? The motive, it appears to me, is man's yearning to control reality fully through his thought and actions. Man's ambition is to control the origin, existence and destiny of all things, subjecting all things to himself. Man keeps trying to break reality down into its smallest, basic elements in order to reconstruct it according to his own power structure.

This fundamental motif was already apparent in the Fall, but it was not until after the Renaissance, during a period strongly influenced by modern humanism, that this drive was reinforced by the energy of modern natural sciences and technologies.

Protagonists of the Renaissance bade farewell to Christianity. They continued to use Christian terminology, but in a thoroughly anthropocentric (man-centered) perspective. Creation was no longer considered to be the handiwork of God, but rather man's own handiwork. The Fall, according to humanists, was not a denial of God, but rather a denial of the self. Redemption is not the restoration of communion with God through Jesus Christ but rather the assertion that man can learn to stand on his own two feet. Belief is not reliance on God in Christ, but rather faith in oneself. And finally, the future is not whatever God places on man's path, but rather the organization of the world according to one's own insights. The spirit of the Renaissance has permeated the thinking of most modern philosophers and scientists – including the Enlightenment, modern philosophy, positivism, Marxism and materialism. The spirit of the Renaissance, featuring self-reliant and self-fulfilling man, has dominated developments in economics, politics, science and technology. Man has become the measure of all things in most sectors of culture, and science is his instrument for controlling reality.

Scientific rationalism has pushed technology to enormous proportions. At the same time, cultural development has been retarded – a sad fact which few people notice. Most people, motivated by materialism, consider non-technological matters unimportant.

Other motifs have sprung from the same root of man's pretended autonomy and self-sufficiency to bolster rationalism. Most importantly, there is the motif "technology for technology's sake." Whatever can be made, must be made, and the bigger the better. And so technical development spins beyond man's control. Man may pretend to be lord and master of technology, but he becomes, in fact, its slave. People become prisoners of their own work when they refuse to think about appropriate norms for technology. Problems of the environment, and the dangers associated with nuclear energy, computer technology and biotechnology warn us that technology is becoming an absolute power which threatens both nature and culture. And its growth seems to be out of control.

A second principle that plays a large role in technical development holds that technology must serve economic power. Technological development becomes totally subjected to the profit motive. Other norms are paid scant attention. One of the painful results has been with widespread environmental pollution. The aberrations brought on by a society dominated solely by economic motives have brought about serious problems. The potential blessing of technology has turned into a curse. Potentially man's friend, technology has become his enemy.

We cannot blame only the philosophers, scientists, engineers and economists. Many people not directly associated with technology are also dominated by a materialistic spirit, so much so that they ascribe to technology messianic power. Blinded by insatiable yearning for material prosperity, modern man idolizes technological development as a means of obtaining ever more consumer goods and material blessings – his kind of happiness.

Thus, both inside and outside the process of scientific-technological development, we find that the current problems and perils are brought on by people who develop and build without norms. The anormativity of their production is the result of their pretension that they, not God, determine the development of science and technology.

As pointed out earlier, modern man has been tricked and trapped by the power science offers.

Science, with its abstraction and reduction, gives knowledge of only a part of reality, not of its whole. The unlimited application of science amounts to a reduction of reality. Tremendous things may be achieved, but the reduction may well lead to the eventual annihilation of reality.

If, in order to solve existing problems, technocrats turn to yet another field of science, the problems may temporarily be suspended. But they will reappear in a more menacing measure later. If all of this technology finally results in the establishment of a global technocratic dictatorship, then there will be no more room for human freedom and responsibility. Man will then become prisoner in a universal concentration camp.

3. Babel Culture

We now come to the question of how best to describe the development of the scientific-technological culture. If Christians can thoroughly understand this culture, they will be better enabled to consider their own responsible involvement in it.

We are confronted with many problems and perils, including the possibility of a global nuclear war. It is becoming increasingly clear that, whether through nuclear war or through accident, a global catastrophe is not at all impossible. We will have to subject that possibility to the Word of God. We face the greatest and most pressing problems of life and death in global proportions. This is a "sign of the time."

How do we perceive that sign and what is our reaction? Perhaps Christians will have to realize that they are facing an apocalypse. Perhaps we still underestimate the potential for destruction, especially concerning nuclear warfare. The consequences of such a war would be practically indescribable. And should any future generations survive, they, too, would suffer terribly.

What kind of culture can tolerate such a threat? What should the Christian's attitude be toward it?

Such a culture can well be called a Babel culture. A Christian's responsibility in such a culture cannot be denied. The Christian must appeal to a return for a cultural life *coram Deo,* to a responsible cultural development. At the same time, the Christian must assess his time prophetically; he must remind the people that indescribable disaster will take place unless there is repentance. Christians will need to proclaim that political solutions cannot alter the current course of our modern culture. They will maintain that beneath our current cultural morass is man's radical religious choice. Christians must assess the spirit of current culture on the basis of God's Word, and on the basis of that Word they must look for properly normed political responses to today's problems. A prophetic assessment and analysis of culture – a task to which the church has been called – will enable Christians, including those active in politics, to exercise influence in the broad developments of culture.

What does the term "Babel culture" mean? The Babel motif, dynamic self-willed ambition, harks back to man's fall into sin, and has often made itself felt since. Currently, however, this motif has reached unprecedented prominence, and for two reasons:

First, we now live in a secularized culture, which no longer concerns itself with God and His commandments.

Second, this secularized culture now has at its disposal tremendous scientific-technological power. Let us look at the combination more closely. We notice that the Babel motif manifests itself in both an integral and global fashion in our current culture; it keeps pulling science, technology, economics and politics into one massive entity. These various sectors of culture reinforce each other's strength as together they head toward godlessness. Has this culture not been described in Revelation 13, which prophesies that the beast of political power will increase its strength toward the end of time by mobilizing the beast from the earth, the beast of the powers of science and technology? Thus science and technology, now in the service of politics, will display power, signs and deceptive miracles (compare 2 Thessalonians 2:10). Within such a culture, material prosperity will be interpreted and even worshiped as progress. Mankind will consciously choose for the things of this earth rather than for the things of heaven.

Development will become unprecedented overdevelopment of science, technology and economics, and overdevelopment will become exploitation. What was alleged to be progress will turn out to be regression, a kind of retrogressive development. The appearance of material prosperity and welfare materialism is a real and deceptive threat. Catastrophes in the environment, the depletion of natural resources, the exhaustion of energy sources, the increasing alienation between people, and the growing chasm between rich and poor nations show that this powerful Babel culture is threatened by impotence and decay from within. Again we see the prophetic truth of Revelation 18. Verses 11–14 make clear that the demise of the Babel culture will coincide with the end of the mineral, vegetable and animal kingdoms as well as with the end of the world of men. Man's self-willed ambition has put culture on the road to doom. That road leads to the kingdom of man, but since man is incapable of ruling justly, it also leads to destruction and death.

As I mentioned earlier, the Babel motif comes from man's wish to regain Paradise lost through his own strength. Men try to make their own legacy on earth, to create eternal rest and the conditions for their own utopia. Cain and Nimrod were the prototypes of such men. The Babel motif recurs repeatedly in Scripture; Sodom and Gomorrah, Egypt, Babylonia and Nineveh all showed the signs of this culture. The Bible often prophesied against these forms of self-willed ambition. Revelation 11 shows us that even Jerusalem, the chosen city of God, could become a form of Babel. This should tell us something: it is precisely in Western, post-Christian culture that man's self-willed ambition has run wild, inspired by the powers of science and technology.

Much could be said about the history of this development. But it is even more important to see what the Bible says about the religious direction of the Babel culture. I am reminded of 2 Thessalonians 2:1–21 and 2 Timothy 3:1–9. If these passages are read against the background of Romans 1:16–32, the religious direction will become clearer. Romans 1 shows us what happens in a pagan, heathen culture and it reveals to us the religious direction and commitment of non-believers. 2 Timothy 3 and 2 Thessalonians 2 describe the religious direction of neo-pagan, secularized, selfish man and the cultural impact this direction has. Romans 1 parallels 2 Thessalonians 2 and 1 Timothy 3. Looking at modern, post-Christian and even anti-Christian culture, we can say that its evils and disasters have come about precisely because of the potentials offered by science and technology. The modern forces of science and technology are enormously threatening, and for that reason demonic. The Bible shows us why they are demonic; it is because our current culture not only rejects the revelation of God as seen in the works of His hands – as took place in the pagan culture described in Romans 1 – but also rejects the revelation given in Christ. And just as the rejection of creation as revelation had disastrous consequences for the pagan world, so the current rejection of both creation and the incarnate Word has multiplied the evil. The escalation of evil is made possible by the unnormed forces of science and technology. Why? Because "they exchanged the truth of God for a lie, and worshiped and served created things rather than the Creator – who is forever praised. Amen" (Romans 1:25; compare 2 Thessalonians 2:10–11).

In the light of this prophetic Word, we see why our culture can be called the Babel culture of the end of time. Our culture combines sin, apostasy, idolatry and lawlessness with science and technology. Therefore we contend with the dangers inherent in the lawless forces of science and technology, and with the destructive corruption of life and society. A Babel culture is always a culture of confusion. Those who see only the lawlessness in life and society while ignoring the lawlessness of science and technology will not recognize the deep spiritual force which unites the spiritual disintegration of our times. But those who take note of that deep spiritual unity, seeing also the revelation of God's wrath on all apostasy and lawlessness, will know that God has surrendered man to decadence and blindness (Romans 1:18 and 28; 2 Timothy 3:9).

Through his Babel culture man tries to erect a counter creation. His efforts glitter like gold, and science and technology promise an escape from God's judgment. However, appearances are deceiving. As counter creation, Babel carries the seeds of its own destruction, and will effect its own judgment and demise.

Within the force field of science, technology, economics and politics, nuclear weapons appear. Defense policy can be understood only in the context of the forces of science, technology and economic interests. When we see the big picture, we come to see that world events are drawing man toward a vortex of evil and death. Who can escape the thought of nuclear war? Demonic powers would then be revealed in all their power (Ephesians 6:12); humankind would face the consequences of the law of sin and death (Romans 8:2)

4. Challenge to Christians

Within this Babel culture which seems hellbent for destruction, Christians must learn to see that the armor of God equips them to fight against dominions and powers, against world forces of darkness and evil spirits (Ephesians 6:12ff.).

Christians may not allow themselves to be lured onto the road of counter-creation. Instead, they must follow the way of renewal. We are on the way to the new Jerusalem, and we must be pulled back to it continually. Removing ourselves from the Babel motif and speaking against it will put the Christian into a position of sojourner and cross-bearer. He will be the kind of witness found in Revelation 11. His witness is a prophecy even as it is an admonition to return to God's norms, which teach and allow true freedom and responsibility.

We cannot, in good faith, avoid the world. The Babel culture is a perversion of the Kingdom of God, a perversion which feeds off the forces of God's

Kingdom. The Bible calls this culture the culture of darkness, and yet this darkness cannot extinguish the light that has burst upon this world with the coming of Christ (John 1:5). Through the coming of the Kingdom of God the Babel culture will be judged. The perspective of renewal, signified by Christ's glorification, is the perspective of responsible thought and action.

If the man of science, technology, economics and politics would again choose a responsible, normative course, science and technology would no longer be threatening forces. They would present fascinating possibilities for researching and disclosing the secrets of creation. To such discovery and unfolding there would be no end. By contrast, the way of the kingdom of man always threatens an end, for it threatens humankind with its destructive and demonic developments.

If man were once again to discover the way of normativity in science, technology, economics and politics, or in other words, to seek the Kingdom of God responsibly, his efforts would generate many problems. The right way of following God's commandments would be problematic simply because so many cultural structures are presently locked into the perspective of the kingdom of man. We have to take this obstacle into account when we seek to pursue our Christian perspective on culture. We would first have to recognize the existing situation and find ways of dealing with it. And we would always be tempted toward accommodating ourselves, such being the nature and power of the Babel culture. Christians find it more and more difficult to live in that Babel culture and yet not be a part of it. To choose responsibility *coram Deo* means resisting the powerful forces of human ambition and will. Responsibility also means rejecting revolution. Revolutionary changes cannot provide solutions for existing states of lawlessness, for they themselves own no law. Christians face the enormous task of beginning with the existing, decadent situation and trying to renew it, using the norms given by God.

If Christians had more insight into what was really going on in the world they would be misled less easily by the world's motive. An understanding of the world should be the basis for their apologetics, especially when they attempt principled action in cultural activities. In our present culture which is tending toward nihilism, it seems that Christians are all too willing to choose either for technocracy or revolution. We often see young Christians take one standpoint, and then exchange it for a completely opposite point of view. There are always convincing arguments on both sides. The winning point of view is then modified and accommodated before it is passed off as Christian. Through such vacillation and compromise, Christian political thought and action in Western European countries has all but lost its central dynamic. And why has this happened? Because Christians have failed to recognize the spiritual conflict inherent in the cultural motifs which determine contemporary development.

In the past, the weapons of apologetics have defended the Christian faith against the spiritual force of paganism. A similar defense should now be set up against neo-paganism, the modern spiritual force and hallmark of secular culture. Such an apologetic would eagerly confront philosophies, ideologies and thought systems which, as false revelations, willed tremendous apostate religious power.

A Christian apologetic should give a united and biblically responsible, biblically normed vision of man, culture and history. Such an apologetic would help Christians test the spirits and would enable the church to remind its members of the privilege of having the Kingdom way of life open to them despite all the anguish and problems that arise from their culture's way of death.

Developing Ethical Awareness

We have dealt with the problems of scientific life and technological culture, pointing out that these problems are signs of man's self-willed ambition to form culture. Escape from this road to death is possible only if one chooses a different road. And we maintain that it is the church's prophetic task to point out the better road.

However, because of the church's rightful limitations, it cannot solve the technological crisis. To this task, there are called Christians, either as individuals or in groups. The church's task is to *warn* against wrong motives. The right way can be travelled only if we allow ourselves to be led by biblical wisdom which, while ancient, always remains new and relevant.

In the biblical dynamic, man is not the centre of reality; he is not a totally self-willed ambitious creature. Scripture shows how man is created according to God's image, and thus emphasizes man's responsibility. As a responsible creature, man is to love God above all else and his neighbour as himself. Practical political results of such love would mean that man gives up his lust for power and seeks to promote justice and righteousness. For the economy, love means that man is no longer driven by an absolutized profit motive but instead exercises responsible stewardship. For science, love means that knowledge is no longer raw power, but that it serves the interest of wisdom. We must come to see that science and technology can be humanity's helpful servants, rather than our tyrannical masters.

We need not deny the great significance of science or technology; however, we should resist the belief in man's self-willed independence that has become part and parcel of science and technology. Through man's apostate faith the development of science and technology have come to determine the course of our cultures; actually their function should be limited to being the roads which lead towards building culture. Both science and technology need to be subjected to responsible thought and action.

We can better understand the service science and technology should provide for culture when we return to the original motif. The Bible teaches us that man is permitted to build upon creation, but only with the intention of preserving creation. Only to preserve creation without developing culture leaves fallen man in the grip of natural forces. On the other hand, to build without thought of preservation is arrogant. Ignoring judicious and discreet preservation will eventually turn into a situation in which natural perils are replaced by cultural perils; menacing technological forces will threaten to bring about total ruin.

Within the context of this harmonious calling both to build and to preserve, man sees himself as the image of God. In both building and preserving, he confirms his love towards his Creator and Redeemer. He will then treat creation with the concern and respect it deserves. Responsible humanity recognizes the need both to develop creation and to resist every form of distortion and chaos. If man allows himself to be empowered by scriptural norms, his cultural endeavor can be a blessing even for the realm of nature. Such was the case in the days of King Solomon (1 Kings 4:33, 34); such can also now be the case if we allow our economics, politics, science and technology to become a harmonious action of building and preserving or, put differently, if they become part of our search for the Kingdom of God.

Admittedly, achieving harmonious interaction between building and preserving is not easy. Many people will strongly resist biblical direction. Even after rejecting the idols of science and technology, many people turn to other idols, such as the idol of revolutionary freedom or the idol of nature. Others try to design a modified strategy for science and technology but continue to rely on a closed view of the world and life which still excludes God.

But God does not permit Himself to be excluded, and so He permits cultural developments within this closed world to run haywire. We recognize His judgment here, and yet at the same time hear His call to return to Him and follow His norms. This gives us hope. The horizon of our hope is the horizon of the Kingdom of God, which will be the fulfilment, reconciliation and renewal of the entire creation. With this perspective, we do not need to consider ourselves sojourners or spectators in the world, but rather citizens of a Kingdom that has once come with Christ and will come again. *We* belong in this world, even though we are exiles in a Babel culture!

Exiles, while not builders of the hostile culture, are also not its slaves. Their relationship to that culture is one of tension. God's love demands that we reject such a culture, but it also demands that we address that culture in love. Thus, a Christian cannot avoid his cultural environment. On the other hand, a Christian cannot expect to see too much fruit come of his Christian mandate within such a hostile culture. For the biblical motif, which features living and working out of God's love and grace, is diametrically opposed to the motif of the Babel culture. The Bible rejects unequivocally man's pretension that, with the use of science and technology, he can build a counter-creation to which he himself can give meaning.

Scripture calls us to live and work with a recognition of creation order, and with the confession that only Christ can give meaning and the expectation of renewal. In Christ the Kingdom of God has already been given to us and will, upon His return, be given to us in its recreated dimensions. This is the true view of history. Nor can this perspective be in any way altered by the pretensions of our Babel culture. The biblical perspective, it seems to me, also gives the resources with which to counter our culture's reduction of the meaning of science and technology, for it addresses our attention to the rich and inexhaustible meaning of science and technology. Our Babel culture has reduced science to a caricature of its true self. As exiles we can witness to its genuine and full meaning by focusing on its original meaning and normativity. This concerns global issues of our time such as nuclear armament, biotechnology, computer technology and energy problems, but also the more personal problems such as abortion, crime, the dissolution of marriage and the family and the increasing decadence of our society, problems having personal roots but which have come to assume huge proportions. Living in the midst of all these perversions and dislocations, Christians must carry themselves responsibly, which means being of service and at the same time witnessing to the values that society must honour. For this reason, the personal cultural endeavours should not be forgotten, for it is in those areas that cultural reformation must begin. We often think in global dimensions and analyse our culture as part of a global culture, but the smaller cultural endeavours is where we must begin. Science and technology can assist us in our endeavours, but we must always be mindful that they are potentially subversive forces that could, if not used properly, once again imprison us.

In this regard it would be helpful to say a few things about politics, which is where people usually seek solutions. Christians should first pursue a different course, beginning with questions about spiritual and historical background matters. Given the religious background of a problem, they will then have to articulate an ethical position before they can proceed to looking for responsible political solutions. To begin and end with politics is a superficial approach whose efficacy will never be more than superficial. I could mention in this connection the current political discussions concerning nuclear weapons, one of the global issues, and, as an example of the more personal issues, the discussions about legalization of abortion.

I suspect that many Christians will consider my approach impractical. Nevertheless, I maintain that accepting responsibility and normativity allows a great variety of possibilities and benefits, and also a stable course. The course of our current Babel culture may appear to be a road to freedom, but man will eventually find himself to be a prisoner of his own misdirection, on a course that offers no future, only fear.

Those who keep in touch only with factual developments run the risk of always accommodating themselves to those developments. By contrast, those who orient themselves to the perspective of the Kingdom of God, given to man in grace, do more than keep up with the facts. They will resist the spirit of godless development, and will also accept their own responsibilities to pursue a biblically normed approach. The Bible gives us the examples of Joseph in Egypt and Daniel in Babylon, and the examples of the two witnesses in Revelation 11 are also encouraging and hopeful.

At the end of these chapters, I conclude that we should first of all learn not to see the Kingdom of God as the final goal of history and of our cultural endeavours. We must constantly remind ourselves that the Kingdom of God is a gift that has been given, is also truly given now, and, finally, will be given to us again in the future. The final renewal will show us the real meaning of our cultural endeavours. Then even Babylon will become Jerusalem. This divine mystery, given to us throughout history, cannot be comprehended, yet it is a life-giving dynamic deserving our respect, devotion, gratitude and sense of responsibility.

A normative view of life in culture which rejects both frenzied expectations of culture and outright avoidance of it is most accurately described by the prophet Jeremiah in his letter to the Babylonian exiles, in words that are serene, yet filled with expectations:

> This is what the Lord Almighty, the God of Israel says to all those I carried into exile from Jerusalem to Babylon: "Build houses and settle down; plant gardens and eat what they produce. Marry and have sons and daughters; find wives for your sons and give your daughters in marriage, so that they too may have sons and daughters. Increase in number there; do not decrease. Also, seek the peace and prosperity of the city to which I have carried you into exile. Pray to the Lord for it, because if it prospers, you too will prosper" (29:4–7).

and:

> "... For I know the plans I have made for you," declares the Lord, "plans to prosper you and not to harm you, plans to give you hope and a future ..." (29:11).

PERSPECTIVES ON TECHNOLOGY AND CULTURE (1995)

1. Basic Questions and Background[41]

1. Technology and the Technological Sciences

Technology is as old as humankind. Originally it included only human handicraft and was part of the work done in the home. As such it was sensitive to immediate and local needs. One worked with modest tools and instruments that were seldom more complicated than a simple loom, millstone, or lathe. The energy required for this labor was supplied by humans or animals. Technological inventions like the wheel, the lathe, the potter's wheel, the water wheel, the smelting furnace, etc. were developed only to the extent that human or animal propulsion allowed.

Technological development did surge here and there, for example, during the rule of Rome, in the execution of public works, or in the waging of war. Only when science began to work closely with technology and the industrial enterprise was there an explosive development of technology.

This book studies the relationship between science and technology. The scientific foundation of technology has many consequences. Science solves technological problems systematically. It analyzes nature with an eye to its possible technological functions. Scientific methodology can solve many problems quickly and efficiently. These solutions are recorded, remain scientifically adaptable, and can there- [2] fore be applied in new and better ways. Electronics, chemical technology, material science, aerospace engineering and technology, maritime technology, mines engineering, or civil engineering and technology, are unimaginable without the influence of science.

Years ago design and implementation were undivided. Due to the influence of science on technology these have become two distinct phases. Especially in the preparatory phases of design, for example, the influence of science is unmistakable.

Literally, "technology" or "technological science" involves the scientific knowledge of *technique*. Technological science and technology are closely interwoven with each other. Given the wide variety of sectors within technological

---

[41] This selection includes its own index (pp. 331 ff. below), for which reason the original pagination is embedded in the text.

science we will also speak of "the technological sciences," each of which corresponds to a certain field of technology. Paragraph 18 is devoted to an overview of the "technological sciences."

2. Philosophy of Technology and the Technological Sciences

Technological science is not the same as technology. Both are, however, inseparably connected to each other. Technological science focuses on how something ought to be made. Technology and the technicians make it. The technical scientist is usually also a technician who applies the science or who further develops the science on the basis of the application. But a sharp distinction between technological science and technology is not possible. New developments in technological science go hand in hand with new technological possibilities. New inventions are significant for technological science and, as we will see later, also for the natural sciences.

Philosophical reflection on technological science is, as a result, also dependent on a philosophical consideration of technology. That is to say: a philosophy of technological science is inconceivable without a philosophy of technology. This book strongly emphasizes the mutual [3] relationship of these two. And yet, distinctions can be made. The remainder of this chapter is primarily devoted to a philosophical consideration of technology. We have to know precisely what "technology" is, how in overview technology has developed, which forces have driven it, the rich diversity of tools and methods that characterize it, what its possibilities and limits are, etc. But because technology, due to the influence of science, has become a pre-eminent cultural power to which we are deeply indebted, philosophical consideration requires that we examine the inner connection of science and technology. A survey of the various currents in that philosophical reflection (in §8 and following) brings home the fact that the possibilities and dangers of modern technology have everything to do with the relationship of science and technology. This survey forms the transition to the philosophy of the technological sciences, which is the topic of Chapter 2.

Because the dangers and threats of modern technology can be traced to an irresponsible connection between science and technology, it is unavoidable that the ethical dimension will surface in the elaboration of various philosophical currents. After a philosophical discussion of the technological sciences in Chapter 2, we will return to the matter of ethics in Chapter 4. There we will want to clarify how science and technology can best be connected with each other so that the problems, threats, and dangers can, if not be avoided completely, at least be held within certain bounds.

3. What is Technology?

Philosophical reflection on technology and technological science must first answer the question, "What is meant by *technology?"* The words "technique" (or "technic") and "technology" are derived from the Greek word *technè,* which has an array of meanings. It can express all human (cultural) activity in distinction from nature as *physis.* But it sig- [4] nifies especially "skill" and "proficiency." It can also connote "cunning." *Technè* also proves to be related to the words *epistèmè,* meaning "understanding" and "(artistic) skill," and *poièsis,* that denotes "doing," "working," "creating," and also once again "skill." Obviously, the Greeks did not clearly differentiate between technology, science, and art. In other words, the etymology of the word "technique" is not going to be of much help in answering the question as to what technology is.

In their common usage "technique" and "technology" have, as well, a variety of meanings. We use the word technique when we want to talk about a certain kind of human activity and its result, but also generally when we have our eye on human proficiency and the methods it involves. Piano technique is an example of the first use; social techniques an example of the second. These ambiguities can hardly provide a basis for theoretical analysis. Any philosophical reflection on technology and technological sciences will have to ascertain with some precision what technology is.

Sound insight as to what we are discussing will prevent misconceptions and erroneous views of development. Unfortunately, philosophers themselves use these terms in a variety of ways. The French philosopher Jacques Ellul speaks about technique as a scientific method of domination. José Ortega y Gasset, the Spanish philosopher of culture, used the terms "technology" and "science" interchangeably. For Martin Heidegger, "technology" includes many cultural expressions, like machines, bureaucracy, information, and organization. The distinction between science and technology is not at all clear, given these uses. And yet the distinction does refer to a decided difference. The aim of science is always to gain knowledge of reality, while the goal of technology is to change reality. In like manner, we speak of "discovery" in science and of "invention" in technology. Science is the power of knowledge; technology is the power of formation. In other words, in science it is a matter of *knowing*, in technology a matter of *making or forming.* [5]

In this study we take *technology* to be *the human formation of nature with the help of tools for human purposes.* We have to remember, however, that the nature of tools has changed substantially in the course of time due to the influence of science (see §5). This is so evident that technology is often simply referred to as *an applied natural science.* While this rightly emphasizes the significance of the natural sciences for technology (see §10), modern technology is itself primarily not a matter of knowing, but, as we said, of forming. Besides that, instead of saying that technology is applied natural science, it could better be said that modern technology is *applied technological science.* We will refer to the scientific, theoretical knowledge about the technological activity of formation as "the science of technology." This science can rightly be called a practical science.

"The science of technology" is usually equated with "technology." We will want to distinguish these two so as to avoid confusion. Technology and the science that deals with technology are not one and the same. That people confuse the two is more than accidental. Science has often been taken to be a technological instrument of domination. As we will see (§§21–22), it is precisely this instrumentalistic view of science, and of technological science in particular, that has given rise to ethical problems.

The power of human formation through the use of tools, with which human beings invent, design, construct, and fabricate new things and processes as the fruit of productive imagination, has changed substantially due to the influence of science. That is why, to distinguish it from the classical technologies of handicraft and guild, we speak of *modern* (science based) *technology.*

Before turning to the differences between guild or handicraft technology and modern technology, we must mention the fact that modern technology, in the sense of the power of formation, has in the recent past developed beyond tools. Biotechnology is one such example, in which [6] scientists control life-processes or manipulate the genetic structure of micro-organisms, plants, animals, and even humans. In order to distinguish this kind of technology from *tool technology, we* will also want to talk about *biotechnology* and the science of the same (see § 19).

Another misunderstanding deserves attention. When we say that technology is the power of formation through the use of tools, we are clearly distinguishing the tool from what is being made. The product of technology can itself be of a technological nature. For example, technology can produce new technological tools. But usually its products have another purpose. An automobile, a house, a church, a rocket: none of these have a technological purpose. All of these are results of technology, but they do not have as their purpose to further the development of technology, at least not directly.

4. History of Technology

To get a correct sense of its development, to gain insight into the influence of science on technology, and especially to be able to direct technological development responsibly, we must know the history of technology, and particularly of the spiritual forces that drive technology.

The history of technology can be subdivided into three main periods. During the first period technology developed very slowly. This phase, the *traditional era,* lasted until about 1500 A.D. We refer to the period from 1500 to 1800 as the practical era. Given the predominant influence of science on technology, we can call the next age the *era of modern technology.*

During the traditional era technological development remained static. This can be attributed to a lack of technological possibilities, but even more to deeper lying reasons. In the traditional era human beings were not motivated to change the practice, activity, and constructions of everyday life, but to *contemplate* nobler things. Inventions were made [7] on paper, by Archimedes for example, but because of a dearth of technological motivation they were seldom implemented. For religious reasons, among which a devaluation of earthly life was predominant, work and technology were not encouraged. That is why the technology of the traditional era is more or less limited to things like the techniques of war, technological products connected with worship, and technology as a kind of intellectual free play.

The rise of the practical era is closely related to the spread of Christianity and the motivation to form culture that accompanies it. People are called to serve God in his creation and to help unfold it. The cultural mandate of Genesis impels humans to develop and value technology positively. This Christian motivation did not immediately develop a culture or dynamically unfold history. That came only after 1500. Some of the obstacles were spiritual. It took time to see what Christendom implied for culture, how it should influence practice. At first this process moved very slowly, and often very superficially, due to the wholesale conversion of entire peoples. Christianity also was often accepted piecemeal. The characteristic influence of the biblical message was often restricted to "supernatural priorities" and the realm of the institutional church. In addition, socio-cultural developments, like the fall of the Roman empire in the fourth century, hindered a prompt flourishing of technology. Later the rise of Islam deterred sea trade and cultural development stagnated.

Social changes and new spiritual currents between 1200 and 1600 prompted the rise of the New Age, the onset of a dynamic period for culture and history, and of the practical era for technology. To understand this development, we should best not think of "cause and effect," but of a multifaceted process of interaction. However, the spiritual motivation, understood in a religious sense, was the fundamental reason for the genesis of the new development.

Circumstances in society, like the upswing in trade due to the Cru- **[8]** sades, the developing craftsmanship in the guilds, and the social emancipation of the common folk, stimulated a developing technology. The growing strength of this liberation movement contributed to the disbanding of the guilds, which in turn contributed to renewing technology.

The history of technology is more complex than is often thought. Christendom itself was of considerable significance. A spiritual renewal and at the same time the secularizing of that same Christendom also supplied a powerful impetus.

From 1300 to 1500 mounting spiritual movements imparted a new attitude that would eventually split in two: on the one hand, the movement of the Renaissance and, on the other, that of the Reformation. Both of them address the place of man in the world. The Reformation focused on the human being as called by God to serve him in all things. For the Renaissance the theme of the self-empowerment of man began to predominate more and more. This motif of the Renaissance, amplified by the movement of humanism, extensively influenced philosophy and science. In the long run, after an initial flowering, the Reformation's influence on philosophy and science, in spite of its historically significant positive view of these pursuits, remained limited. The Reformation set its stamp on commerce, agriculture, and the development of trades. The spiritual movement of the Renaissance, intensified by the Enlightenment, became the moving force in the development of science and technology.

The practical era was ushered in by new developments in the technology used by craftsmen. Many inventions came into play, often one right after another. Even academics, like Leonardo da Vinci and Simon Stevin, joined the ranks of inventive craftsmen.

Inventions like gun powder, typography, compass, the mechanical timepiece, and many of the instruments needed for scientific research helped turn the page around 1800 from the practical technology of [9] craftsmen to the era of *modern* scientifically based *technology* (Forbes, 1963).

From the eighteenth century on, the spirit of the Enlightenment shaped science and modern technology. In organizing society it let science lead. As a result, and with the rise of modern technology, culture became secularized.

An interesting detail deserves mention. More than one scholar has pointed out that the later motif of the Enlightenment had begun to develop already at the end of the Middle Ages. The Enlightenment made the claim that mankind, independently empowered by the light of reason and with the help of science, would be able to control the future. Some said the influence of magic on developing technology was the precursor of this Enlightenment pretension. People turned to alchemy as life's cure-all, hoping that it might even be able to keep mortals from death. The same spirit is evident in its search for that primal material from which gold could be fashioned and one's financial independence guaranteed. Attempts to transcend the bounds of space and time and to conquer the law of cause and effect likewise manifest a motif that, modified by the admixture of science, became even more preponderant in the Enlightenment.

In summary, the spiritual forces shaping the development of modern technology are derived from the Reformation and from the Renaissance, later transposed into the Enlightenment. Both movements confirm the influence of science on technology. However, the direction of each is different. Only now that we are confronted with a diversity of problems, tensions, and threats within and around modern technology are we more and more becoming aware of this divergence in direction. That is also the reason why unanimity is lacking today as to the road to follow in the future. As a result we can speak of a crisis in the development of technology. For that reason, too, ethical reflection is inescapable.

5. Technology Then and Now

[10] The history of technology teaches very clearly that technology has to do primarily with *tools* and their use. The history of tools discloses the gradual growth of human formative mastery; a growth that today has resulted in the imposing power of puny people – puny compared to the means at their disposal.

This development is strikingly evident when we compare the classical technology of the tradesman and craftsman with modern technology. In making the *comparison* we can leave the actual development out of the picture for the moment. We talked about that in the previous paragraph. We will limit ourselves, point for point, to differences in *milieu,* materials, energy, skill, the sequence of the technological operations, cooperation within technology, work methods, mass production, and the nature of technological development.

a. In the era of classical technology man was by and large surrounded by a natural milieu. He was basically limited to what was allowed and supplied by nature. There are only incidental examples of an ability to distance oneself from nature. In contrast, people in the modern era live in a milieu already formed by technology. Highly developed tools are readily available and the distance from nature is on the increase.

b. It used to be that people had access only to *materials* that were readily available in nature. In the modern era the desire is to depend on natural materials as little as possible. Natural materials of the same kind still differ too much in particulars and are usually mixed with natural "impurities." In modern technology materials are expected to be fully refined and have a uniform consistency. An extreme example of this development, in which the technician dictates the desired characteristics, is commonplace in materials technology. One can think, for example, of very light materials that can at the same time withstand very in- [11] tense forces, of nonabrasive materials subject to little or no wear, and of modern synthetic materials, like plastics, that display a whole new set of characteristics.

c. In the previous era the *energy* required for technological formation, for construction or fabrication, depended on the push and pull of animal and human muscle power. In modern technology energy is drawn directly from nature, from wind or water, or indirectly through the burning of coal and oil or even by splitting atoms. The direct use of the sun's energy via solar cells and panels is presently a reality.

d. For centuries human *skills* in the use of simple tools defined the extent of *technological formation.* In modern technology these have been replaced by machines. Machines not only refine and enhance these skills. They also give rise to new possibilities. Electronics and computer technology are particularly obvious examples.

e. As far as *technological operations* are concerned, craftsmen and tradesmen had the lead from beginning to end. In modern technology this process has been taken over by automation. The modern tool is an independently working, technological operator.

f. In technology of the practical era there was a direct relationship between *producer* and *consumer.* Modern technology is characterized by a strongly differentiated *cooperation* between engineer, builder, and employee. In addition, production is usually interwoven with an industrial enterprise. There is no longer direct contact between producer and consumer.

g. Pre-modern technology was, as it were, enfolded within nature. Modern technology is characterized by a *distance* with respect to nature because of the *scientific approach* of modern technology. By distancing itself from everyday life, the plan and its execution have become divorced from each other. The workers' place has become a scientifically calculated station. When technological formation and its execution become self-supporting in the process of automation, the worker is left [12] without work. At the same time, with the scientific approach, the process of design is brought to a higher level, creating the possibility for new constructions and new inventions.

h. When technological formation is automated, we have *mass production.* Products of the craftsman, in contrast, are unique and of a singular character.

i. Earlier, technology was limited to the natural possibilities of man, in the sense of the creationally given measure of bodily exertion and capacity. It remained within parameters defined by the human hand and sense organs and was based on everyday knowledge. This is what makes the technology of the craftsman "natural." In this kind of technology there is little or no talk of development. The experienced rhythm of the routine was passed on from father to son, from master to apprentice. *Pre-modern technology is undifferentiated and static in character.*

Due to its scientific basis, *modern technology* has experienced a massive proliferation. It is *highly differentiated and dynamic in character.* In addition, the breadth of modern culture bears the stamp of technological development. A technological culture is usually vivacious. This cultural dynamism especially typifies the recent past. If we could condense the entire length of technological development into a period of twenty-four hours, then the invention of the steam engine would have taken place about thirty seconds ago. The most recent seconds would have contained tumultuous developments that unimaginably change the very countenance of the earth with lightening speed. In a material sense, life is enriched enormously. At the same time it becomes increasingly apparent that these changes are accompanied by colossal disadvantages. The extent to which science is to blame for this remains to be seen (see §21).

6. From Tool to Robot

To get a handle on the change that has taken place in *tools* over the course of time, we can best focus on the changing relationship between [13] humans and their tools. What becomes apparent is that human beings gradually recede into the background and that tools more and more take over what used to be human tasks. Originally humans made or formed everything using tools. Later they directed and controlled the tools that did the making. Finally, today, tools have become technological operators, working on their own, with humans determining the process at a distance through design and programming.

This changing relation between human beings and their tools has been an on-going topic of discussion among philosophers of technology. Ernst Kapp, said to be the first philosopher of technology, introduced the subject already in 1877. He considered the development of technology to be an unconscious extension of human organs. Though unaware of the fact, we tend to make tools that amplify natural skills and abilities like seeing, hearing, grasping, and lifting.

Human beings with tools are in a position to do more than they are without them. The power and achievement of human labor increases as these tools become more advanced. Although there are exceptions, we can in a sense agree with Kapp that people have objectified or projected certain human functions in their tools. In doing so, the burden on human beings decreases while the human function in effect increases. Because of the influence of science, particularly of the natural sciences, information science, and the technological sciences, tool development is still experiencing a stage of rapid growth. An increase in tools that do their work with little or no human intervention has been the result.

Originally human beings supplied the energy required for using tools. Their skill, or lack of it, in using these tools determined the outcome. When tools become *instruments,* human skill is defined by predetermined limits. The instrument's construction allows human energy to achieve more than ever before. At the same time, the actual human energy exerted decreases to the extent that energy can be drawn from nonhuman sources: horsepower, then wind and water power, and after that [14] a long gradation of energy sources that reaches its limit with atomic energy. Though exerting less, humans today can do more. This is especially the case where, with the help of the natural sciences, energy is produced by technological formation.

During the next stage human beings still often have to service, steer, and supervise their tools. But in the process of turning to other sources of energy, the skill and proficiency are increasingly transferred to the *machine,* with generally positive results. Ultimately, using the principles of cybernetics, the regulation and supervision of the production process is also transferred to machines.

These last developments move us beyond the phases of *material technology* and *energy technology* and into the period of *information technology.* With the appearance of *computers* and *automatons* (e.g. robots) we stand on the threshold of a new era.

The age of information technology is characterized especially by the introduction of the *computer.* It is a fascinating technological achievement which, at the same time, breaks new ground. In the last thirty years, developments in computer technology have been almost overwhelming. We have moved from mammoth, unreliable, expensive, main-frame computers to small, dependable, relatively inexpensive, user-friendly machines that can process mounds of information very quickly.

With respect to technology, the rise of the computer generated a giant step toward complete automation. In cases of total automation the computer is completely integrated in the production process. Acronyms abound: CAD computer aided design; CAM computer aided manufacturing; CIM computer integrated manufacturing. Their presence pervades, for example, the tool and die industry and modern chemical processing companies. We see analogous developments in, for example, aerospace, communication and medical technology, in telecommunications, and in automation in the service sector. Developments in [15] *micro-electronics,* including micro-computers and micro-chips, have accelerated this process. However, as a result of automation, increased unemployment is no longer limited to technicians, but now includes all fields of cultural labor that are systemized.

In all of these cases the computer is present as a *steering or control mechanism.* In the case of science the computer is used as a *tool for thought.*

More than any other tool, the computer has shown that science and technology reciprocally enrich each other. The computer as technological result serves science, while science is a perpetual impetus for technological development. The computer is not only important for technological science. Because of the applicability of logic and mathematics as scientific aids for (quantitative) analysis, the general use of the computer has become commonplace in scientific research. In addition, computers are used in fields like operations research, decision-making, and praxeology – the study of human action and conduct.

Emphasizing that the computer, too, is a tool can keep one from falling prey to all kinds of speculation with respect to "thinking machines," as though a computer could ever exist or work independent of human beings. To believe that, is to become the dupe of the computer. Conjectures about machines that can think too often forget that when it comes right down to it the computer is and will remain a tool of human beings. Computers can process very large quantities of information quickly and without error. They perform mathematical operations that are so involved that humans could not even begin to duplicate their calculations. And they generate answers that are more inclusive than humans, given their limited ability to survey all the available information, could ever produce without the computer. Nevertheless, one ought not to forget that the information stored in its memory and the programs it uses are where they are because humans put them there. In this sense the computer remains subservient to human beings, who in turn are [16] responsible for the outcome. Even though computer results may be startling and their makers or users incapable of ever achieving as well, if at all, similar results without a computer, that does not mean that computers are not dependent on human beings. Computers are there to serve human beings. They help to analyze, search, recognize, calculate, sort, and, when it comes to technology, to control (Schuurman, 1982).

But now back to the continuing development of technology. In the development of industrial as well as nonindustrial *robots* the attempt is being made, with increasing success, to transfer the remaining functions of human beings to the technological operator. We can speak of a robot when the internal heart of the technological operator is a computer, the information available comes from external sensors as well as from the computer's program, and the external effects of the information processed run via artificial limbs. The robot should also be able to learn from its own "experience" and in so doing be able to better itself. It is understandable that where this combination of "sensitive" and "mental" functions presents itself people speak of a *robot with artificial intelligence.* However, some who use the term "artificial intelligence" are referring to *expert systems.*

In noting the place these systems occupy in the production process, we can correctly say that the history of technology shows that human tools have gone from things we can easily get a handle on and oversee to a *technological system* that is beyond our circumspection (Ellul, 1980).

Having come to the present in this long historical process, the dynamic development that characterizes modern technology is clear. And as though that were not enough, it still has many surprises in store for us. Robots are in a position to do even more than most can even imagine.

As a result, both optimists and pessimists speculate as to how technology will yet develop. On the one hand, the use of robots provokes very intense premonitions concerning the future: they will be used in [17] ways presently unheard-of; wonderful promises are on their way to becoming reality. On the other hand, unanticipated results incite fear and anxiety. Will we be able to continue to oversee and control these new developments? Is the product going to consume the producer? Is science being misapplied or does the blame lie with science itself? How could that have happened?

7. Problems in the "Technological Culture"

Modern technology, whose activities and results are rooted in science, commences with James Watt in the second half of the eighteenth century. The industrial revolution follows in the nineteenth century, and in the twentieth century, at least according to Norbert Wiener, there emerges, riding on the shoulders of information technology, a second technological revolution.

Machine construction, chemical technology, electronics, transportation technology (automobiles and airplanes), materials technology, atomic energy, computer technology, aerospace technology, telecommunication, and biotechnology are developed one after the other and influenced and stimulated the one by the other. Education in the technological sciences and related disciplines skyrocketed as well. Some of the industrial arts that had been taught in high school were now given academic status at the college level, while university level institutes for technology were being founded as early as 1860.

Given the fact that we find ourselves right in the middle of the development of modern technology, we know not only of the advantages it has brought to man and society, but also of the disadvantages that accompany it. Modern technology, it would seem, has a questionable character.

For a long time the advantages of modern technology contributed to an uncritical attitude with respect to it. People spoke about the "triumph of technology," the "century of technology," even about the "miracles [18] of technology." Many expected that the combination of science and technology would be the beginning of the end to humanity's problems. They had their eye on an increased life expectancy, overcoming most illness and transmissible defects, improving memory and intelligence through symbiosis with the computer, the automation of most activities, growth in the material standard of living, and the artificial generation of more suitable plants, organisms, and animals.

Needless to say, many of these are now reality. There is an uncanny dynamic to recent human history. Human beings are more powerful today than ever before. Technology is shoring up a level of material prosperity that is unparalleled in history. Moreover, leisure time is on the increase, such that many can liberate themselves and, for example, further themselves in many more ways than ever before.

Modern technology has put its stamp on culture. At the same time it functions as a basis for a phenomenal integration of humankind. An integral technological world culture is upon us.

However, it is time to attend as well to the problems, tensions, and threats that characterize the other side of the coin. People perceive these burdens to be the result of the growing technological power that has taken hold of the history of humankind at its core, imparting to it a fragility that forces us to ask the question whether present developments find us on the right road. In connection with this conundrum, there is even talk of a *crisis*. That is to say, there is no unanimity as to the solutions of these problems, which in the meantime are increasing.

I will focus briefly on a number of these problems. They have to do with (a) the relation between people and technology, (b) the relationship of people, technology, and nature, (c) the socio-economic context in which technology is developed, (d) questions concerning human responsibility, and (e) the relationship of technology and religion. We will see how much these problems coincide with *modern* technology and hence are connected to the influence exerted by science on tech- [19] nology. That will become particularly clear when, after reviewing these problems, we discuss different philosophical conceptions of the "technological culture."

a. The problem area that looms foremost concerns *the position of human beings in modern technology* as compared to their place previously. In the technology of the craftsman, human achievement is determined by the eye and hand, by a feel for the material, by imagination and a formative knack. Physical and mental abilities are brought to bear directly. People's work and their products are marked by quality, as well as by being distinct, one of a kind, and nonrepetitive.

In modern technology the situation is completely different. It is no longer a matter of a simple, personal, comprehensible technology, with few, if any, unforeseen consequences. Modern technology, in contrast, is very involved and characterized by an unoverseeable interaction, with, in many cases, unanticipated, radical repercussions. Within the arena of modern technology almost everything is quantifiable and characterized by an increase in tempo and range. Human labor has become a quantified business whose intensity is being stepped up continually. At the same time, the laborer is being taken up into an objective, impersonal, and dispassionate process. What was previously personal becomes matter of fact and vacuous. That helps to explain the callous attitude and indifference of many with respect to their work. People feel as if they are being sucked into an insurmountable, massive, mechanistic process; a process in which their energy is being exhausted. When this process becomes fully automated they are simply eliminated and can only look on, unemployed.

The designer of this technology, the *engineer,* experiences something very similar. In designing he has, with the computer, the most modern technological options at his disposal, and yet he often feels strapped. As a specialist he is increasingly dependent on other specialists. Given this dependence on others as well as on the computer, the individual engineer lacks the insight and capability to help direct the development of the technology.

[20] For the laborer and engineer, as well as for the "outsider," the cutting edge of modern technology arouses apprehension. For example, the development of nuclear reactors promises a secure supply that can meet the increasing demand for energy, but we are at the same time oppressed by the many as yet unresolved technological and political risks that accompany these facilities, which can hardly be surveyed in their entirety. In addition to difficulties in controlling the technology – think of Three Mile Island and Chernobyl – problems concerning radioactive wastes continue to vex the minds of many. It remains to be seen whether the elementary forces of matter that are being released can be controlled sufficiently.

Something similar holds true for computer technology. It does not take too much to see that the computer works quickly and accurately and that what it generates will never fall outside of the programmed parameters. And yet we fear becoming increasingly dependent on the computer. Its results – albeit within certain limits (Schuurman, 1980) – have something dazzling about them. This is enhanced because the user and the programmer need not be one and the same person. As the users change, they are no longer cognizant of the criteria according to which the computer works, with the result that one can only trust that what it does it does well. This problematic situation is going to become more pressing as the capacity of computers increases and as machines are introduced that can learn or reproduce themselves. When one banks on so-called expert-systems and on robots with "artificial intelligence" (see §6), stagnation can arise because these systems can never completely adapt themselves to the dynamic change of reality (see §21).

A *computerocracy* seems to be just around the comer when these machines not only find their place within the production process, but also settle into the areas of economics, politics, and the like. Especially the manipulating of personal information stored in data banks, with its consequent manipulating of people, can be dangerous.

[21] The perils of modern technology are manifestly evident when we note the dangers that are connected with the newest techniques of waging war, the pit of which is nuclear suicide.

The development of recombinant DNA technology – also called "genetic engineering"–is just one facet of biotechnology that next to its many surprises has as many imminent dangers in the offing (see §21).

b. A second problem area has to do with *how human beings relate to nature.* More than one scholar has pointed out that the earlier organistic view of nature has been replaced by a mechanistic conception. Nature is seen in terms of the physical categories of number, space, and causality. Measuring, weighing, and counting determines what is nature. The remainder is neglected. This mechanistic view of nature invites pervasive technological control in service of goals and desires set by men. Nature no longer seems to have its own value or meaning.

The means by which and the manner in which nature is controlled are given little or no attention in this utilitarian framework (see §21).

Early on, small-scale technology was still characterized by an integration with nature. That is no longer the case for modern technology. A rift has arisen between man and nature. Aided by modern technology human beings are intruding on nature on a large scale and with increasing rapidity. In doing so they undo the manifest coherence in nature. What is more, nature is not even given the opportunity to recover from the damage inflicted. All the while, unfavorable side effects of modern industrial technology, modern transportation, and what all too often approaches factory farming, continue to reinforce each other. The imminent collapse of the biosphere is one of the dangers. We use more from nature than it can produce and we leave more behind than it can decompose – not to mention what cannot be disintegrated. Because of the disruption that precipitates, nature when controlled technologically itself becomes a threat to humankind. That is evident from the greenhouse effect, changes in climate, a rising sea level, ozone depletion, acid **[22]** rain, and desertification. All of these are the result of irresponsible procedures that have risked the coherence in nature and whose consequences can no longer be completely undone.

c. Technology is interwoven with *economic and political forces.* In addition to the difficulty of surveying the structure of modern technology, this intertwinement explains why the development of technology is experienced as a self-governing, *autonomous process* (Winner, 1977). The research and development of new technologies has often taken place (secretly) in multinational conglomerates and within the so-called military industrial complex. When government begins to play an ever-increasing part in this process, experience shows that technological development embarks on an upward spiral. This growing *technocracy* is coupled with specialization and mass production. Technology responds, in the meantime, not only more to present needs and desires, but also artificially creates new ones. An inability to survey the whole, on the one hand, and redundancy, on the other hand, make it such that technological means not only have a devastating influence on nature, but at the same time constitute a direct threat to humanity. Because people more and more are becoming components of the technological system, philosophers like Karl Jaspers and Jacques Ellul have emphasized the inhuman character of modern technology. People are uprooted from their tradition and alienated from their culture. The gravity of the situation is often concealed by the numbing effects that emanate from a sweeping material prosperity.

d. Given what has been said about the previous problems, it follows that the question of *responsibility* within the context of technology presents great difficulties. That is true for those involved in the production process as well as for those in design. Because of the scientifically defined task, the worker, to the extent he still has a part to play in production, must completely disengage his freedom and responsibility. Or better put: these must *be* eliminated.

[23] The responsibilities of the engineer, given how technology affects culture, have become heavier. At the same time he finds, possibly due to his one-sided background, but especially because of the situation in the workplace, significant obstacles in carrying out that responsibility. In addition, due to the intricate relations of cooperation, responsibility has more and more become a matter of ever-increasing *communal* responsibility. We become even more conscious of the extent of this responsibility when we realize that we are often confronted with new technologies with which we have no experience. We do not have the advantage of being able to draw any lessons from the past so as to avoid old and new calamities. For example, many have the sinking sense that while recombinant DNA technology (see §19) will proffer unforeseen possibilities, it will also bring with it new hazards. Ignorance, innocence, a lack of experience, and increased risks go hand in hand (Jonas, 1984).

Clearly, our responsibility has become *greater* and *more difficult* to bear. We can give a number of reasons for this. First of all, we are uncertain with respect to almost all sectors of technological development – precisely *how* we can carry out our responsibility when it comes to modern technology? The operative range of the engineer, as we saw, is expanding, while at the same time the nature of his work keeps him from a comprehensive purview. Of increasingly little he knows increasingly more. He will have to submit himself in trust to his cohorts who work in an equally specialized manner. But it is precisely this specialization that weakens the community called to bear a greater responsibility. Moreover, even where there is sufficient insight to bear the personal and communal responsibility, these efforts often are being frustrated by imposing scientific, technological, organizational, economic, or political allegiances. It is very difficult to break through an existing, dynamic trend, especially when it pervades the present culture. [24]

e. The problematic situation with respect to responsibility deepens with the realization that the perplexities of modern technology are not incidental problems with one or more components, but that they all hang together. That is why partial solutions are also inadequate. An integral approach is required. But that in turn presents additional problems because our "technological culture" is at the same time a *spiritually disintegrated culture.* The Spanish philosopher of culture Ortega y Gasset put it succinctly when he wrote: "Our time, being the most intensely technical, is also the emptiest in all human history."[42] There is no longer a unified vision of what it means to be human, of history, culture, and the future. The answer to the question of responsibility requires a communal response to the question of meaning and a communal sense of normativity. And that in no way is present any longer. That is why Jonas rightly speaks of an *ethical vacuum.*

8. Philosophical Perspectives

Both the advantages and disadvantages of modern technology have given cause for philosophical reflection, especially when considered in light of the development of the sciences. These reflections, however, have historically been highly divergent. Since the rise of modern technology and its problems, Western philosophers wrestling with the background of science and technology increasingly disagree.

Surveying that philosophical reflection reveals the spiritual abyss of our scientific-technological culture. Such an overview can also make clear that the problems we have mentioned have a deep spiritual-historical background and

---

[42] *History as a System*, p. 151.

usually have to do with overestimating the potential of science with respect to technology. On the one hand, we find philosophers who evaluate modern technology positively. Other philosophers reject modern technology, seeing a growing conflict between humankind and modern technology. While the one takes modern technology to be an affirmation of human power and supremacy and [25] of cultural progress through the advance of scientific-technological control, the other sees people alienated from nature and their own freedom as they become the victims of science and modern technology. In a word, a number of philosophers see modern technology in combination with science as a meaningful avenue towards a future that will the liberate us from many evils. Others see in this combination of science and technology a much more somber picture.

There are, nevertheless, so many shades of nuance within both categories as well as points of similarity between both manners of philosophizing that further elaboration is required. The following brief survey of philosophical viewpoints will show how central the human role is in these perspectives and how fundamental and significant one's view of science and its relation to technology is.

We will take a look at six different philosophical traditions: a positivistic-pragmatic position, a Marxistic one, a systems analysis approach, existentialism, neo-Marxism, and a counter-cultural view. Given the context it is not possible to be exhaustive. We limit ourselves to the main points.

a. The *positivistic-pragmatic view* of modern technology is very prevalent. This philosophical attitude is more often than not present in the minds and actions of natural scientists and engineers. Pronounced representatives are Karl Steinbuch and Norbert Wiener (Schuurman, 1980), both of whom were strongly influenced by the American pragmatism of William James. Their estimation of natural science and modern technology is very positive. Technology for them is the engine of progress in culture; growth in scientific knowledge is its fuel. They have put their hope especially in the development of the computer.

In the past, the so-called scientific design method (see §12) proved sensible and successful in controlling "dead" matter. If we are going to continue, extend, and strengthen this successful progress, so they argue, we are also going to have to use this method, as a method of control, for [26] areas that are not specifically technological. We must analyze and form humankind as well as society and its future by means of this method.

When problems arise along the way, people blindly appeal to science, be it an established discipline or a new one, for the solution. They have their eye on a comprehensive, scientific-technological control of society. Such control can only be realized when economic and political forces cooperate. In other words, they favor a technocracy. They see anything that resists such an approach as possibly causing failure; it must be opposed. This includes all those philosophies that put any less trust in science, like existentialism for example. But they are equally opposed to religion, particularly the Christian religion. They argue that religion relativizes science and technology or sometimes even challenges their validity. For positivists and pragmatists, the closed, seientific-technological world picture is a *conditio sine qua non.*

b. The development of modern technology is also a central tenet of *orthodox Marxism.* For the Marxist Georg Klaus, for example (Schuurman, 1980), the person is pre-eminently the technological person who in and through the development of technology not only achieves full potential and freedom, but also undoes all oppression in achieving the goal of an emancipated society.

In contrast to the positivists and pragmatists, Marxists recognize that the development of technology goes hand in hand with many problems, threats, and dangers. Placing the development of technology in the hands of capitalists will result in the increased alienation and coercion of the labor force. At the same time, modern technology will bring an end to the class struggle. Marxists maintain that ongoing socio-economic development that proceeds on the basis of technological proliferation will wipe out the lengthening shadow of alienation and coercion and that an age of freedom will break forth with the dawning of the new day. In that day no one individual or class will have their hands on the reins of power, but humankind collectively will be the lord and master [27] of the works of their hands. Along with the previous tradition, Marxists have a high regard for science as the means to get things under control. But because of their basic assumptions about the nature of society, how this will work itself out is quite different.

Marxists do *not* proceed from the freedom of individuals and from a free corporate means of production within which technology will come to fruition. Their assumption is that technology can facilitate revolution and ultimate emancipation only if it is orchestrated in a centralistic fashion and the use-value of the goods produced, over against their possible exchange-value, remains definitive. Which is to say, they proceed from the idea of a central technocracy.

c. The third group of thinkers are representatives, like Ervin Laszlo, of *systems analysis* (Schuurman, 1982). Systems science is about forty years old. This philosophy drew a good deal of attention because of the Club of Rome publications at the beginning of the 1970s. This Club of Rome made clear that present cultural developments in which science and technology play a dominant role are soon going to be confronted with overwhelming problems and perils. According to Laszlo, systems analysis is the only possible way both to analyze the dangers and problems of a growing world culture, with the use of models, and to turn these risks around and to resolve these problems.

Systems theory looks for the cause of cultural problems in an overly one-sided, hyper-specialistic, scientific approach to things. This philosophy is a proponent of a single science that will overarch all the separate sciences. Often a particular special science has a reductionistic influence on practical life because of the application or instrumental use of this science. Systems theory does not want to orient itself to one component, but to the whole of reality. For these philosophers the whole is more than the sum of the parts. And it is this whole that they are trying to get a bead on in their systems analysis approach.

[28] An integral scientific approach via the system is meant to lead to integrated control. This kind of management is made possible by the newest of technological developments, like information and computer technology. From this we see that the system may not be equated with the *given* whole, but must be seen as an *artificial construction.*

Systems philosophy is trying to provide an answer to the cultural crisis. It is concerned about the dangers of technological development, about the degradation of nature, and about many other social, economic, and political problems. The assumption is that these problems have been elicited by the predominance of a *reductionistic* kind of science. This general systems theory is out to resolve the problems of what could be called first level science with a meta-level, more "holistic" science and with its correlate, a meta-level technology, namely, systems (or computer) technology.

The integrated control of reality that systems theory is aiming at, even on a worldwide scale, coincides with a fundamental shift. The prime objective is no longer the *progress* that positivists and pragmatists yearn for, but rather the *survival* of world culture. This major shift need not imply that the newest science and technology cannot be considered as guides toward a meaning filled future. On that score, we have to say with respect to systems science, that next to a change in orientation, we still are looking at a bolstering of the influence of (the newest) science and technology.

In passing, it should be noted that representatives of systems analysis have emphasized more than once that, on account of the central position of an eventual world government to control global culture and its future, Marxist and non-Marxist societies are approaching each other and will eventually work together.

d. Compared to the philosophical perspectives discussed above, *existentialists* are more aware of the problems of technological development (Schuurman, 1980). An important existentialist is Martin **[29]** Heidegger. He claims that humanity more and more is surrendering to technology and that in doing so is alienated from the origin of technology (Being) and from his own essence (freedom). "Technological culture" for him is one in which Being has been abandoned and forgotten. Though surrendering itself to technology and seeing to its advancement, humanity in fact finds itself on the road to madness. Other existentialists, like Karl Jaspers and Hermann J. Meyer, say that the development of technological culture threatens the human subject, particularly, attacking what makes the human personality unique, namely, freedom and individuality. They turned themselves against science and technology, which they see as autonomous, anonymous forces. It is not that they turn against the abnormal outgrowths of a technological society, but against science and technology as such. For them the scientific method operative in technology is a method of suppressive control.

Their protest is often impressive. But given their assumption that they are dealing with an autonomous, self-governing originating force, one can only ask whether they can provide a meaningful perspective for the future. They resign themselves to the necessity of the present situation, have a longing for the past, or want to flee or transcend contemporary culture by seeking refuge in a freedom beyond science and technology that is itself perpetually vulnerable. Through thought they try to rise above the thinking of scientific-technological control and to enter freedom's space of internality. This turn to the interior must repeatedly counter the ever present and proliferate domination of externality.

e. There is a good deal of similarity between *neo-Marxist revolutionaries,* like Herbert Marcuse (Schuurman, 1982), and existentialists in their evaluation of the position of man in contemporary society. However, they do not consider science and technology to be an autonomous power. Science and its relation to technology are not discussed in a negative manner, at least not primarily. Their critique is especially directed to economic and political power-lords, to an elite that takes science and technology into its service.

[30] These neo-Marxists are inspired by an alternative vision for the future, a utopia in which everyone is free and happy, and try in a revolutionary fashion to transform today's society according to the model of that utopia. A permanent conversion of society will increasingly liberate it from the bondage of this present darkness.

The realization of their revolutionary ideas has extensive consequences for the development of science and technology and for the significance of that science and technology within economics and politics. Achieving practical *goals,* always in a domination-free manner (Habermas, 1970), must precede the resolving of practical *problems.* They rebel against the ideology of technocracy, to help human beings become authentic in daily life. In practice human beings will be less the worker and more the player. In that way there will be more room to satisfy what being human means, to enjoy life and to live out vital urges that now all too often are being suppressed.

f. Thinkers of the *counterculture,* like Theodor Roszak and Charles Reich (Roszak, 1969; Schuurman, 1980), argue as do the existentialists for the renewal of authentic thinking. They do not want to leave it at that, however. Authentic thinking must be confirmed in culture. Doing so need not imply the ways and means of the revolutionaries, who have a societal revolution in mind, after which also the way in which science and technology function will have to change. Representatives of the counterculture favor a spiritual revolution, an inner revolution of human consciousness. The results of such a reorientation would then undermine the present social situation and give rise to a new culture, the counterculture, characterized by another science and technology. As do proponents of general systems theory, counterculture thinkers oppose reductionistic science and its universal application. Their preference, however, is not a reinforced or extended scientific-technological world picture, but its fundamental alteration. They want to return to a sense of the sacred, to the visionary consciousness that rests not in reason, but in [31] feeling. They wish to be liberated from the total alienation of a scientific-technological culture and wish to attend to the non-intellectual faculties of the human personality, for abilities that refresh themselves on the splendor of visions. On the path back from expertise towards wisdom they are seeking for the good, the true, and the beautiful.

The values and the norms of the counterculture clearly differ from those on which Western culture has been based since the scientific revolution of the seventeenth century. Cultural progress is resisted. Science and technology must be "adapted" so that survival once again is possible. The counterculture will have to be characterized by a small, human scale. Instead of everything being standardized, culture will have to be variegated, also in its societal forms. It will have to be more organic than mechanical; simplicity and frugality must replace profusion; varied, joyful work will have to come in the place of labor that alienates and is evaluated only in terms of productivity.

These suggestions of countercultural thinkers have greatly influenced many protest movements that oppose large scale modern technology (Schumacher, 1973 and 1977). But a consistent elaboration of these suggestions has not been forthcoming. As is often the case, counter-movements often linger on as parasites to the present situation, all the while expressing a basic discontent.

9. The Knot of Science and Technology

When we survey the various philosophical views of modern culture, which were themselves shaped by modern technology, we conclude that many experience modern technology as very problematic. In addition, there is little consensus among the diagnoses of the cultural crisis, let alone on the therapy needed to resolve this crisis. Clearly modern technology raises many problems rooted in basic motives, or background forces, that bond science and technology.

[32] In connection with the configuration of technology and science that gave rise to *modern* technology, it would be good to look at the general parameters of science's influence on technology. The French philosopher Jacques Ellul can help us here. In two of his publications, *The Technological Society* (1964) and *The Technological System* (1980), he analyzes modern technology culturally and philosophically. Some of the things he says come close to the position of the existentialists discussed above. What is worth noting and also very interesting is that Ellul also analyzes the characteristics of modern technology – features that

can only be understood in light of the influence of science on technology. He makes the point that its scientific basis and the scientific method (see §12) are the reasons why modern technology is clearly different from the technology of the craftsman. An overestimation of science's potential in technology also suggests why so many problems (see §7) arise in contemporary culture because of technology or along with it.

Ellul says that modern technology is characterized by rationality, artificiality, automation, self-reinforcement, monism, universality, and autonomy. We will briefly discuss each of these.

a. The *rationality* of science shapes modern technology in such a way that its influence on culture is defined by this rationality. There increasingly appears to be less room for fantasy, beauty, and the irrational, says Ellul. By systemizing tasks, dividing labor, and dictating standard criteria for products and their production, less and less room is left for spontaneity and personal creativity. As each new technology is introduced, the schema of logic is strengthened in culture. Technological development follows the ironhanded laws of logic.

b. While technology was once imbedded in nature, it has now brought forth a thoroughly abstract, *artificial* world. Modern technology exhibits an accumulation of technological means that together begin to dominate and even eliminate nature. [33]

c. The *automatism* of technology implies that technology more or less steers itself. Once scientific discoveries have been made, the technological development they make possible seems to happen of itself. In spite of all our deliberations, modern technology is not to be swayed.

d. The *self-reinforcing* character of modern technology is readily apparent in this process. As technology allows the science of physics to penetrate deeper and deeper into matter, its influence on technology increases in turn. This is clear from recent developments in nuclear research and technology. This self-sustaining feature of technology also reaches to education, economics, agriculture, politics, administration, and so forth.

The influence of the computer serves as a prime example. According to Ellul, the self-proficient nature of technological development also implies that it cannot be reversed. This irreversibility holds for technological development as such as well as for the influence that technology has on culture in its entirety. A new technological development often has rippling effects on all branches of technology and every sector of culture.

e. According to Ellul, one result of the automatism and self-reinforcing character of technology is that people have less and less to say about technological development. The phenomenon of technology, though blind to the future, nevertheless increases in intensity. People are affected by technology more and more, but in the process have less and less effect on it. That is what Ellul calls the *monism* of technological development. Weighing the advantages of technological development against the disadvantages of technological development is out of the question. There is also no place for moral considerations. Modern technology becomes totally separated from ethical deliberations. In fact, it is more often the case that ethics is being forced to conform to technology. Now that technology has become a system, the ethics that fits the need is a systems ethics, which does not critique technology as a system, but is there to support it as such. This does not mean, Ellul cautions, that tech- [34] nological development has put us on the road to certain catastrophe. The inner logic of technology, it seems, sees to it that the harmful effects are expunged. In other words, given the influence of science on technology, technology is in a position to correct itself.

This orderliness of technology in Ellul's estimation is devastating for human freedom. Technological order, he claims in typical existentialist fashion, eliminates human freedom.

f. Ellul sees the connection between the self-reinforcement and the monism of technology in the *plan,* taken in the sense of a definitive technological construct. What he especially has in mind here is another influence that science exerts on technology, namely, *universality.* By means of trade, wars, technological assistance, and the influence of telecommunication networks, technology's sweep spans the globe. Technology is disintegrating the barriers that exist between cultures.

The universality of technology is not just geographic. Ellul also has in mind a qualitative universality. Science focuses on what is universal, on what is true everywhere. In providing modern technology with a basis and a method, technology too has become universal in a qualitative sense. That is to say, technology, mirroring science's influence on it, is in turn imposing a uniformity on culture.

g. All the characteristics mentioned above can be traced back to the influence of science on technology. They find their unity ultimately in the *autonomy* of technology. For Ellul this means that technology becomes a world of its own, subordinating everything to itself and putting its mark on it. This all-pervasive technology will become so influential that people will subject themselves to it with the reverence of a religion. Not only ethics, but even religion will be influenced by technology and even defined by it. The society of the future is the technological society. The ethics of the future will follow the dictates of systems analysis. The religion of the future is a technological eschatology. People trust, admire, and worship technology for these kinds of reasons. But they will also come to fear technology's agents as though they were [35] gods. The technological society is constructed by technology, for technology, and is exclusively technological.

As an existentialist, Ellul's critique concentrates on these characteristics of technology. But as far as the influence that science exerts on technology is concerned, certainly where that influence is unwarranted, it cannot be denied that Ellul's characterizations hit the nail on the head. In his later work (1980) as well he stays by this analysis. There Ellul avails himself of the general systems approach, which assumes that the whole is more than the sum of the parts, and analyzes the phenomenon of technology more as an all-inclusive, coalescent system. Technology's totalitarian traits are apparent from its being a system in which all the technological means are incorporated into a common context.

In Chapter 3 we will consider the ethical questions surrounding modern technology. We will see there, in contrast to Ellul's claim that ethics is banished, that because of the advancing influence of science on technology and its consequences, ethics is being given more attention. The big problem with respect to modern technology is that the influence of science is something people now take for granted. In what follows we will see that the characteristics of science, for example, its universality and abstraction(s), are in tension with the fullness of reality. In reflecting on these matters ethically, the central question will be whether technological science, instead of dominating technology, can also seek to serve that technology.

The next chapter analyzes the structure of the relationship between science and technology to discover what makes modern technology unique. By taking a closer look, particularly at the scientific method of which technology avails itself, we will be in a position in the fourth chapter to deal legitimately with the philosophical-ethical questions and problems surrounding technology and the technological sciences. In the end, it is only within that context that we can perceive all of culture meaningfully.

2. Technological Sciences

10. The Foundational Sciences

[37] The immediate foundations of modern technology are the technical sciences. Given the history and differentiation of technology, from materials and energy technology to information and biotechnology, there is sufficient reason to speak about technological sciences in the plural. In what follows, however, I will refer simply to "technological science" and mean with this term the wide diversity of technological sciences (see §18). Each of these technological sciences reflects the scientific knowledge of the respective sectors of modern technology.

Technological science has to do with knowledge about technology, taken here in the plain sense of forming and making. It should be noted that there exists more scientific knowledge about what is formed and made (e.g., technological things and processes) than what is here referred to as technological scientific knowledge. An obvious example to illustrate this would be physics. In technology we try, for example, to give form to inorganic or "dead" materials like stone or wood. Scientific knowledge about these things from physics serves as a basis for technological science.

The distinction between natural science and technology, as theory on the one hand and technological praxis on the other, loses its sharpness because technology is being defined more and more by the natural sciences. Many technological things and processes have become so com- [38] plex that only scientific knowledge can provide us with insight about them. It is becoming increasingly clear that technological science bridges the gap between natural science and technology.

Physics and technological science are closely related, as is readily apparent when one compares these sciences. Why closely related? Physics seeks knowledge about the physical facets of things, while in technology physical subjects are the starting point for technological formation. As a result, technological scientists must know about the physical features of things before they design anything.

To the extent that humans are no longer needed in technological execution, physics and the science of technology look alike. Terminology in both sciences often sounds the same, though the notation in technological science meets the needs of the concrete technological scene. In other words, in technological science one prefers to choose symbols that conform with (automated) technological formation or execution. Drawings, graphics, diagrams, nomograms, functions, and descriptions, depict the technological order. Symbols in physics, on the other hand, are more abstract because they are directed to the physical structure of things.

The points of similarity increase the more technological science deals with fundamentals. Similarities decrease the closer technological science gets to technological practice, because there it primarily prescribes. Some examples are designing a bridge that will require less material, building a dam that is safer, or devising faster train travel. When technological science is so involved with concrete technological practice, it is also more limited. Scientific-technological research ends as soon as a solution has been found. Research can begin again if there is a desire to improve products or if socio-economic circumstances change, or if improvements in a related technology make new developments possible. But it can also be that development in a technological science comes to a halt because no new technological possibilities [39] appear on the scene. So, the more technological science deals with practical matters, the more it differs from physics (see § 18).

One should also see how physics and basic technological science differ. Physics has to do with knowledge about nature. Technological science is based on general knowledge of how technology is formed and how it uses physical subjects. This is specific knowledge, namely, knowledge about forming technology.

Physics has a more universal focus than does technological science. Physics is after universal or generic knowledge about the physical dimension of reality. The science of technology attends more to individualized knowledge about the technological formation of physical things and processes. From this point of view mathematical physics belongs to the basis of technological science.

Finally, concerning their connection and difference, one ought to keep in mind that physics in its pursuit of knowledge increasingly uses technology and the insights that accrue to technological science, for example, through experimentation. Physics, technological science, and technology interact (Van Meisen,

1961; Schuurman, 1980). On the one hand, physics, together with other foundational sciences, lays the foundation for technology, while on the other, technology and insights gleaned from technological science create the possibility for further development in the foundational sciences. An example of this would be when observations and experiments in which answers to questions concerning scientific knowledge of the physical facet of reality can be gained with the aid of technological mechanisms. In that case the technological apparatus serves the process of theory formation in physics. This technological support can lead to new knowledge in physics that in turn can enhance technological development. A scientific discovery can be the condition or catalyst for an invention, because with a scientific discovery the frame of reference within which invention occurs has [40] been expanded. The invention can in turn benefit the further development of physics (Van Meisen, 1961; see §14).

That technology permeates physics is quite obvious from the apparatus used in research laboratories. But it is also clear that the most advanced of instruments, for example the particle accelerator, result from the newest scientific developments. And yet the distinction remains a clear one: natural science describes the material world. It does not change that world. In research, the technological apparatus serves knowledge; it is not itself the goal of knowledge.

The scientific basis of technology and technological science, as was said above, is not only physics. Because what is known in physics is increasingly assimilated in the form of mathematics, we often deal with mathematical physics. In addition, this knowledge has most recently been modeled according to the rules of mathematical, information theory, so that information science too must count as a foundational science (see §6). Other foundational sciences, besides mathematical physics, are biophysics, chemistry, biochemistry, and, to the extent we are dealing with biotechnology, also biology (see §18).

We will see later that biology and ecology should actually also be included with the foundational sciences because technology always takes place in an environment interrelated with the living natural world. We live in an age of environmental pollution. We see nature degraded and even destroyed. We see that the technical schools often include only those foundational sciences that help us continue to expand technological functioning, without regarding its consequences for nature and the environment. Fortunately, many are beginning to realize that the damage this has caused must now be undone and as much as possible be prevented in the future.

Because technology functions significantly in society today, and because, for example, we must ensure the quality of work, it is a good thing that in the last twenty years the social sciences have been accepted [41] as the context within which technological science should be pursued. In this study, however, we will not be able to discuss how important this is for the development of technology.

11. Development of the Technological Sciences

We have mentioned in passing that in modern technology, *design* has disengaged itself in place and time from *implementation.* The connection between both of these sectors of modern technology is formed by the objective construct or design. It results from productive imagination and functions as the prescription for the implementation. The new construct provides the criterion for the technological execution. As knowledge about that technological process, it belongs to the field of technological science.

Because preparation and execution are usually separate in modern technology, these two areas can also be distinguished within technological science. On the one hand, based on present technological development, we know about the technological process. On the other hand, part of technological science involves knowing about how engineers go about designing. This knowledge in turn leads to results having to do with the first sector. Seen in this way, the engineer is the architect and builder of technological science. He is both the scientist and the technician.

Because of the computer, the first sector – actually implementing and forming – will become completely automated. Human beings will actively participate less and less in production.

In the second sector of technological science, the engineer plays the major role. He does not simply follow his intuition or proceed on past personal experience. Instead, given his scientific knowledge (see §10), the engineer distances himself in a typically scientific manner from the immediate environment within which the technological problem has [42] arisen. Design is based on science and is characterized by a scientific attitude or method.

The very first thing an engineer has to do is state the technological problem. Then he asks whether what technological science knows at present provides sufficient grounds for him to resolve one or more of these problems. Usually it does not. So the engineer proceeds from scientific knowledge to analyze the problem, abstracting and isolating it from the immediate context, breaking it down into modules or unit problems, and then posing it in general or universal terms. This is what we call the *scientific design method* (see §12). The engineer, using his technological imagination, tries to solve the subproblems. To solve a concrete technological problem, then, he or she coordinates the various module solutions and integrates them into a whole: *the design.*

While designing, the engineer must check repeatedly to see whether the design is suitable to the technological problem and to the context in which that problem arose. With complicated problems this feedback is especially imperative. Technological knowledge and imagination alone cannot produce a good design. Performing experiments or making prototypes is required to answer questions that technological science cannot (yet) resolve (see §17).

The ultimate intent of technological science, in contrast to physics, is not to collect knowledge, but to develop technology. We could also say that the science of physics goes its own way, while technological science always follows a detour to promote technological development. Physics continues to aim at universal knowledge, while the goal of technological science is new technology. The hypothesis, anticipating a future physical theory, is for that reason analogous to the innovation or invention anticipating a new design (see §14). Here lies the similarity. The difference is found in the direction both sciences are headed (see Figure 1). [43]

A very common problem for engineers today is that the technological knowledge of the individual engineer does not increase proportionately with the tempo of technological development and hence always lags behind the breadth of relevant knowledge. Specialization and teamwork are supposed to alleviate the effects of this problem. Through this specialization the engineer increasingly loses contact with the context in modern technology. Specialization can weaken technological development. Working in teams is one attempt to prevent this. This encumbers the task of the engineer, but the advantage is that more can be achieved in a technological sense.

<table>
<tr><th>Science</th><th>Technology</th></tr>
<tr><td>existing knowledge</td><td>existing product or process</td></tr>
<tr><td>|</td><td>|</td></tr>
<tr><td>via hypothesis and reflection</td><td>via recognition of needs or market research</td></tr>
<tr><td>⇩</td><td>⇩</td></tr>
<tr><td>hypothesis</td><td>new innovation or invention</td></tr>
<tr><td>|</td><td>|</td></tr>
<tr><td>via logic and mathematics</td><td>via feasibility study (technological science)</td></tr>
<tr><td>⇩</td><td>⇩</td></tr>
<tr><td>falsifiable deductions</td><td>adequacy of the design; testing</td></tr>
<tr><td>|</td><td>|</td></tr>
<tr><td>via experiment</td><td>via prototype/development</td></tr>
<tr><td>⇩</td><td>⇩</td></tr>
<tr><td>confirmation</td><td>production</td></tr>
<tr><td>|</td><td>|</td></tr>
<tr><td>via communication</td><td>via public acceptance</td></tr>
<tr><td>⇩</td><td>⇩</td></tr>
<tr><td>new knowledge</td><td>new product or process</td></tr>
</table>

Figure 1: Development in science and technology (see Rogers, 1983)

A special point deserves attention here. In contrast to the foundational sciences we have discussed and classified with the natural sciences [44] (see §10), technological science is a *social science* because it tries to formulate knowledge about *human activity* in a technological sense. The free and creative human being plays a decisive role in its field of investigation, that is, in the entire arena of technology. This is especially true when it comes to designing and inventing, but also to the extent humans have a place that as yet cannot be automated in the process of technological production. If automation becomes fully realized, this would mean that the limits for the technological scientific control of the production process will hold as the limits for a particular field within technological science (see §12). In separate paragraphs along the way we will analyze technological science as a social science and the special problems this presents.

After scrutinizing the scientific design method and its limitations we will turn to invention. Inventing breaks open this method because in and through the process of inventing we find solutions hitherto unknown. Inventions can obviously differ in kind. What they have in common, however, is that they always bring with them new possibilities for the development of technology. But they often also bring new problems.

12. The Scientific Design Method

As was said, the engineer uses the scientific method. This method is like that of the natural sciences but is technological because it focuses on the design of technological things or processes. This scientific design method characterizes modern technology. As a result, the features or characteristics of science are projected in technology. Seeing this close intertwinement of technology and science also helps us see the characteristics of modern technology. In the next paragraph we will see that there are limits to this method and that understanding these limits also helps us see the limits of modern technology. But we will begin by investigating the scientific method. [45]

With the scientific method of the natural sciences we attempt to discover knowledge of reality. With the scientific design method we intend to formulate technological designs both for the product and for producing it. The ultimate aim is *a temporally and spatially remote, scientific control* of the technological process of forming or producing. But its result also belongs to the domain of technological science because the designs furnish us with knowledge regarding technology.

Some of the characteristics of natural scientific knowledge and of its method are reflected in a technological kind of way in technological science. This is in large part due to the scientific basis of design. These characteristics can also be found in the technological production process and in its results.

Analysis, abstraction, and synthesis constitute the method of science. That is to say, in science we consciously turn away from and seek to avoid the influence of the immediacies of the concrete individual, situation, or environment. As a result, scientific knowledge is universal, enduring, coherent, and functional knowledge.

The engineer, too, follows the same procedure. The technical question is broken down into parts which in turn are posed as universal (sub)problems. Universal modular solutions are sought both for the components of the thing to be produced and for steps of the production process. The division of functions continues until one has arrived at atomized, isolated, autonomous functions.

Analysis puts technological modular functions at one's disposal. Abstraction presents each of these functions on its own, bracketing out the others, so as to find a technological solution for that subfunction. The result consists of basic building blocks that have a standardized neutrality of purpose about them and, hence, can have a universal application in technology, as is the case for example with propellers, rivets, or dovetail joints. In this universal applicability, a result of analysis and abstraction, we find science's general or universal knowledge projected [46] in technology. In other words, the universal suitability of the solution of a subfunction is an analogy of scientific knowledge in technology.

In addition to the neutralizing division of functions, the scientific design method, as was said, also aims at a spatially and temporally remote control of the process. Therefore, the tool or technical operator must be constant over the course of time. The control of technological formation at a distance and over time is an analogy of science as enduring abstract knowledge of reality.

This remote control, with respect to the whole, is further made possible by the integration of the neutral modular solutions. Depending on the modules that are merged, this integration of functions can represent many forms of individuality. As such, it is a technological analogy of synthesis in science. *The scientific design method aims then, on the one hand, to standardize how functions are divided and, on the other, to anticipate how these functions will be integrated for various specific ends.* The production process that results is likewise an analogy of the *coherency* of scientific knowledge, with the product of mass production showing something akin to the *universal* character of scientific knowledge.

In like manner, an analogy of scientific abstraction can be noted in the integration of the solutions from each module. These problem sets were originally isolated by means of abstraction. After a standardized solution is found for each subset, the integration of these separate solutions requires that technological formation take place elsewhere, in isolation. Pertinent measures are taken with that in mind. Technological formation has an enduring character about it because influence from the surroundings, like temperature and humidity, are precluded as much as possible.

This scientific design method determines the basic structure of modern technology. Technological science and technology are virtually interwoven. An example can help to elucidate their relationship. [47]

A technical operator, as we said, is a tool that works automatically. Given such an operator we can distinguish the thing and the process of the transformation of energy. We can note the influence of the method on this thing, the machine. The parts of such a machine have a general or universal purpose and are as such technologically neutral. Together they make up this particular, concrete machine. The neutral division of functions, then, is allied to an individualizing integration of functions.

The energy exchange that drives the machine has undergone the same effect. The flow of energy is in fact individually directed for this particular machine, for example, a lathe, but in previous steps of the production process is dependent on energy of a more universal structure, that is, one in which the energy flow must have as little individuality as possible. Universal neutrality in a technological sense is always accompanied by individualizing integration.

When developments in science lead to a new method, as was the case with systems analysis, then this method recurs analogously at the technological level. That is why general systems theory and computer technology are complementary.

The method of systems analysis (Laszlo, 1974) chooses not to proceed from analysis and abstraction, because that would fragment and reduce reality. Systems thinking attends to the whole of reality. According to Laszlo, that is why its method is expansionistic and synthetic.

Without going into great depth about this method, I must point out that it provides a new scientific prospect towards advancing the cause of modern technology. At the same time, one must be warned of a possible misunderstanding. Although the systems analysis approach differs, it is nonetheless a scientific method, with the same promise of possibilities and potential for problems for technology that come with scientific methodologies. Its possibilities and problems are simply different from those of the method of analysis and abstraction. [48]

The most striking difference lies in the shift in scientific focus. Whereas earlier attention was given to a component, an aspect, or a function of reality, in systems theory the *whole* remains central *as a system. That whole is more than the sum of the parts.* Closer consideration shows the whole to be the sum of the parts plus the interaction between those parts. The interaction is controlled by additional information, communication, feedback, equifinality, self-stabilization, and self-organization. This approach also attends to the system and its environment in terms of the input and output of material, energy, and information. We see from this that although the method is new when compared to previous methods, the result leads to an enhanced continuation of modern technology in information technology, computer technology, and the technology of integrated systems.

13. Limits to the Scientific Design Method

That the scientific design method we are describing is technological, and not something of the natural sciences, appears from the fact that reality resists the realization of this method. Owing to its scientific character this method has a universal pretense. But because it is also directed to a concrete technological application, this method must conclude with individual designs or techniques. That is why applying this method in the practice of technology has its limits. Its possibilities are in principle limited because science is not *congruent* with concrete reality. The universal, abstract, and enduring or resolute character of science pulls in a direction other than the fundamental features of reality, where everything is singular, hangs together with everything else, and changes. People try to suppress these features of reality with the scientific-technological method, but they are never entirely successful. Concrete reality, in which technological forming takes place, sets limits to the scientific design method. For example, entirely universal products cannot be made, the technological process can also not be fully isolated from its [**49**] environment, and finally, technological things that are altogether resistant to wear are out of the question. It is the resistance of reality that prevents an all-encompassing scientific control of modern technological formation. A scientifically recalcitrant remnant of reality will always, in one way or another, continue to make its demands and to confront the realization of remotely controlled, total automation with singular problems.

The resistance of reality manifests itself in many ways: in occurrent variations in natural materials and energy sources; in perpetually changing factors in the technological milieu, like temperature, humidity, pressure, and so forth; in changes in the technological operator due to wear, corrosion, aging, breakage, fatigue, fluctuating temperature, and the like; in the individuality and freedom of the workers, to the extent they must still necessarily be incorporated in the technological process of formation; and in the fluctuating tastes of the consumers of the products to be made.

These variances present problems when we try to control things scientifically from a distance. However, these problems are not always insurmountable. With the introduction of the computer, for example, digital control has made great strides in satisfying the individual desires and wishes of the consumer while maintaining remote control (i.e., complete automation).

Although a complete scientific formulation of technological formation is not possible due to these variances, it is interesting that this does not always mean that remote control is out of the question. We see here the fruit of general systems theory. Certain degrees of variation owing to natural inconsistencies can be corrected with the help of a "trick." At the point where a set standard is not met, an automatic switch is triggered and on the basis of the cybernetic principle of feedback, the deviation can be nullified. The occurrence and frequency of these variances cannot be theoretically predicted. [50]

Such limits are due to the laws of nature. The success of the scientific design method easily tempts people to absolutize this method and to apply it where it ought not be used. They forget that technological science is a social science. The relevance of this design method is overestimated when, for example, the worker in the production process becomes an object of scientific-technological control. The situation becomes very serious when the scientific flow-chart dictates that, due to a lack of technological operators, workers have to repeat monotonously a single subfunction. The individuality and freedom of workers has been eliminated because these disturb the degree of scientific control. They are expected to work as mechanical parts.

A good deal of labor's drudgery, for example, the tedium of work on the assembly line, has disappeared with the increased application of the micro-computer. But that has often brought with it the problem of unemployment. The complete automation of the production process requires highly qualified workers. Each step of the process once performed by human beings is now projected or objectified in the technological operator. But each of its steps, whether that used to be head or hand work, expediting or controlling, can after all only be worthwhile if there is, at its side, a person who remains physically and mentally the master of its work (Van Meisen, 1961).

Overestimation of the scientific design method is also often evident in mass production. Mass production is not so bad for making component parts. But when the final mass-produced product is supposed to serve the consumer, many problems surface. Van Meisen rightly points out that a mass-produced product can have an unmistaken leveling effect on the user. The statistically average end user, as the cross section of many users with a diversity of choices and desires, ultimately becomes the actual user. Every consumer eventually becomes a representative of the average user. People begin to adapt themselves to what technology prescribes: standardized goods for standardized users, that [51] is, for the masses. In all honesty, we have to say that these mass-produced wares are cheap and are available to many who otherwise would have to do without.

The objections and dangers are greater when the scientific method becomes the absolute in technology and is given regal status (see §21 and what we will call "technicism"). When "technologize" is the answer to every question and "technologically possible" dictates the agenda for tomorrow, the limits of the scientific design method are being or will soon be transgressed. It was especially this absolutization of the scientific design method as well as its imperialism that Ellul had in mind (see §9) and that constitutes a problem for the ethics of technological science (see §§21 and 22).

14. Invention

The pivotal point for human work in technology, then, will be in the design phase, in preparing for the process. To that end, people will need to learn the foundational sciences of technology, technological science, and the scientific design method.

Even though designing is dominated by this method and it guarantees the continuity in technological development, we must attend to breaks in this continuity. We are talking here about *inventions.* Technology continues to develop in the context of scientific consultation and formulation. In ways that are often unexpected and cannot be simulated, productive imagination can put people on the track of technological innovation. These innovations or inventions can interrupt the developmental process of technology.

Where invention begins is difficult to ascertain. Usually a number of factors like need, coincidence, want, playfulness, imagination, together or individually, play a role. But that is not all. The unthought-of dimension, which makes invention what it is, is not sufficiently explained by these factors. [52]

In modern technology, invention takes place at a higher level than in classical technology. Today inventions build on mathematics, physics, information theory, and technological science. This also means that when inventors are onto something new in technology, they try to develop it scientifically and to check it out with precision instruments to complete the invention. But at the decisive moment of inventing, theory plays no role. The invention and its effects will later belong to the field of technological science, which also served as its basis. This being so, the invention contributes an additional path to those presently available for new technological possibilities.

Among inventions we can distinguish primary inventions, which have to do with an entirely new technological unit, and inventions that improve on components. Dessauer speaks in this context of "pioneer inventions" and "developmental inventions" (1956). In this connection, however, we should add that because it is becoming very difficult for one person to invent something new, the inventor more and more must work with a team of innovators.

It is important to note that although technological science affords a more advanced basis or context for invention, there is a danger that this basis for invention or innovation can stifle the development of modern technology. The sheer growth of theoretical means and their ready access threatens to sterilize the productive imagination. On the other hand, invention can be an opportunity to break through an unfavorable or infertile development and to open up new perspectives. This is especially true in developing computers, where even playing around can spark new inventions.

Raby was correct in pointing out that the technological realization of inventions often depends not only on the scientific basis, but also on the material capability of technology itself. Already in 1823 Babbage had worked out a calculator in theory, that is, on paper, but its realization [53] met with many difficulties. Apart from financial constraints, the technical ability was not yet available.

Remarkably, the reverse situation presents itself at the moment in computer technology. Computers offer many more possibilities than there are applications – the prospects precede the novel technological ideas. The human imagination is at the moment not in a position to oversee and take advantage of all the possibilities provided by computer technology.

Inventions in technology spring from "objective" elements as well. (The same can be said analogously of discoveries in the sciences.) More than once inventions have occurred simultaneously in different places. Stork speaks in this context of the "births of time" (1977). Given the scientific foundation of modern technology we see this phenomenon more in our day than previously. Because of the scientific basis, the phases of preparing, incubation, illumination, and verifying or testing, have become clearly distinct and inseparably connected with the process of invention. Though imposing to the individual inventor, they at the same time actually invite many to become inventors. This situation gives rise to the idea that inventions can be more or less planned. That is why research laboratories try to program for future inventions and innovations.

Once inventions become known, variations are quick to follow. In connection with patents, it is important to note the legal requirements placed on an invention. They must include a new, precisely defined, distinct, and integral assembly of technological things or processes. Inventors must orient themselves in two directions. First of all, they have to determine what is and what is not included in the invention. What defines its individuality? What makes it peculiar? Secondly, moving in the opposite direction, they have to answer the question whether their invention is not a specialized application of a more general order of technology that made this new invention possible. [54]

In summary, inventing, because of human creativity, cannot be fully analyzed scientifically. Psychological studies, however, do show that creativity, as an expression of human freedom, can be nurtured with special education, through "brainstorming," and by learning to deal with critique. These three completely coincide with the following traits of an inventor: flexible thinking; a playful, young, inquisitive mind; the ability to see what everyone else does differently and, hence, to suggest nontraditional solutions; a vibrant intuition; a desire to perfect a technique; a bit of an introvert; great intellectual prowess paired with courage and a reticent social attitude. These are all conditional factors for invention. When it comes right down to it, however, inventing eludes scientific inquiry (Schuurman, 1980).

Finally, modern inventions are founded on science. Inventions can form a prologue to the rise of completely new scientific knowledge. At the same time, scientific discoveries can likewise be the catalyst for new inventions. As these things happen science and technology interact and the complexity in technological science increases, requiring a clear analysis and definition of place.

In previous paragraphs we saw that the scientific design method is an analogy of the natural scientific method. Inventions and innovations can be compared with discoveries and hypotheses informing scientific theories (see Figure 1 above [p. 236]).

15. Technological Science as Social Science

Technological science encompasses any scientific knowledge that has to do with designing, inventing, and technological execution. In other words, technological science also investigates human beings. That is why technological science, in contrast to physics for example, is a *social science.*

Technological science, as a social science, is the first science in the transition from the natural sciences to the social sciences. That is why **[55]** there is much in technological science that is similar to the natural sciences. We can see this from another angle as well. Technology, we said, is the tool-aided formation of the natural side of reality. The preparation for and execution of technology belong as well to technological science's field of investigation. When it focuses on fully automated technology in the production process, technological science is dealing with a determinate reality. Obviously, similarities with the natural sciences will be the greatest in that part of technological science. To the extent it deals with the engineer as designer and inventor, the weight will fall much more on technological science as a social science.

People quite rightly speak of technological science as a *normative science.* The engineer and the inventor are obliged to seek normative principles for technology, especially for how technology should relate to human beings and to nature. A clear example of this, mentioned already in §13, is that in a production process that is not fully automated, workers may not be dealt with as though they were technological objects and be expected to work as though they were components of a machine. When this happens in technological science, human freedom, creativity, and responsibility are undercut. That is why technological science, as a social science, must give heed to the quality of the work that the process requires as it seeks to combat and circumvent drudgery. The engineer and inventor must also follow normative principles when developing and renovating technology. Designating a direction for technology is a normed, cultural business (see §24).

We find then in technological science two distinct sectors. Producing and executing, whether it is fully automated or not, is the first sector. The second sector has to do with the designing, inventing, theorizing, and experimenting that accompany these (see § 17). There is interaction between both sectors. The design sector has to solve problems that arise in technological production. These solutions at the level of technological science serve, in turn, to instruct us in forming technology. **[56]**

This scientific control of the production process has its limits. The surroundings in which production takes place cannot be scientifically determined in their entirety. The same holds for the aging and wear to which the tools, or technological operators, used in production are subject. Products and the raw materials used to make them also display idiosyncrasies that can be factored in and controlled beforehand only within certain limits. In addition, at each point along the way there is always the possibility of mistakes or breakdown due to severe, unexpected irregularities. If these difficulties are not covered by automatic adjustments or switches, remote scientific control has to leave room for the responsibility of the workers in the production process. This kind of responsibility also appears in the final stage of manufacturing an individual product, although digital computer monitoring does offer possibilities here.

In general, we can say that the influence of technological science is such that people will want as much as possible to automate production or technological formation according to a scientific design. The inherent inclination here will be to rid it of the human presence as much as possible. Consequently, any attention within technology for human beings in their freedom and responsibility has come to focus more and more on the design phase, although there too scientific operations can be automated with the help of the computer (see §17).

16. Problems in Technological Science

The development phase of technological science is defined by a number of constraints. The problems with which the engineer is confronted arise in a specific context and are defined by historical factors and, further, by industrial conditions. The immediate context is determined in a historical sense by the amount of reliable technological knowledge about production processes and products fabricated to date. [57]

Technological science is also conditioned by the actual state of technology within a particular industrial plant.

A separate problem is that technological knowledge grows by leaps and bounds and few designers can keep up with all the developments in technological science. This has become one of the principal problems in modern technology. The obvious solution to this problem is to step up specialization. But this, too, has its limits. The more specialized people are, the less they know about life beyond their specialty. That is why the dangers of specialization are offset with teamwork. For the rest, steps must be taken to make the newest developments in technological science available to designers in a way that is both quick and efficient.

Although technological problems arise in a defined, particular context, it is important for the technological scientist to be able to pose and to resolve these problems in a way that is as general or universal as possible. This can result, however, in a tension between the universal solution and the individual context in which the problem surfaced. To the extent the outcome varies from what the original problem required, the original context will have to adjust itself to fit the solution. The issue here is once again the tendency to neutralize individuality and the functional integration that accompanies it.

On the other hand, we have to keep in mind that the scientific treatment of technological problems often arrives at solutions that have a wide-spread significance. Actually, one finds a broad scale of possibilities between the extremes of the most universal and most individualized solution. Illustrations of the first category would be, for example, the transformation of energy from one kind into another (photoelectric, thermochemical, electro-mechanical, and so forth) and within one energy group (e.g., lenses bending light rays, the damping of energy, mechanical transmission, etc.). When people try to solve broad problems for all time, they tend to design neutral building blocks (like [58] photo-electric cells, capacitors, gear wheels, and electron tubes) which, when integrated, as we saw, can lead to an individual final product.

Thus we see that technological science is busy with problems of universal consequence and with problems of individual specificity. The first category belongs to the foundational technological sciences, the second to the more practical or applied technological sciences. (See § 17 for a survey of the technological sciences.) If engineers would limit themselves to the applied technological sciences, they are going to lose contact with the breadth of technology. That is also why students in technological science more and more feel frustrated when they lack a solid grasp of the technological sciences in general. At the same time they realize that their inability to take insights from elsewhere and constructively apply them directly to their own field is the price they will have to pay for this deficiency. Think, for example, of the generalization of vibration systems for the different kinds of energy processes. In addition to the specialist's knowledge of the applied technological sciences, students still need to know the foundational technological sciences.

From a scientific point of view specialization is understandable. But it has its drawbacks as far as the development of technology is concerned. The curriculum for scientific, technological training is going to have to strike a good balance between the foundational, scientific knowledge of technology and knowledge about the practical face of technology. This tension between universal and individualized scientific-technological knowledge may not be resolved on pragmatic grounds by emphasizing the latter. Universal knowledge about things present throughout technology, such as measurement, connectors and regulators, and energy transfer, is an indispensable prerequisite for applying this knowledge in individual cases. (See Figure 2 on page 254.)

17. Scientific Operations

The engineer appeals to technological science and its possibilities both when beginning to design as well as when finishing. At the start, [59] design has to do with a scientific formulation of the technological problem. When finished, the design, as the law or order for the technological solution, is placed at the disposal of technological science. In the meantime, scientific operations are carried out in the designing that continually emphasizes the relation between designing and technological science.

The scientific operations are set up according to the scientific method, the purpose of which is to come to a symbolic objectification of the technological structure or order. We can distinguish two groups of operations: (a) *theoretical* or *deductive operations* and (b) *experimental* or *inductive operations.* Given their significance in technological science we will briefly discuss each of these operations.

(a) *Theoretical operations,* in the first place, make it possible to work with technological questions and problems. They also are aimed at deriving new designs for technological problems given the present state of the technological sciences. This usually involves moving from what holds in general to what holds for a particular, concrete, technological question.

The symbols used in these operations are usually borrowed from concrete technology. There are two types. Symbols that represent things denote the relative durability of technological operators or tools. The other symbols refer to changes in the technological processes. When scientists use arithmetic symbols, they refer to them respectively as *constants* and *variables.*

The symbols for constant and variable magnitudes can in turn relate to the physical, spatial, and numeric aspect of tools and processes. A few examples of symbols for constant magnitudes, that is, for the lasting characteristics of things, from the technologically unfolded physical aspect would be these: self-induction, resistance, mass, resonance, but also the elastic moduli, permeability, and the dielectric constant; from the spatial aspect: the (technological) moment of inertia and the cross [60] section of a conductor; from the numeric aspect: the number of coils of a spool. Symbols for variable magnitudes for the physical aspect of technological processes are, for example, the strength of the current and the field, speed, and the amount of energy; for the spatial aspect, the volume of the cylinder between piston and head; for the numeric aspect, the frequency and revolutions per minute.

Among scientific operations we can distinguish the *argumentation,* the *depiction,* and the *calculation.* Each has its own function and as such, supplements each other. Their use depends on how straightforward or complicated the technological problems are. For a simple problem, calculation is often sufficient. For a more complicated one, argumentation or demonstration is called for. In general, argumentation is less effective, certainly than calculation, which is the most effective. One cannot, however, always use calculations. When people play an important role in forming or producing technology, one has to revert to diagrams and sometimes simply to a verbal rationale. On the other hand, it often happens that one begins with a verbal presentation. Argumentation can then, after necessarily discounting certain factors, give way to depiction and calculation. The limits of the scientific, technological method, discussed previously (§12), play a part in leaving these factors out of account. They have to do with the tension, also mentioned previously, between science and the individuality, change, and coherence of reality.

Related to this tension is the fact that calculating can be applied most easily at the most general level of technological science. The more the design is individualized, the more difficult calculations become and the more will depiction and argument take over.

A very prevalent approach, that again makes calculation meaningful once certain factors are disregarded, is determining the safety coefficient of stressed technological constructions. By methodically leaving **[61]** things to the side as one proceeds further, the technological focus remains. As a result, these calculations may not be considered simply a matter of doing mathematics. An example may help. When calculating for a truss bridge one represents beam junctions as though they were hinges while in fact that is not the case. Because of this simplification one must assume a large safety margin so as to once again make the calculation possible.

Important types of theoretical, deductive operations include *calculations,* the graphic formulation and representation of the order or structure of technological things in *blueprints* and *drawings,* the graphic formulation and representation of the order or structure of processes in *diagrams* and *graphs,* and *technological instructions.* Theoretical operations can be applied in technological science just as they can in physics. The only difference is that in physics these operations have a universal scope, while in the technological sciences, which ultimately have to do with concrete technological things and processes, they will be more or less individualized.

Calculations have to do with relations between arithmetic symbols which stand for characteristics of technological magnitudes. One carries out mathematical operations on these relations, and in doing so, gains insight into the structure of technology, insight not gained through argumentation and demonstrative statements.

Graphics focus on the structure of technological things and leave the implicit energy processes out of the picture. The design is objectified in the form of geometric symbols. Arithmetic symbols and brief descriptions may supplement the drawings. The geometric operation takes place as the drawing is being made and usually goes hand in hand with calculations.

Graphs and diagrams, in contrast to drawings and schema, always depict a technological process. [62]

Instructions can be either a description of a technological thing or technological process. As such, they connect drawings and schema, on the one hand, and graphs and diagrams, on the other.

(b) In technological science one might shift to *experimental operations* if theoretical ones, and particularly calculations, are no longer possible either because a good starting point is not available or they will not withstand the course of time. Chemical technology, on the other hand, which has been around for long time, cannot do without experimenting. Qualitative chemical analysis and classical organic chemistry are prime examples. In technological science, however, one usually experiments to confirm or to correct a theory.

An important characteristic of experimentation is that one arranges reality in such a way that one gets a clear answer to the question being posed. The answer is received via *instrument measurement* in the form of arithmetic or geometric symbols and then assimilated into knowledge about the technological structure of things and or processes.

When following the deductive method of theoretical operations, one usually proceeds from the general or universal and moves toward the particular or individual. The experimental method, however, requires that one move in the opposite direction, namely, inductively. One proceeds from the individual results of the experiment – though many individual features already have been eliminated by the arrangement of the experiment – with an eye to arriving at knowledge or a point of view that is as general or universal as possible. This tension is the same as occurs in theoretical operations. In that case, the procedure of methodically leaving things out of the picture is necessary because of the individuality, coherence, and change of technological reality that science cannot control. In experimenting one tries to eliminate these factors by concentrating more on the question the experiment is meant to answer. Despite all this, one will *not* find much in the final calculations that actually belongs to the technological structure, and all kinds of things [63] will *implicitly* be incorporated in the results of an experiment, such as the influence of the environment and the interrelations of the measuring circuit, that were *not* intended in the question.

Both of these operations interact with and supplement the other. They can guide and stimulate the productive imagination of the designer, and at higher levels, help develop technology.

A special point deserves mention. Today scientists use computers in the operations of technological science. This is a prime example of how much science and technology have infiltrated each other. Computers speeded immensely the development of the technological sciences. The engineer can now do much more than ever before and in less time. Theoretical questions today can be so complicated and wide-ranging that engineers alone, either individually or even with cooperative effort, cannot solve them. The computer affords a remedy. The same is true, more or less, for experimenting. The experiments called for may be such that, humanly speaking, they can no longer be carried out. It could be that the preparation time is prohibitive or that the number of factors that have to be reckoned with in the experiment has become so large that a reasonable amount of insight into the outcome is almost out of the question. Some experiments may be too dangerous or simply impossible. Today, these kinds of experiments can often be simulated with the help of the computer. The computer's calculations take the place of the experiment and generate accurate answers to the questions posed.

Because the computer, in principle, can take over most everything that is determinate in the design process, the emphasis comes to He more and more on what is typically human in design: freedom in the context of responsibility, the normative dimension of technology, and its translation into the creative or productive imagination. We speak then of computer aided design (CAD). [64]

18. An Overview of the Technological Sciences

When we look at the preceding paragraphs of this chapter, the interrelation and the difference between natural science and technological science have been clearly addressed. Both sciences display analogies in their development, but the intent of each moves them in opposite directions. The natural sciences are directed toward knowledge of physical reality in a universal sense, while the technological sciences aim at concrete, individual, technological realization. Natural scientific claims must, with an eye to attaining universality, be falsifiable. The innovations and discoveries of the technological sciences have to be tested in practice, using specific designs, to see if they are actually going to be fruitful for technological practice.

That is not to say that there are no differences between the various technological sciences. Technological science does form the bridge between the natural sciences and technology (§10). Some technological sciences stand closer to the natural sciences. These we have called the foundational technological sciences. Others focus directly on more tangible matters like materials technology, product technology, and production technology, sometimes also referred to as the applied technological sciences (Teichmann, in Rapp, 1974). Figure 2 provides a partial listing of the basic sciences and many of the technological sciences. It indicates the relation between these basic sciences and the technological sciences as well as the degree to which some of the technological sciences are tied to concrete technology.

The sciences that form the basis of technology are the most general. Then come the foundational technological sciences. Building on these are the sciences dealing with materials, natural resources, and semimanufactures. Finally come the sciences having to do with concrete technological products or processes, the applied technological sciences. Figure 2 does not include philosophy and the social sciences. Given their place at technological universities, they could be considered as contextual disciplines for technology and the technological disciplines. **[65]**

19. Biotechnology

The development of the technology discussed up to this point has moved from materials via energy to information technology. More and **[66]** more this technological development has become involved with, and supplied the basis for, the development of biotechnology. Tools are no longer the means used in biotechnical formation. In biotechnology, life forms are active in such a way, possibly via manipulation, that they change according to the desires of the technician and subsequently deliver the desired products.

Applied technological sciences:	
knowledge of technological products and production technology:	
machine tools	steam turbines
gas turbines	coke furnaces
nuclear reactors	generators
automobiles	aircraft
radio and television	telecommunication
presses	computers
bridges	ships
agricultural machinery	
knowledge of technological materials (including semi-manufactures) or *materials technology:*	
iron and steel	aluminum
oil	uranium
gas	rubber
paper	ceramics
glass	plastics
concrete	textiles
foodstuffs	chemicals
Foundational technological sciences:	
engineering thermodynamics	surveying
mechanics	statics
materials science	metallurgy
calculus	machine mechanics
fluid mechanics	circuit analysis
electronics	kinematics
weights and measurement	
Basic sciences:	
Physics, biophysics	information science
biology	mathematics
Chemistry, biochemistry	ecology

Figure 2: Overview of the technological sciences (adapted from Rogers 1983; the lists are not all-inclusive)

Biotechnology uses microbes, tissue cultures taken originally from plants or animals, and manipulated micro-organisms, plants and animals. It is a mistake to think that biotechnology is of recent vintage. That is simply not true. People have used the biotic processes of microorganisms for a long time already. Think of baker's yeast, yogurt, cheese, and wine. How these products first arose is often unclear. Through centuries of experience, people have made this technology their own.

About a hundred years ago, the systematic, scientific study of biotic processes was initiated. The results of that investigation, gleaned from life sciences like molecular biology and genetics, lead to the incorporation of biotechnology in fields as diverse as industry (especially in the processing of food), agriculture (particularly bioindustry), health care, environmental protection, and energy production.

The development of biotechnology in the last decades has moved very quickly. New possibilities are the order of the day. One of the most astounding discoveries in biology has been the possibility to play around with heredity. Little pieces of hereditary material, called genes, found in the carrier of hereditary information, the DNA molecule, can be transferred from one organism to another such that cells in the second organism form having new characteristics. With the help of enzymes, pieces of one DNA molecule can be spliced and then recombined in another DNA molecule. Spliced genes, either individually or in groupings, can be inserted in other cells without those cells dying. In this way artificially new cells are made that in turn are able to reproduce **[67]** themselves. This process is called genetic engineering or recombinant DNA technology and is one application of the same scientific method we discussed previously (§12). The point can then be made that recombinant DNA technology is a counterpart to biotechnology just as modern engineering is a counterpart to technological science.

In 1974, at a conference on molecular biology in Asilomar, California, Paul Berg and Stanley Cohen announced that they were conducting experiments to transfer genes from one organism to another. They related at the same time that they had no idea as to the possible ramifications of their experiments. Scientists themselves became deeply worried about the potential catastrophic consequences, such as the spread of dangerous viruses and bacteria. At that same conference they decided to curtail all research, worldwide, and only under very special conditions to allow further experimentation. From that time on experiments were performed in isolation under very strictly controlled physical and biological conditions. Physical barriers consisted of air locks, filters, and even constructing a research facility on an island. Biological containment required that micro-organisms be "crippled" with the help of recombinant DNA technology such that should these microbes escape the laboratory environment they would disintegrate and destroy themselves completely.

This moratorium on DNA research, in other words, arose within the scientific community itself. Never before had that happened. These circumstances led to a discussion between science and the populace such that society became informed about this new area of research.

As research continued it gradually became apparent that the risks and dangers were not as great as were originally thought. The reins were loosened a bit. Throughout the world today there are set guidelines for quarantined experimentation. In addition, precautionary measures are in place to see to it that genetically manipulated organisms are not introduced outside the laboratory without permission. **[68]**

The development in biotechnology brings with it unexplored vistas. It is as though people can command and program nature, manipulating it genetically to do their bidding. Crops can be made to resist drought or heat, milk production can be increased, using recombined bacteria in bio-reactors, vaccines can be developed against a sickness like foot-and-mouth disease, a polluted environment can be cleaned up using microbes that are recombined in such a way that they eat up the pollution. New medicines can be made. For example, by inserting the human insulin gene – the gene responsible for the production of insulin – in harmless bacteria found in the human intestine *(Escherichia coli)* these recombined bacteria will produce insulin. This is a great improvement over the only other commercial source of insulin, namely, the pancreas of either cattle or pigs. The same can be said for the production of interferon that is needed for fighting viral infections.

This all sounds rather fantastic and in large part it is. But it is becoming increasingly clear that next to the many advantages, there also can be significant disadvantages to manipulating plant and animal life. Genetic engineering can result in an enormous increase in crop yields. But these crops cannot reproduce themselves, so that new hybrid seed will always be required. The farmer, and here one should think especially of those in the third world, becomes dependent for a large part on the industrial powers. Moreover, the resulting surplus can bring with it considerable economic problems.

Success can also be had in the field of animal breeding and livestock production. But here too there are accompanying ethical problems. How far is too far when it comes to genetic manipulation? Where do the limits lie? May we do with animals simply what we find useful for ourselves? Could it be that animals produced by this technology would suffer under the consequences of manipulation? For example, because of obesity or a disproportionate or weak skeleton? Furthermore, is it pos- **[69]** sible that by transplanting or cloning embryos we are introducing a technology that depreciates the very nature of these animals?

The ethical questions will obviously be even greater if genetic manipulation would expand to include humans. On the one hand, we may expect that within the context of a responsible medical ethics, medicine will be able to advance, as is already the case with the genetic manipulation of certain organs. On the other hand, human beings can also become the victims of this technology.

With this new technology it is possible via prenatal or even preimplantation diagnosis to determine beforehand any hereditary deformities. This clearly raises many ethical and social problems. What are the implications of this kind of research? Would knowledge of this kind of diagnosis not put insurance companies in a position to exert inappropriate pressure? The problem would be multiplied with a complete catalog of a person's genome. This would include the total complement of genes in an individual, even though the particular significance of each of these genes might not be known. The function of many genes is already known, but for a good number, that is still not the case. That is the agenda of continuing research. The increased ability to predict more and more diseases that will result, will in tum have societal implications for the relations between employee and employers and between client and insurance companies.

In short, given the new possibilities that biotechnology brings, particularly in genetic engineering, we have arrived at a stage in the development of technology that is thoroughly fascinating, but that can also become extremely hazardous. Technology began developing when human beings mastered dead material. But scientific developments have now made human beings themselves the object of technology. A few of the many problems and ethical questions connected with this last phase will be addressed in the fourth chapter.

3. Philosophical and Ethical Foundations

20. Ethics and Technological Science

[71] In our culture, science and modern technology influence all kinds of areas. Compared to the past, materially speaking, our culture is rich. Our age has seen an unparalleled increase in human power and freedom. At the same time, it cannot be denied that its problems are large and on the rise. For example, because of the predominance of scientific-technological power, agriculture has been disconnected from its ecological and biotic context and the strain between agriculture and nature is intensifying. A rich diversity of flora and fauna are being wiped out, damaging the environment. The pollution of earth, air, and water seriously infringes on the healthful fundamentals of the biosphere. The power of modern technology and the ensuing dynamics of history prove to have a flip side in the threat to nature and man by degrading the environment and destroying our health. The big question that will not go away is whether or not the power of technology is going to get out of hand. Or as Bruno Guardini posed the question: how is our power over the power of technology going to be maintained or reinstated?

In a variety of contexts people are talking today about a cultural crisis. This is due particularly to the fact that our culture has come under the grip of an *overly extended* scientific-technological control. The word "crisis" suggests the fact that we are confronted by an accumulation of problems upon whose solutions few can agree. [72]

There has been a good deal of reflection from a variety of philosophical positions on ethical issues related to technological science. In our survey of philosophical perspectives (§§8 and 9), for example, we saw the extent to which cultural philosophical discussion has focused not only on philosophical questions concerning technological science, but on ethical problems as well. In this chapter I will want to contribute to that philosophical, ethical reflection. In doing so I will be discussing an approach that is more radical and more integral than is usually proposed.

In that context, I will analyze the influence and pivotal position of technological science. This is necessary because previous discussions have shown that within technology, tension exists between science, on the one hand, which by nature is always universal, constant, abstract, and logically coherent, and reality, on the other, which is singular, continually changing, replete, and everything but rational. Given the extensive influence of the scientific-technological control of culture, we cannot exclude the possibility that causes for the crisis can be found in this tension.

Problems indicated in Chapter I (§5) could very well be related to an undisclosed prescientific or extra-scientific dynamic and view of science and to its structural consequences in the repercussions of technology's scientific control and its influence in so many areas of culture, such as agriculture, economics, health care, and politics. Through scientific-technological control, science is having more and more influence on society.

Moreover, it certainly seems that there is a blind spot when it comes to considering these preliminary questions. The silence must be broken regarding the philosophical and ethical foundations having to do with what most deeply moves human beings.

I want to emphasize that the philosophical and ethical approach I am taking proceeds from the perspective of reformational philosophy [73] (Dooyeweerd, 1953; Kalsbeek, 1975). In addition to its analyses of the structure of reality, particular attention has been given in this philosophical tradition to the deeper driving forces, or motives, in the cultural activity of human beings. After discussing some of these we will also want to look at whether reformational philosophy's theory of the nature of things contains a normative context for scientific-technological control which might suggest a way to understand the problems of technological culture.

Before proceeding I want to clarify what I understand ethics to be. Ethics can be defined as the science or discipline that deals with what is good or responsible human activity. It also involves making explicit the body of norms and values that make an activity "good" or "responsible," as well as the motivation that lies at the basis of that activity. In this chapter I will speak of ethics in this last sense. I will be talking about setting an ethically good, and hence responsible, meaningful direction for developing culture and, in so doing, for developing both technological science and modern technology in its relation to the various sectors of society. We have to seek out a normative context within which modern technology and technological science, to which modern technology is attuned, can develop responsibly; a normative context that opposes the overestimation, or absolutization, and blanket imposition of the scientific design method.

21. Technicism, Utilitarianism, and Instrumentalism

Human power has increased tremendously, but human beings do not seem able to handle this power. Using science and technology, we have conjured up very serious problems which now seem beyond our grasp. We stand powerless over against the deterioration of our inner cities, the devastating character of nuclear war, the problem of aid to third world countries, the degradation of nature and the pollution of the environment, the many problems tied to unemployment, the many irrational [74] reactions to these demoralizing problems, for example, in the use of drugs, the growing loneliness of people and their increasing sense of alienation, and the list goes on.

In our quest for the causes of the present cultural crisis, we are going to begin with a discussion of the spiritual and historical roots of the development of science and of the technology to which it has given rise.

Following more or less the path of a growing number of philosophers, like Heidegger, Ellul, Sachsse, Staudinger, Horkheimer, and Ihde, many have come to recognize *technicism* as the *predominate* spiritual force in the development of science and technology. It is important to note, however, that this driving force, while predominant, has not been the only one. The fact that there have been others makes it possible for us to indicate another, more acceptable route for the future.

Technicism is the human pretension to be able to autonomously bend all of reality to our will through the use of scientific-technological control and in doing so to solve any problem that arises and hence to guarantee material progress. The assurance is that even the problems technicism creates can in turn be mastered with new or proven scientific-technological solutions. Spengler had this conviction in mind when he spoke about a "faith in technology" as the stimulus for technological development.

To avoid any misunderstanding: my intent in discussing "technicism" is certainly not to dismiss technological science and modern technology. My critique of technicism is aimed at rejecting the *predominance* of technological science, while promoting its ability to be of *service;* to dissuade people from *overestimating* technological power, while encouraging *responsible, meaningful* technology.

Technicism has come to dominate more and more the ethos, the guiding principles of human knowledge and activity, in science and technology. In talking about technicism, then, we are really talking about a religious conviction that manifests itself in what many have referred to [75] as a "faith in progress." That confidence, as a faith in material advancement, is the driving force behind the expansive and absolutized growth of science and technology. As far as this faith is concerned, all of reality is just "stuff" awaiting the formative and creative power of humankind. Under the influence of technicism, the successful scientific design method of modern technology (see §11) is converted into an absolute. Furthermore, there seems to be something imperialistic about this method as it also begins to make its presence felt in areas of culture not directly related to technology. Considered an absolute, it proceeds to impose its irrefutable mark on everything under its sway, instead of rendering a service within the various sectors of society.

The influence of technicism is also readily apparent in the misconceived notion that one need not distinguish between natural science, technological science, and technology. This misrepresentation quietly harbors the view that science is the *instrument* for technological control. In this instrumentalistic conception of natural and technological science is hidden a *utilitarian* ethos. What can be made, must be made, with an eye to enhancing the utility of what is made, for as many people as is possible.

Let us take a closer look at each of these three.

a. The influence of technicism comes to expression in Western culture for the first time in the spiritual movement of the Renaissance. The sweep of technicism broadens from the moment that the Renaissance put its stamp on Western philosophy and the development of science. Later, due to the influence of the Enlightenment, of positivism, and of pragmatism, technicism became dominant in our culture.

The influence of technicism has not always been so great. Through the influence of Christian movements like the Reformation, the Counter-Reformation, and the Great Awakening, but also owing to the influence of Romanticism and, more recently, of the Counter-culture movement, the promulgation and application of technicism has met [76] with some resistance. As for the question whether Christendom is culpable for technicism and its consequences, much has been written. Historically, the influence of the Reformation on the development of the natural sciences is unmistakable. But yet it was philosophy, which became predominantly technicistic under the influence of the Renaissance and later humanism, that more and more put a stamp on the development of science and technology.

The father of this technicism is in a certain sense René Descartes. He claimed that the rules and laws of mechanics are the same as the rules and laws of nature. Basic to his philosophy of nature is the paradigm of the automaton, that is, of the machine. "Nature is a machine, as easy to understand as clocks and automata, when one but investigates it carefully enough," says Descartes. What this eventually implies is that nature can be figured out and controlled. After all, man is "the master and owner of nature." That is why Descartes, already, considered plants and animals as things to be manipulated.

Under the leadership of later philosophers, this technicism became stronger and its influence began to spread into society and politics. Hobbes raised ideas about a technocratic state and Comte wanted to develop techniques to control societal processes. In our day, with genetic engineering, we realize that humankind itself can become the object of this technicism (see §28).

b. Technicism expresses an ethical conviction. This means that under the influence of technicism the development of science and technology has been set on a specific course. This direction corresponds to what in ethics is called *utilitarianism.* In utilitarianism those goals of science and technology are considered good which are useful to as many people as possible. For example, in the case of scientific-technological control, people pursue those goals that comport well with technicism. Utilitarianism has an eye only for material prosperity, long life, happiness, and the like. Again to avoid any misunderstanding: these goals [77] in themselves are certainly valuable. The ethical problem concealed in technicism, and its bedfellow utilitarianism, is that these aspirations are converted into absolutes and that these absolutized goals have become definitive for scientific and technological activity. That is to say, the manner in which and the means by which these ends are achieved through scientific-technological control are not given any, or at least not enough, attention. At the same time, it is the ends that sanction the means to scientific-technological control.

Utilitarianism's narrow vision of a "good" and "responsible" technological science, and the technology that comes with it, has led to many of the problems we have mentioned (see §3). The ethical attitude of utilitarianism results in a one-sided technology that is harmful to people, nature, the environment, and society. By means of scientific-technological control, nature is reduced to the use humans can make of it. What we have to do to use it as we wish and how we use it are of little concern. This course proves to be destructive. Anything that does not fit the parameter of technicism is misjudged or disregarded. Nature no longer seems to have a value or meaning of its own and human beings are too often reduced to machine components or to a complex of consumers.

The outcome of an unlimited scientific-technological control of inanimate and animate nature can lead to the dehumanizing of labor, the atomizing of society, and the degradation of nature. The biosphere in which we live forms a highly complex, singular whole. When this is not taken into account the result will be the gradual reduction of plant and animal species.

But destruction is not the only consequence. Due to limitless industrialization, to modern scientific-technological practices in agriculture, and, not to be forgotten, due to modern means of transportation, earth, water, and air pollution are increasing at an alarming rate. And as nature becomes less stable, the desire for technological control grows, generating a dangerous and vicious circle. [78]

c. To better understand the consequences of technicism and utilitarianism we must examine the characteristics of natural science and the *scientific design method* (see §11). Technicism and utilitarianism project, via the *instrumentalistic* maneuvering of science and method, their characteristics onto the breadth of reality, such that it is remodeled, reduced, and disfigured.

Technicism, we saw, is the attempt to create a world that is molded to the human hand. When science and the scientific method are absolutized, we can refer to this method as one of *destruction and reconstruction.* Without taking into account the given order of reality, its structural diversity, and the mutual interrelationships within that reality, everything, including their smallest components, is broken down or taken apart with strict consistency and then, using these elements, put back together again as we see fit. Converting the method of systems theory into an absolute leads to the same result (see §11). This philosophic and scientific "creative" renovation of reality gives rise to a mechanized world picture (Dijksterhuis, 1986). Later this inventive reconstruction results in the practise of scientific-technological control and a corresponding technological world picture.

It is only during and after the Industrial Revolution that technicism comes to cultural fruition. With the rise of industrialization all of reality is placed within the grasp of an overextended, scientific-technological control.

The substance and consequences of technicism in the guise of instrumentalism can best be approached from the spiritual attitude of the Enlightenment which arose in the eighteenth century. In this movement, the spirit of the Renaissance, to wit, an all-out trust in the human ability to renew life, is connected with the development of the natural sciences. The arrogance of human autonomy, man as Prometheus, someone who needs nothing more than himself, attaches itself to science. Inspired by the thriving development of natural science, the super- [79] human of the Enlightenment thinks he can solve all problems and with the help of the natural sciences, and later technological science, bring new life to society.

Owing to the influence of the Enlightenment, and via that of positivism and pragmatism, the ethos of many scientists today is characterized by technicism and utilitarianism. The essence of science is interpreted technicistically. Science becomes an instrument of control and is used to resolve any presently pressing problems. Utilitarianism affirms and reinforces this process with the promise of unlimited material prosperity. Because people no longer see in this process the limits and limitations of science, the way is open for a limitless scientific-technological manipulation of reality (Staudinger, 1976).

The preponderant role of scientific utilitarian thinking means that values and norms that conflict with people's wants and desires are simply dismissed. That is why reality's pregiven meaning and coherence are disregarded more and more in science.

Within that spiritual climate it was easy for positivism and pragmatism to stamp out any resistance to instrumentalism, the unbounded scientific-technological control of reality. Everything there is, is considered as material to be controlled. The worldview that fits with this assumption we call materialism.

Consequences of Technicism, Utilitarianism, and Instrumentalism

Initially the results of this development are impressive. Actually, they fascinate many still today and are often perceived as a means to unprecedented progress and material prosperity (Von Weizsäcker, 1964). Yet, it is important to realize that from the very beginning of this development there has been talk of a latent lack of direction. Only later did some begin to listen and take note. [80]

On the one hand, human beings are assumed to be the only authority to which one can appeal, and as such, the supreme beings within reality. But on the other hand, seen from the side of science, human beings are the result of a material development that has been determined by science. Man is "lord and master." But as the product of evolutionary coincidence, he is also victim. Because human beings do not want to draw this conclusion of being victims of chance, the quality of mastery remains dominant. Nevertheless, in the face of looming problems, it cannot be denied that pessimism and uncertainty are present and growing as things continue to develop. The lack of direction mentioned above is becoming manifest.

That was hardly the case in the eighteenth century for a materialist like Baron d'Holbach, to mention just one example. He says that the world shows itself to be nothing more than matter and motion, connected in an infinite chain of causes and effects, and that our scientific knowledge of the same provides humankind the key it needs to be able to control reality. In the early nineteenth century, the mathematician Pierre-Simon Laplace was convinced that we would be able to get both reality and the future under control by means of science. Many representatives of the Enlightenment, including Immanuel Kant, thought the same way. Although he was not a materialist, he was still strongly influenced by technicism. With an eye to controlling history, he called for a Kepler or a Newton who would discover the natural laws of history. The ultimate goal that Kant, as well as d'Holbach, had in mind was a congruence of physics and morality. This would imply that by means of instrumentalization natural science would come to function as the standard for practical activity.

When this process is affirmed culturally and begins to spread, and when, as a result, problems and threats, like the degradation of nature and environmental pollution, increase, and when people begin to realize that they cannot simply appeal to science and to scientific-techno- [81] logical control for a solution, then, as they come to see their predicament as the result of their own handiwork, they will be left without a clue as to what to do. Today, our culture's lack of direction manifests itself in the confusion that reigns among suggestions for a way out of the crisis in which we presently find ourselves.

We have to recognize that under the influence of technicism the world around us has been forced increasingly to conform to the characteristics of science. The instrumental use of science that technicism encourages, to wit, *instrumentalism,* has tried its best to fine-tune reality according to the sounds of science.

To understand the effects of unlimited, presumptuous scientific-technological control we must examine the structure of science.

One hallmark of scientific knowledge is that it is abstract. When scientific knowledge, due to the influence of the Enlightenment, is elevated to an instrument of control and people no longer have an eye for the reductions implicit to the abstractions of science, these abstractions, through the irresponsible use of science, begin to mark a culture. That is to say, because of their inherent reductions, the wholesale use and resolute instrumental employment of the abstractions of science can lead to the degradation of reality and, by mutilating its meaning, even ultimately destroy it.

Which *abstractions* characterize science? To come to scientific knowledge one must follow the path of *analysis and abstraction.* Along that path (which is what "method" means literally) more is going on than analyzing and abstracting. For example, we form hypotheses and we experiment. But analyzing and abstracting are certainly the most prominent activities. The scientist distinguishes within a many-faceted reality among a great diversity of functions or aspects, for example, the physical, biotic, and economic aspects. To investigate this diversity scientifically, the scientist first abstracts from this coherence of aspects just one aspect or function. The second abstraction is that the scientist, with- **[82]** in the field of the initial, functional abstraction, brackets out the particular and attends to what is general or universal in that field. The third abstraction is that scientists distance themselves from sensible, perceptible reality and turn to the *laws* that *hold* for this reality. That is why science is often, and correctly, compared to a "map." One arrives, for example, at the formulation of a natural law with which the researched phenomena comply. A fourth and final abstraction is that the scientist is supposed to ignore any benefit to or interests of oneself or others. Which is to say that "science must be disinterested" (Van Meisen, 1970). Failing to honor this abstraction as the standard or norm for scientific research is to give tacit consent to utilitarianism and implicit approval to having science simply benefit human interest through scientific-technological control.

Within the context of these abstractions, the scientist is seeking *to gain theoretic knowledge.*

Given the above, we can say that science follows the route of these four abstractions. To arrive at good scientific results, every science must learn to follow this route. Exaggerated slightly, this means that the scientist proceeds wearing blinders. This is inherent to science. The scientist ought to be conscious of this fact. He brackets out a great deal to which he gives no further attention and concentrates on what remains within those brackets. The scientific knowledge gained along the route of analysis and abstraction is then incorporated – using logic – into a *system* of scientific knowledge. That is also why we say that scientific knowledge is *logically coherent* knowledge. When we transpose this scientific knowledge into mathematical relations (what we could call a second level abstraction), we can manipulate it by using information technology, which in turn positions us to manipulate the reality that one originally set out to study.

To get a good view of the features and characteristics of scientific knowledge, we need to set these four abstractions in a row. Scientific [83] knowledge is universal, functional, and, as knowledge of the law for whatever we are studying, valid. These properties, as Van Riessen (1970) has rightly noted, are *incongruent* with the fundamental characteristics of the full reality we experience. In *that* reality, everything is different, interrelated with everything else, and ever-changing.

When we do not recognize that science distances itself by means of abstractions from the fullness of experienced reality, as happens under the influence of the Enlightenment, then a good deal of the fullness of that reality will be lost with the instrumental use of scientific knowledge.

A simple example can make this clear. In theory we know how to distribute four apples among four children. But in reality doing so is not always so easy. Not all apples are created equal and the preferences of children are not only different, but change continually. Our calculation did not take the size, color, or condition of the apples into account. But in reality, it is usually these factors that determine the choice of children. Obviously, applying the division of four by four has no earthshaking repercussions. But it is different when through the "instrumentalization" of science, for example, in agriculture, industry, or health care, the fullness of reality is converted into the abstract, qualified context of science. Influenced by technicism, people have claimed that science can fully grasp reality. They are convinced that science can solve any problem that arises. In other words, science's "map" is equated with reality. Aided by science, all of reality is molded to the human will with an eye to increasing human power and generating unprecedented and unlimited prosperity. Science no longer serves everyday life but is beginning to influence and even to dominate daily life. This especially happens when the scientific design method becomes a will-less pawn of economic and political powers. These powers exploit the scientific-technological power of control to bolster themselves and the materialism present in a culture. [84]

But has modern systems analysis not seen and sought to remedy many of these dangers? In one sense this is true. But giving attention to the whole as a system does not take away from the fact that it is still a constructed system, a *simulated* whole, and, as such, abstract. Systems analysis does resist the world picture nurtured by technicism. But however important and valuable its revisions have been, systems thinking itself remains tied to a technological world picture (Ellul, 1980). Invoking systems analysis, as the new way of thinking to resolve the cultural problems of our day, may possibly bring temporary relief. But remembering the structure of systems thinking and the limitations inherent to it (see §§8 and 12), this uncritical appeal will, in the long run, confront us in a more intense manner with both old and new problems. Ultimately, we model reality according to the standard of the system we have constructed. As we nullify earlier reductions, we see new ones. An apparent solution ends up producing a convoluted complexity without prospect. Furthermore, systems thinking does not consistently halt growing materialism or faith in progress. To the extent that the theme of progress is replaced by that of survival, the cultural dynamic can indeed be slowed down. But, even then, the question is whether this deceleration is possible given the trends of mass culture or whether it will be acceptable to those who have yet to enjoy the fruits of modern technology and scientific-technological control. Therefore, for the moment it seems that, all things considered, a systems analysis approach advances a technological culture more than it slows it down or changes its direction.

22. Making Everything Technological: Technologizing

We noted above that science and technology are taken up into an interrelated complex in which economics and politics also play an important part.

Technicism is the driving force in this complex. Bolstered by utilitarianism, technicism, as we saw, has turned science into [85] an instrument of control. In doing so the possibilities of technology are overestimated, so that now, influenced by the imperative of technological perfection, everything that can be made is made. The eventual result is that everything becomes technologized.

Technicism, which wants to guarantee material happiness and prosperity, also leaves its mark on economics and politics. Economics becomes the economics of material progress, in which the power of money and material gain dominate (Goudzwaard, 1980). Materialistic politics enhances this process and in so doing, plays into technicism. Politics, as many have noted, is becoming increasingly technocratic, meaning that the method of scientific-technological control also influences politics.

People outside these cultural sectors, given an insatiable desire for material prosperity, eagerly buy into this development. They are convinced that they can only gain by it. Other sectors of society, like agriculture, health care, and a diversity of organizations, have to learn to put up with the influence of a clustering of cultural powers. Although the opposite was promised, here too the convergence of powers proves a disruptive influence and accelerates the crisis within culture. For example, farmers tend to overproduce, on the one hand, and to impoverish the soil, on the other.

When this development began on a small scale, the ramifications were hardly serious. In fact, short term effects promised nothing but success. That is also what makes it so appealing. Building a kind of "alter-creation" in which human beings would be lord and master over everything and be certain of untold material wealth, blinds one to what is actually going on here.

As the process of technologizing increases in extent and magnitude, something that reflects the preponderance of science, negative effects become apparent and threaten to predominate. Reality is shaped according to a reduced, logically coherent network. The abstract framework [86] becomes so dominant that the fullness of reality is broken into atomized and functional units or reduced to artificially constructed systems. The absolutizing of the artificial, from flowers to intelligence, makes reality more matter of fact, maims its meaning character, and rips reality into component parts. The problems of a technological paradise are great.

This process of making everything technological is evident most everywhere: in the dehumanization of labor, in the devastation of nature, in the pollution of the environment, in problems related to nuclear energy, in making society information-compatible, in problems related to farming and meat production, and in problems surrounding the genetic engineering of humans. In a certain sense, the magnitude of these problems increases as we move down this list. Fortunately, we have not yet been confronted as such with the full potential of many of these problems; in part because the practical expansion of biotechnology and genetic engineering techniques is still subject to clear limitations. However, we must weigh more fully the eventual consequences of their unrestricted application.

Two examples can help acquaint us with the significance and influence of technicism and its consequences in the technologizing process. In order to also make clear that in rejecting technicism and the attempt to technologize everything we are not dismissing the value of technological science and technology, we will examine briefly a responsible development. The first example has to do with the development of nuclear energy; the second with the information age.

a. Within a context defined by technicism nothing is allowed to hamper the development of nuclear energy. In contrast, the large scale of the technology used to generate energy seems to be the source of so many advantages and to guarantee increasing prosperity. It is this materialism that initially sees to it that nuclear energy must be had as cheaply as possible.

[87] Since Three Mile Island and Chernobyl, it is obvious that grim catastrophes can happen with atomic energy and that their repercussions are immeasurable. Scientific-technological control is not nearly as safe as many have claimed. If the control mechanisms should fail at a certain point, the aftermath could be devastating for a long time and for a large area. When we also take into account the unresolved problem of storing radioactive nuclear waste, it is clear that we have yet to arrive at a responsible final phase as far as this technology is concerned.

The waste products of nuclear reactors pose significant health problems for people today as well as for those in the (distant) future. That is why this radioactive refuse must be able to be stored securely for a long period of time, often for a good number of centuries. Does this responsibility not go beyond what man can handle? And even more primary is the question whether producing these hazardous wastes is permissible. Every cultural activity comes with a price. But can the phenomenon of radioactive waste as such, to which so many objections and concerns are tied, be reconciled with a correctly *normed*, ethically acceptable, technological development? In other words, can we speak of a good and proper technology if its long-term products are hazardous? If they remain hazardous for so long that it is simply impossible for the designers of this technology, the technological scientists, to bear responsibility for them?

This peril in the development of nuclear energy is only accentuated when we consider that the development of atomic energy goes hand in hand with possibly producing nuclear arms. This, by itself, can have significant detrimental consequences. In addition, we have to figure that, politically, strong protective measures will have to be taken. The political realm, confronted with the drawbacks of technicism and the desire to make everything technological, is becoming more and more technocratic and, as a result, is often at odds with democracy. This development has elicited a number of strong reactions. For example, [88] virtually every new nuclear power plant brought on line or even only contemplated in the last fifteen to twenty years has met with demonstrations of protest.

In order to constrain a growing cultural adversity to large scale technology, energy policy makers must find an alternate route. Instead of technologizing everything, they will have to consider conserving energy and developing alternative energy sources, like solar and geothermal energy or energy derived from wind- and waterpower or biomass. These forms of energy make a culture less dependent on one particular source and hence more stable. With a more varied development of energy, fewer risks in space and time are taken than is the case with the unhindered development of nuclear energy.

Only after these initiatives have been realized could it perhaps be possible to talk about a limited and stringently safeguarded development of atomic energy. The suggestion of C. F. von Weizsäcker, advisor for the German government regarding energy policy, to build maximum security nuclear reactors underground deserves further attention.

b. Many of the social sciences seek to emulate the natural sciences in method and outcome. When technicism's procedure of brutish deconstruction and headstrong reconstruction of given reality extends by way of the natural science to the social sciences and then to *society,* the disastrous consequences of that method will also begin to affect society itself. This disruption will manifest itself in fragmented communities and isolated individuals who have been alienated from their original place in society. Was that not often, even still today, the reason for the dehumanizing of labor in modern factories? Is not the same thing happening to many today who work the land? Many farmers are being robbed of their freedom and responsibility. Some are being forced into early retirement. Others are coerced to change careers. Many feel they have been alienated from their land and animals and from nature.

[89] Modern society is entering the information age. Interactive television, computerized education, and electronic banking, news reporting, and mail order shopping, all supported by electronic information systems, will soon be considered commonplace. It is all just one more step in the process of technologizing society. However, the hope that the newest technology is ushering in a new future stands in the way of any critical distance with respect to information technology. Given their conviction that if it can be made it should be made, combined with a faith in progress, many uncritically assume that information technology will be the panacea for all of our past and future problems, facilitating a perfect, technological streamlining of society. But the other side of the coin will be human alienation, a feeling of disintegration, and a lack of direction.

The desire to make everything technological is evident in education, politics, futurology, and in the modern workplace (see §26). What follows will focus on the workplace.

Human labor is multifaceted. Because of the influence of technicism, work has become more and more related to business and technology. The production sector, and the services that support it, tend to be considered the only kind of valuable work. That is to say, imaginative fields, work in the social service sector, and works of compassion are depreciated. Vocations that are devoted to the transmission of culture, of a culture's norms and values, also receive less attention. As a result, secondary education is experiencing a shift toward fields related to business and technology, while other educational fields are left in the lurch. For culture this means a significant loss.

The unrestrained effort to technologize work will eventually pose a significant problem, certainly in the production sector. In principle, this kind of work can for the most part be automated. When this happens many will become unemployed. Unfortunately, most of these people will not understand that other valuable kinds of work exist. C. J. Dippel quite rightly speaks in this connection about the missed chances of [90] modern technology. His point is that technology offers so many new possibilities for all kinds of work, but that a narrow view of technological development is yielding just the opposite.

Although it should be clear from the above, I want to emphasize again that protesting against the move to make everything technological does not imply an objection to technology as such. The point is rather to defend a responsible technological development that does justice to its meaning and relative importance. People are going to have to recognize more and more that technological science must serve humanity. Its place is to support and not to suppress (see §§24 and 25).

In summary: the dubious process of making everything technological, of which we have given two telling examples, is based on absolutizing scientific-technological possibilities, allowing them to rule life.

Influenced by the materialism of producer and consumer, the move to technologize has been pushed by economic and political forces that themselves put their stock in technological power. In agriculture, for example, technology contributes to overproduction. To keep ahead of progress, the renewal of scientific-technological control finds itself in an upwards spiral that both is world-wide and, given the effects of technologizing, can become catastrophic. Global contrasts, as well, become the more biting. While forty percent of the world's population lives in poverty and suffers from malnutrition, people in the United States, for example, spend fifteen billion dollars per year on losing weight, and underscoring the sharp contrast, an extra twenty-two billion on beauty aids.

In the meantime, economic and political powers, aided by biotechnology and the newest information technology, are allowed to increase in scale and intensity. As they do so, one's ability to scrutinize them decreases. Together with the tangible problems of technologizing, they are defining the crisis within culture and setting a course that knows no direction. [91]

23. Science and Wisdom

Up to this point we have seen that technicism influenced science to become an instrument of control. This instrumentalism has been very successful in bringing about material prosperity. But at the same time, its large-scale application, absolutization, and imperialism have led to the problems of technologizing.

The science that plays such an important role in modern technology need not be renounced. Modern technology in its entirety would be impossible without that science. What we need is a different view of science, one in which science is not allowed to dominate, but is nurtured with an eye to service. The place of science should be to assist and bolster. We have to let go of the mistaken notion that with science at our side we, as "lord and master" (Descartes), can mold reality as we see fit. In contrast, I would maintain that as stewards of creation we are obliged to insist that science be of service in its dealings with reality and, likewise, in technology.

Within that context science can contribute to a growth in wisdom. That is to say, we have to recognize that scientific knowledge is carried by experiential knowledge, which is full and concrete. Science ought once again to stand in a correct relation to reality in its fullness. We should integrate it into the fullness of the reality we experience every day. Doing so will deepen that experience. In this way science contributes to an increasingly extensive understanding. That is what *growing in wisdom* is all about. Science will no longer function as the main thoroughfare in cultural activity, or even when it comes to technology, but as a helpful service road. The difference this would make in technology is that it would not only be a matter of building, and then building some more, following the technological mandate that what can be made must be made, but it would be just as much a matter of preserving and protecting nature. Wisdom, in the sense of a growing, compre- **[92]** hensive insight, will also lead to creative and circumspect activity in technology.

In order for us to grow in wisdom and for science to remain of service, we have to acknowledge that we must integrate the abstractions discussed previously into prescientific, or extra-scientific, experience, so that the limitations of scientific knowing can be lifted. When we integrate the scientific knowledge we bracketed out along the scientific path, we must once again take it into account. That holds first of all for the abstraction of a functional field with respect to the interrelatedness of reality. We must do justice to all of reality and it exhibits more aspects than the aspect investigated by one particular science. The approach of any one science is then insufficient for solving some problem, certainly today, when so many complex problems exist. Problems require a *multidisciplinary* approach. And yet multidisciplinary collaboration, as the sum total of the knowledge coming from the various special sciences, is still no guarantee that justice will be done in a concrete situation. We understand this fact only when, in contrasting the generalizing approach of multidisciplinary collaboration with the approach of specialists in their discipline, we realize that scientific knowledge is characterized by more than one abstraction. Within its functional field, a science focuses only on what is universal. This abstraction, and the universal scientific knowledge that ensues, we often mistake for knowledge of concrete reality. This reality consists of only particular, discrete, everyday entities. Restoring the abstraction of what is universal will ultimately contribute to an enriched knowledge of the singular reality of everyday. But the richness of reality may not be relegated to abstracted universality nor may science claim that it can grasp what is unique and may control what it can grasp. Scientific knowledge may not be blinded by its instrumental use, but must seek to serve our knowledge of the particular things around us and what we do with them. When we recognize, for example, the distinctiveness of plants and animals, we will want to **[93]** set limits on our desire to submit them to our scientific-technological control.

The map that science provides may not be equated with reality. As map, however, science can provide us with a better sense of orientation. This orientation is pre-scientific. The full knowledge of everyday life will be enriched by science and help us become more responsible people.

In this connection we should attend separately to the view that a different scientific approach, for example that of an integrated systems approach, solves the problems we have been addressing. This new approach would have to see to it that both technology's development and the study of societal (cultural, economic, political) consequences are investigated in relation to each other. This new science would have to nurture our knowledge and ability as well as the required socio-practical insight. The advantage of doing so is that we can pose questions concerning norms and ethics clearly.

Many, including someone like the German philosopher Jürgen Habermas, assume that we can discuss normative and ethical questions rationally. But there lies a problem in this hope for communal, rational communication. There is simply no way around the fact that a fuller or broader rationality that includes more sides and more areas, a rationality of social reflection that replaces the present rationality of the scientific-technological dimension, will still remain dependent on rationality. The only answer that can be given as to the basis or grounds for this rationality is, ultimately, rationality itself.

To dispel the pretended self-sufficiency of rationality of whatever kind, reformational philosophy acknowledges that reason is rooted in a given, normative structure of reality that is neither subjective nor arbitrary. This normative structure precedes any discussion regarding rationality and therefore is crucial for determining the place of reason as well [94] as that of science. It also indicates the subordinate place of science and its significance for scientific-technological control and for technology.

It is actually the given, normative structure of reality that points us in a responsible direction and that will help us out of the crisis of the technological culture without having to abandon science, technological science, scientific control, and technology. They will have to conform to the normative limits, which together give a meaningful direction for culture.

24. Driving Forces and an Integral Framework of Norms

What can be said about this given, normative structure? This structure, which the Christian philosopher Johan Mekkes often referred to as the "dynamic of creation," consists of a large number of distinct but interrelated norm principles to which all human beings are obliged to *respond*. The direction of this *guide* for life is "horizontal" (or immanent) as well as "vertical" (or transcendental). Human beings are required to work out these norm principles into norms that can function as signposts for responsible cultural activity.

This *response structure* ought also to help technological science serve the development of modern technology. Usually, that is not the case. Moved by the spirit of technicism and a naive view of progress, what can be made, is made. Few seem to notice what is actually happening. The result is that raw materials are wasted, ecosystems are damaged, jobs are dehumanized, and no account is taken of future generations.

When one recognizes that human beings are meant to do their work in culture *coram Deo,* before the face of God, and be guided in doing so by the normative character of the dynamics of creation and one acknowledges that human beings are not, as in technicism, the center of reality, but must in cultural activities deny themselves, loving God and neighbor, then the intent of their various cultural activities receives a different content. Instead of the central motive of power, in which [95] Driving Forces and an Integral Framework of Norms everything we do revolves around ourselves, we have the central motive of love, which produces divergence in our various cultural activities.

Setting our sights by the normative structure, implies a broad passage and, at the same time, a fixed course. This path begins with accepting the *motif,* mentioned above, that must pervade science and technology, namely, that of *wisdom* and *building and preservation.* "Wisdom" and "preservation" have often been forgotten. Were that not so, biology and ecology would have been accepted as foundational sciences for technological science (see §10) a long time ago. With "preservation" we are talking, for example, about preserving the integrity of nature and maintaining a healthy biosphere. Contemporary developments in technology often lead to the technologizing of nature. All too often there is no sense that technology needs a more comprehensive scientific basis and that this basis, as a contributing factor to a growth in wisdom, leads to creative and judicious activities within modern technology.

Wisdom and the motif of build and preserve dictates that modern technology *ought* to involve itself with the immediate situation in which people find themselves. This includes attending to nature, the environment, and even, for example, the landscape. Modern technology ought to be *adaptable* and *ecologically responsible technology.* Where disruption has occurred, we must make the greatest possible effort to restore things. That in no way means reverting to the technology of the craftsman. Compared to present developments, an adaptable technology ought to be expanded. This differentiation in technological development clearly will also require a cultural dimension. Technology ought not to conflict with the state of cultural development, and the rich variation within it, but to tie into that development in such a way that technology enhances a culture. Unfortunately, the opposite is often true in third world countries. But even in industrial countries there are serious problems. Sometimes technological wastes harm the environment and human beings. We must realize, and that holds in the first place for the [96] engineer as technological scientist and technician, that hazardous wastes cannot be accepted in responsible technology. The engineer's task includes responsibly disposing of technological waste. Sometimes that proves to be very successful. One example: for a long time people in the ice cream and cheese industries had no solution for the obnoxious smelling and sewage contaminating watery by-product called "whey." Later they found that what was first considered waste could be transformed into excellent feed for hogs. This was, of course, both ecologically and economically advantageous.

If technology proceeds from the starting blocks mentioned above, modern technology and alternative technologies, what E. F. Schumacher (1973 and 1978) calls intermediary technologies, should not oppose each other, but can complement each other. Presently these approaches are mutually exclusive. On the one hand, some over-estimate the effects of scientific-technological control and the scientific design method; on the other hand, others, motivated by a romantic view of nature, plead, often as a reaction to present hazards, for alternative technologies. The same tension is evident, for example, between a scientific-technological approach to agriculture and the movement towards organic farming. This tension, caused by scientific-technological control and the problems it raises, elicits reactions. This tension can be avoided and abated when the development of technology and technological control simultaneously satisfy the standards set by many interrelated norm principles.

We will briefly review these norm principles and their articulation in norms and discuss a few key points about their significance for modern technology.

The *cultural-historical norm* is the standard of differentiation and integration, of continuity and discontinuity, of centralization and decentralization, of large scale and small scale, of uniformity and pluriformity. The various components of this norm may not be taken as **[97]** opposites. Setting the sights of the scientific design method and technology by one side of this norm, will put them on a one-sided and eventually dangerous course. When the focus falls on large-scale integration, continuity, centralization, and uniformity, the problems of technologizing will by and large be the result. In addition, the technological development that follows this skewed target will, in turn, result in overproduction and redundancy.

When both components of the cultural-historical norm are met, technology develops more steadily, leaving room for creativity and innovation in new inventions. When we satisfy the cultural-historical norm, technology will be adapted to the existing culture, renovation will become possible, and a stimulatingly diverse technology can arise. For example, rather than expecting nuclear reactors to answer all our energy problems, people will attend to as many different energy resources as possible (see §22). But we must also observe additional norm principles and articulate them responsibly.

Technology unfolds, and its meaning deepens, when we obey the lingual and social norms. The lingual norm is that of *information* or openness. This means that we must clearly inform society about technological renovations. Only then can those working with the technology, or purchasing its products, responsibly evaluate things and make decisions. That is why this norm is closely connected to the social norm, namely, to the norm of *communication* or interaction. Without open communication, it is impossible for those who participate in technology to fulfill their communal and individual responsibilities. Those who develop technology have the responsibility to take the time to inform and communicate with the public. Quite obviously, people in the technological sciences, in particular, have to honor these norms of information and communication for these are the ones who stand, as it were, at the cradle of every new technological development.

The economic norm of *stewardship* must also be honored, but not to [98] the exclusion of the others. As it applies to technology we are talking here about the matter of *efficiency*. Its presently one-sided application is due especially to the prevailing influence of economic theory within industry. Influenced by technicism, industry has focussed far too narrowly on a natural scientific sense of efficiency, in which only those goods that can be expressed in monetary terms determine what has value. We must integrate the economic norm into an integral framework of norms. We must apply it not only to production, but to the use of raw materials, energy, nature, the environment, landscapes, animals, and even of the people involved in developing technology. Problems arise when we limit the economic norm exclusively to techniques of production. Technology and the economics of industry develop into a jumble. But when we apply the economic norm, in conjunction with the other norms, simultaneously across the board, we can prevent a kind of *overdevelopment* in producing surpluses and restore a kind of *underdevelopment* in being stewardly in dealing with nature. We will begin to attend more to nature and the environment, to the scarcity of raw materials and energy, and will come to see that people are much more than their ability to function economically within a technological context. We must acknowledge the responsibility of employer and employee.

The normative development of technology is advanced when we abide by the norm of *harmony*. Non-essential surpluses, the decadence of the materialistic way of life, and the degradation of nature, make it abundantly clear that this norm is not being followed. If it were, given the other norms we have discussed, people would realize that technology ought to be developed in a balanced manner. This norm also requires that we introduce and incorporate new technology in a revolutionary fashion. Doing so can bring with it societal unrest and a loss of communal support. We must also consider this norm of harmony in the multifaceted interrelationships between nature, people, culture, and technology. Technology ought to adapt to people, not people to technology. For **[99]** example, it is not without reason that we appreciate user-friendly tools. That is the way all tools should function. When that is the case those who operate them will also have greater pleasure in doing so.

In honoring the norm of *justice* we oppose any and all injustice that the development of technology may bring about. Engineers, instructors, and employees must all ask themselves whether their contribution to technology does justice to the plant and animal kingdoms, to our sources for raw materials, to consumers, to society, to culture, to third world countries, and the like. This norm of justice is an intrinsic part of technology. When it is disregarded, the government must take specific measures to restore justice. It is worth emphasizing that the positive influence of technological developments that obey this norm will also be felt in the many sectors of society in which technology is playing an increasingly important role. That is especially true, for example, in modern agricultural practices and in health care. The crises in these areas can take a turn for the better when we maintain this normative sense of justice consistently.

All of the norms indicated above are opened up and deepened when people hold themselves to the ethical norm of *care and love.* We are called upon to nurture a concern and compassion for everything that has to do with technology. This, of course, includes caring for and loving our neighbors, far off and close by, but also protecting the great diversity of God's other creatures. When love binds itself only to scientific-technological control, people become so obsessed with control that it is frightening. When this norm is not honored on all sides, people become more and more alienated from their work, for example, the farmer from the land, nature, and his animals.

The last norm to which scientific-technological development is subject is the *pistical norm,* that is, the norm of faith. The norm for faith, in a restrictive sense, is trust. Those who use the tools of technology have to be able to trust that these tools work and are safe. When this norm is [100] met, they will be. Or more broadly, when people accept and operate according to the framework of norms sketched above, technology will be safe. That does not necessarily mean that all of technology's problems and hazards will be completely abated, but that these problems will be kept within bounds. We have to remember that with all cultural work there is a price to pay. No cultural work is a "path of roses." "Thoms and thistles" will continue to accompany the works of our hands. Within the integral framework of norms that we have sketched, however, cultural problems will not become intolerable or insurmountable. That is due to the transcendental direction of the norm of faith, namely, the acknowledgement that nothing in the world, as creature, stands on its own, but is dependent upon the Creator and ought to be directed to Him. This direction is the meaning and fulfillment of everything in the Kingdom of God. When people who trust in technicism and believe in the scientific-technological mastery of all our problems, self-confidently think to put themselves in the driver's seat, the ultimate effect will be the very opposite of what they expect. The problems will expand into dangerous, intolerable tensions, some of which could easily erupt into catastrophes. This faithful allegiance to technology is the background for the problems that technologizing calls forth (see §22).

Before concluding this section I must mention the context in which technological science and modern technology is developing. The normative framework sketched above holds as well for a "free market economy," assuming one takes free in the sense of freedom within responsibility. In reality, however, economic forces often bolster technicism and the process of technologizing. That is why many turn to politics. But a materialistic political program will advance the same process. And yet, because the workings of government affect everyone and legislation can place restrictions on business and industry, the political arena is the place to address and deal with developments that have gone awry. This political activity must be attuned to the normative frame- [101] work in general and, given the nature of the political realm, focus particularly on questions of fairness and public justice. Presently pressing problems will require that we take steps to contain the process of technologizing. It is possible to make a political choice for a normed technology that has a more sensitive attitude toward nature, animals, and the environment and still does not decrease employment. We have to realize that this kind of national political program can be effective only if this sense of equity and public justice also finds political support at the international level.

25. Technological Science Serving the Meaning of Technology

Technicism is one expression of the professed autonomy of humankind. Supplemented by modern science, it indisputably permeates our "technological culture" and, with that, so many different sectors of society. Industry, agriculture, and health care, to name just a few, all experience the pressure to technologize their ways. It is here that the problems and perils of modern culture are concentrated.

Technicism and technologizing, along with the reductionism that accompanies it, can be overcome through a renewed orientation toward the authentic dynamic of culture and, within that context, by explicitly posing the question as to the meaning of things. Technological science ought not to dominate modern technology. Technological science ought to be a productive detour that helps to nourish the wisdom of the engineer. With wisdom's discernment and circumspective insight, the engineer will be able to work cautiously and creatively, both as a scientist and as a technician, to advance the cause of a responsible technological science and technology.

Technological science and the scientific design method ought to satisfy the standards set by the normative framework sketched above.

[102] Following this ethically acceptable path will contribute to a meaningful perspective for the development of technological science. Technological science will then serve the meaning of technology.

The meaning of technology in the perspective we are sketching is exceptionally rich and can hardly be articulated comprehensively. But it does include at least the following. Technology can alleviate in part the bind in which humankind naturally finds itself. Technology can increase life's possibilities, decrease physical burdens and difficulties at work, and free people from routine activities while opening the door to all kinds of mental and creative labor. Natural disasters can be averted, illnesses overcome, and, in a certain sense, with the aid of electronics and microprocessors, the deaf can hear again, the blind see, and the lame can walk again. Technological development can provide houses and food, supply a degree of social security, and increase available information so as to extend and deepen communication. Greater harmony between technology and nature is possible. Through all of this the responsibility of humankind grows as well. Material prosperity will not have a stranglehold or gain the upper hand, if it keeps in step with mental and spiritual well-being. The many gifts and diverse qualities of individuals and peoples will have a chance within technology and by its means. When it is situated within the perspective of an integral framework of norms that holds for all cultural activity and its hazards are kept within bounds, technology will make room for re-creative activities and a rich cultural involvement that are in balance with a conscientious stewardship of nature.

In this meaning-filled perspective technological science, the scientific design method, and technology will be able to contribute meaningfully to other cultural sectors, such as economics, agriculture, and health care. [103]

4. Current Issues

26. Information Technology

George Orwell, writing in 1948, shared his glimpse of the future in his fascinating novel *1984*. His projections uncannily resemble our information age. "Big brother is watching you!" could easily become a reality in an age defined by the permeating presence of communication and information technologies. So many things, like production processes, social services, and even political activities, have changed drastically through the use of the computer. Will an "electronic shadow" soon accompany us from cradle to grave? By connecting new and existing information systems, we can monitor, control, and possibly even manipulate everyone registered in the system.

Dissatisfaction with this development is growing. Computer prowess has caused a sense of impotence among human beings. Genetic engineering, as well, opens a vista of imponderables. Technology's attempt to streamline society perfectly is accompanied by human alienation and societal disintegration and aimlessness. This chapter will focus on these current issues.

a. *Optimism.* Curiously, whenever computer experts discuss the meaning of computer technology for society, they present an optimistic and generally simplified view of the future. The standard version has it that we are entering an era in which all human activities will be con- [104] trolled by electronic information systems. This will result in the following changes: two-way television, electronic banking systems, computerized education, electronic news media, electronic mail service, shopping by computer, and the like. The industrial society, in which people depend on the production of materials, will change to an information society that will satisfy the economic and social needs of all. While current society may be characterized by the division between rich and poor, between the privileged and underprivileged, some claim that the new information society will obliterate all these differences. Everyone will have instant access to enormous quantities of valuable information. Whereas other areas of life are plagued by limitations to growth, electronic technology and telecommunications have no such restrictions. Access to these services is becoming increasingly cheaper and more manageable. There is virtually no limit to our consuming and producing information. As people are liberated from the need to work, some claim that alienation will vanish and be replaced by a more genuinely human environment that promises cultural growth. For the first time in history, concentrations of power and social hierarchies will disappear. A resurrection of democratic ideals, better techniques for controlling the future, computerized education, and a global boost in human creativity will fulfill the lofty promises of a global village.

b. *The Technology of Hope.* Why this optimism? After all, the problems and faults of industrial technology have actually caused widespread pessimism. Nevertheless, belief in the idea of progress through modern technology is still pronounced. Whenever new developments are introduced, such as DNA technology, laser technology, or information technology, people's hopes for technological redemption are rekindled. The experts always promote their new technologies as solutions for age-old problems and as well as for those problems that resulted from the now largely obsolete industrial technology. [105]

Obviously, this worship of technology is really nothing but a repristination of the old idea of progress. New inventions often have this effect on many people, so much so in fact that the new information technology is, with no proof, widely believed to be capable of producing a better world.

Minsky, a prominent computer expert, when discussing fifth generation supercomputers, remarked that the new freedom will enable everyone to organize his life exactly as he wishes. Apparently, self-realization is now within our reach. A religious renaissance is upon us that features, instead of faith in a supernatural God, faith in collective human wisdom and the collective human spirit. At long last the earth will become a place of peace and rest, if only we buttress our attempts to develop information technology and make it available to all. The hope that the newest technology will usher in a glorious future, together with the continued blind belief in the idea of progress via technology, would like to keep us from looking critically at this technology. The old lesson that unnormed developments are nearly always accompanied by unfavorable side-effects and wholly unexpected problems is completely ignored. The experts are by and large blind to the real dangers associated with information technology. As a result, they excuse themselves from seeking responsible solutions to the most pressing problems.

c. *Computer democracy.* The computer expert Licklider supposes that by using information and communication technology can rekindle democracy. Such was not possible in so-called mass cultures. But today, says Licklider, by applying the newest techniques, millions of people will be able to discuss issues with each other and then decide democratically by simply pushing a button. The dreams of many social reformers will finally be realized. Since the available information will be universally accessible, everyone will have equal political influence and power. In other words, true democracy will be fulfilled in a computer democracy. [106]

The first flaw in Licklider's view is that he sees democracy as first of all a matter of distributing information. In fact, the information will always be interpreted and selected on criteria or principles not necessarily shared by all. Because programmers use different criteria than the users, certain information may be withheld. Moreover, unlimited information may not raise political awareness; the "technologizing of information" can have an alienating effect. An overabundance of information can result in incoherence and disorientation. Unable to see the woods because of the trees, many people will be tempted to give up in apathy.

And what about the plans that are presented to the voters in the first place? They originate with a select group of researchers and planners of the future who determine the format and content of these plans and thus retain a good deal of control. Thus the people's freedom of choice will be restricted to the preconceived options of the experts. This is all the more true when communication media present a particular proposal as self-evident.

The delusion of computer democracy becomes even more transparent when experts tell us that decisions made through such a direct system of democracy can be predicted by data analysis and can thus be incorporated into the new plans for the future being proposed. Computer simulation will predict psychological and social reactions to proposed social changes. With this information in hand, computer technology will supposedly be all the more capable of designing and administering a better future. Current technocratic tendencies are thus reinforced by this "computerocracy."

d. *A Computerized Future.* Computer literature commonly suggests the notion that computer technology can better engineer and control the future than can human beings. They are considered incapable of absorbing all the information necessary to control the future. However, with a computer, different kinds of models of the future can be simulated and predicted precisely, allowing one to choose from among the various [107] models thus constructed. The working out of such a model is then left to the devices of the expert. The basic idea is that we can shape a future that is in line with our wishes, solve all the eventual problems, and avoid all future threats.

However, attempts to create a global model of the future that integrates every possible factor have not proven to be very valuable. The models designed by the Club of Rome two decades ago have proven to be obsolete. The model called "Global 2000," which appeared in 1980, now seems to be largely irrelevant and needs to be supplanted by another model, "Global 2000 Revised."

World history cannot be programmed into a computer model. We cannot presume to have a complete overview of the future. Insofar as we do have some insight into the factors that influence the future, we still do not understand the scope and degree of the interaction among the various factors. The future is really inscrutable and full of surprises and upheavals. But if we continue to pretend that information technology can conquer the problems of the future, we will only find those problems reappearing in a more intense form.

Some experts will seek to carry the continuity of the past through the present into the future. But this tactic will only reinforce the problems of the past, especially humankind's loss of freedom. In reaction to this loss of freedom, other people will actively seek to oppose computer technology. They will seek to liberate their peers from the computer through revolution. As a result, computer technology will aggravate the old tension between power and freedom. An information society without tensions will not happen.

To avoid misunderstanding, I should point out that computer simulations and models of the future can help map out courses of development. But such a mapping-out has principal limitations. No one should assume that these projections can serve as blueprints for the future. Such maps are for reconnaissance purposes only, not for plotting a [108] course for humankind. There must be room for human freedom and human responsibility. Computer simulations should serve, not curb, such human responsibility. Information technology must assist people in their quest for responsible courses of action in the future.

e. *Computerized Education.* Publications about computers in education are voluminous. Experts cherish high hopes for computer aided education. Predictably, commercial enterprises overestimate the computer's value in education. That education will improve through using information systems is not a foregone conclusion. Such supposedly self-evident conclusions are dangerous. Whatever comes to us via information systems results from scientific analysis and the scientific-technological control of information. But science works through abstractions. The methodology of science, as we saw, reduces reality. This means that the scientific information programmed in computer systems is always incomplete. Moreover, since scientific knowledge is of a general, universal nature, the information is also universal. In a very real sense, information technology presents a reduced, generalized image of the world.

When the computer dominates education, student users may adapt themselves uncritically to the computer's reduced worldview. Whatever does not make it to the computer screen is then simply not considered by the student. Computer programs contain only information that has to do with measurements, sizes, and statistics. Whatever falls beyond those measurable quantities – for example, faith, hope, and love – are considered unreal because they do not occur in the computer programming. Thus Minsky errs tremendously when he equates computer information with wisdom. Computer information represents a specific form of human knowledge: it is scientific knowledge programmable in mathematical-logical formulas. It can never contain all forms of human knowledge, such as a knowledge of love, a knowledge of practical real- [109] ity, or a knowledge of trust. True wisdom, which begins with the fear of the Lord, cannot be discovered by a computer program.

The type of education that seeks to transmit culture and to prepare young people to accept their responsibility for culture can certainly use the computer as a teaching aid. As a teaching aid, the computer can save an enormous amount of time. But the free time that results must not be filled up with other computerized activities. This would set students on a course of narrow specialization. The extra time should be filled by forms of education that cannot be accommodated by the computer. Thus, in addition to the growing specialization promoted by the computer, students must reflect more generally on reality. By pursuing both avenues, students may use the computer responsibly and wisely.

f. *Computerized Labor.* Many utopias sketch a future in which people no longer have to work. The world of our dreams seems to be a land in which we have to do nothing. And yet, now that we have liberated many people from their labor through automation and robots, we are beginning to realize how large a problem unemployment is. Recent economic recessions may be in part to blame, but that does not take away from the fact that when industry turns to contemplate new investments, it usually thinks in terms of automating production. This means that even if the economy picks up, technological developments and their new automated and robotic potentials will not necessarily create more jobs. The "utopia" of not working will follow.

Having nothing to do will not bring happiness because of the increased social problems and unemployment costs resulting from continued automation. Even if those problems could be solved, the general emptiness, boredom, and loss of human creativity and applied skills would continue to undermine utopian ideals because of the reduced worldview of scientific technology. If we see everything only in terms of science and technology, in the hope that it will somehow improve our lives, then the solutions can only be of a scientific-technological nature. [110]

That in itself is an enormous reduction of what constitutes human happiness. People may often seem to be content to live by bread alone, but from a spiritual point of view, they will be impoverished. In the end, their material security will also be affected by science's reductions.

In short, the reduced worldview of scientific-technological control common to the information society confronts us with a caricature of the most beautiful promises ever uttered. The origin of that caricature is not to be found in information technology as such, but rather in humankind's materialistic inclinations, which have already betrayed their emptiness and hopelessness.

g. *Demythologizing the Information Technology.* As with many revolutions, it can also be said of the computer revolution that it devours its own young. If we expect everything from information technology, we will be confronted with serious tensions and an appalling void. It would be easy, but equally dangerous, to take a reactionary stance. What we have to do is identify the proper place of computer technology. Instead of allowing it to render our responsibility superfluous, we have to realize that it increases our responsibility. Without human responsibility, an information society will develop almost automatically. Computer systems will evolve into more complex systems which reduce the user to a slave of the system.

To maintain the *service function* of the computer and to exercise *effective control* over it, we need to analyze the structure of information technology. For the moment, however, I will underscore the limitations of information technology and define responsible potentials within those limitations. This task could be described as demythologizing the myth of information technology.

This demythologizing begins by asking what is meant by information. Information means *news* within a specific situation, something that was previously unknown. This new information is something people [111] will have to consider in their own situation. Every bit of information is related to their responsibility.

But the word "information" in the phrase "information technology" is in a certain sense misleading. The "information" of computer technology amounts to a *reduction* of information, as pointed out earlier. Though we must conform to conventional terminology, we must also realize that there really is no such thing as information technology. For what happens in computer technology? By analyzing and abstracting, we formalize the human information process so that we can program. We program data in a computer according to a specific method, in a certain way. The procedure is referred to as an algorithm. The computer has a technically shaped physical structure that absorbs data previously processed by mathematical, logical analysis.

What happens in the computer may be called *signal absorption.* The computer absorbs technically rated signals that have come about through a mathematical, logical programming of human information processes. The signal absorption of the computer reduces reality enormously. The myth in current information technology lies in the fact that experts simply will not recognize this reduction. But if we remain aware of this reduction, we can see both the potentials and limitations of information technology. Its potentials include the speed with which a computer can absorb and reconfigure data, a very helpful feature. The limitations of information technology should discourage people from seeing the computer anthropomorphically, as a kind of mechanical human being.

Strictly speaking, computers cannot even perform simple mathematical functions, such as adding and subtracting. It is people who use the numbers. Fully human means of thought will always remain somewhat of a mystery to strictly scientific analysis. Admittedly, scientific analysis can convert human calculations to computer programs, but such programs are technically rated signals that only represent numbers. [112]

Similar considerations also lead us to question the idea that computers think or possess some sort of artificial intelligence. In a technical sense, the process of signal absorption has lately been expanded and improved greatly. We should not take lightly the potential of what we call information technology. But computers will never replace human beings. They may be able to help us solve our problems better. When used responsibly, the computer can positively contribute to politics, education, and business. Within the context of human responsibility, which is ultimately a matter of living *coram Deo,* we should recognize the dangers of a narrow, scientific-technological image of the world. With a sense of responsibility we can use meaningfully the limited – and therefore potentially far-reaching – possibilities offered by computer technology.

h. *Responsible Use of the Computer.* Earlier I noted that the current trend in culture proceeds from systems analysis by way of systems designs to system control. In this way, the computer is incorporated whenever possible, and technocracy is perfected in computerocracy. Modern culture becomes a *Babel culture* (see Schuurman, 1987) with a *universal* language. It may appear that the problems of technocracy can be solved through systems philosophy and computer technology, but in fact they are only temporarily disguised and will shortly manifest themselves more intensely. This is true because as new avenues continue to be opened up to systems analysis and computer technology, these new avenues are also controlled by the idea that consistent application of these procedures is necessarily good. This process has contributed to increased alienation which in turn fosters such phenomena as unemployment, loneliness, isolation and standardization. For if computers are used wherever they can be used, many social contacts, human interrelationships, and opportunities for mutual sympathy and creativity are lost.o If, in addition, we attempt to solve human and social problems by applying systems philosophy and computer technology, we will see [113] those problems reduced to simple, technical definitions to be attacked with equally simple, technical solutions.

But it should be clear by now that the search for technological solutions to human and social problems will only intensify these problems. Dehumanization will result. In the meantime, computer technology appears to have become a nearly autonomous force. Enormous unmanageable stores of information are combined into a strictly centralized system. Information systems now operate autonomously selecting and interpreting whatever relevant information is necessary to make important decisions. One result of all of this is that people feel that their own responsibility is greatly diminished.

I have emphasized our need to relativize this systems thinking with respect to the original, primary world of experience and our need to emphasize human, creative responsibility in the relationship between abstract theory and full, concrete, practical reality. It is clear that computers should be used critically and cautiously. Wherever computers are used, it is essential to discover what kind of world-and-life view lies behind their use, which hidden philosophies are prevalent, and which viewpoints on science and technology are operative. The findings will determine whether we use or do not use computer technology in a given situation. We must also remember that constructive computer use can itself decline over time. For example, the collection of personal data for social care could in time begin to invite invasion of privacy.

Computers can be used responsibly in many ways. The Love Command requires that we explore computer possibilities for helping the handicapped. Furthermore, after admitting the computer's limitations, we should also use it in education. And we can use computers wherever the norm of efficiency is the highest consideration. It certainly has a place in space travel, but also in heavy, dirty, hazardous, and boring work. Computers can warn us about environmental pollution; [114] they can also help us use energy resources and raw materials more economically. In many current cases, however, it is not at all certain whether using a computer is indeed suitable. In such situations, we must consider the permanent responsibility of the person who introduces and uses the computer. In all cases, the user should be able to explain why a computer is necessary and for what purposes it is to be used. In a certain sense, this calls for a sense of social responsibility, the parameters of which should be defined by the government. Furthermore, incomprehensible computer programs should be made comprehensible to prevent users from transferring their responsibility to the computer. And just as a single user should not transfer responsibility to computers, the same holds for other individuals or groups of people. They, too, may not be robbed of their responsibility.

Only if people are prepared to live responsibly, remembering the purpose of computers and rejecting the temptation to exploit the computer in various kinds of technological power trips, will it be possible to expect a meaningful future with unprecedented and challenging possibilities. To recapitulate: systems thinking and computer technology do not result in their own unavoidable, autonomous development. Within normative boundaries, computer technology, together with an adequate sense of personal and social responsibility, can contribute much to a decentralized, pluriform, stable, healthy, and enriched culture. Certainly, much work still needs to be done. That work can be accomplished only if undergirded by a conversion of the scientific and technological models of life and culture. In those models, the real meaning of science and technology has been compromised. To see again the meaning of creation as the Kingdom of God is a challenging mandate. That Kingdom perspective does not exclude the computer, but instead makes it serve Kingdom activities. [115]

27. What is Permissible?
Thoughts on Genetic Engineering

A legitimate question in technology is whether it is permissible to do everything that is possible. This question is especially crucial in modern technology, based as it is in the natural sciences. This question was also asked when the first steam locomotive came upon the scene. Many people protested strongly, for they thought that the noise caused by the locomotive would prevent their chickens from laying eggs and their cows from producing milk.

This example has been used frequently to show that our ethical reflections on such matters can be flawed. The implied conclusion is that any new scientific knowledge permits its application. Yet the ethical problems can never be completely suppressed. One obvious example is the practical application of nuclear physics. Both the development of nuclear weapons and the technology of atoms for peace beg the question of our responsibility in managing the new scientific possibilities and technological achievements. Generally, however, ethical considerations are thought to be something like an afterbirth, and just as irrelevant. Space explorations continue unabated, and few people wonder whether the money spent on these developments is money spent responsibly, especially when one contemplates the enormous global problems of hunger and poverty.

Of late, ethical questions have once again moved to center stage, in the wake of the rapid advance of new technologies. Yet at the same time, the development continues unabated. The gynecologist Dr. Arie Haspels recently observed that the time for ethical reflection on test-tube baby technology is long past. Since the development is now going full steam ahead, he concludes, it can no longer be stopped.

I wish to raise the matter of responsibility concerning genetic engineering. Of special importance is the recombination of genetic information – the so-called recombinant DNA – a matter related to test- [116] tube babies and artificial insemination. The new discoveries of recombinant DNA technology increase the possibilities for artificial insemination.

a. *The Experiment.* At present, genetic engineering, the exchange of genetic information among various species, is still largely in the experimental stage. Many scientific researchers feel that their research has nothing to do with ethical problems. Ethical questions, they claim, would arise only when this new form of biotechnology is applied to industry, agriculture, animal husbandry, or medicine. Basically, they argue, scientific research aims at obtaining new knowledge; ethical issues belong only in applied science.

But the nature of the experiments in such pure scientific research shows that this approach is flawed. Besides, pure scientific research is the first step, whether large or small, to concrete application. Thus ethical questions must be raised at this point, if only to avoid the later excuse that ethical questions are no longer relevant since this research has already begun anyway.

First, we need to look at the nature of the experiment. Ever since Galileo, the experiment has served as a powerful tool for obtaining scientific knowledge. By experimenting, we find answers to scientific questions. The experiment actually simulates a small piece of reality. In doing the experiment, the researcher starts with his knowledge about given reality, but then proceeds to change it to the extent that the experiment's conclusions will provide him with uniform answers to his abstract, scientific hypotheses.

Most of us have had some experience with this form of experimentation in high school. There are few objections to this. The picture changes, however, when we begin to carry out our experiments on animals. Many people consider vivisection to be ethically irresponsible. In fact, legislation now restricts experimentation with animals to very specific conditions. Oddly, there is no similar legislation restricting exper- [117] imentation with people, even though the introduction of such legislation is expected soon. The fact that such legislation is necessary indicates that experimenting with people is on the increase. In any case, it is evident that biology and medicine raise more ethical problems than physics does. Many people consider experimentation simply necessary to promote the high ideal of new scientific knowledge. It may, at first, be quite harmless. However, advanced scientific research requires more precise and more incisive experiments to address the increasingly more complex scientific questions. Not even physics escapes this dilemma. For example, when experiments are conducted with radioactive material, the ethical problems can no longer be ignored. But when biology begins to experiment on living beings, the ethical problems are immediate and should be addressed at once.

This can be illustrated from the field of molecular biology. To gain a greater understanding of how living beings function, scientists now conduct large-scale investigations into the structure of the DNA molecule, the basic building block of information in all living organisms. Recently, molecular biologists have developed a specific experimental method to find out more about the structure and function of the DNA molecule. Sections of a DNA molecule from a particular species are introduced into the DNA molecule of a bacteria. These experiments do not involve pathogenic bacteria, but rather bacteria that are harmless to humans, animals, and the environment. Bacteria are used because of their rapid growth: they are capable of almost instantly producing the desired changes.

As was noted previously (§19), American scientists in 1974 decided to halt such experiments because they could produce bacteria which, although not immediately pathogenic to the carrier, could readily become so. The freeze did not last long. Subsequent experiments were subjected to a number of security regulations, but since experience proved the immediate concern unfounded, the rules governing these [118] experiments were gradually relaxed. However, caution is still in order because it is impossible to determine beforehand whether such experiments with bacteria are entirely harmless.

Needless to say, the new knowledge obtained from genetic research has been used. This knowledge enables scientists to rearrange the information in genes or to modify and manipulate it. Seldom do they see serious ethical problems with combining components of genetic information of a higher species with those of a lower species, such as bacteria or spores, provided the proper precautions are observed.

Combining the DNA of a certain plant species with the DNA of certain animal species offers us many opportunities, I believe, as long as the possible negative effects are considered right from the start. In fact, I think it would be irresponsible to disregard the potential benefits of this development. It could solve many of our present problems. To reject this technology could be more harmful than to use it judiciously. Outright rejection of this technology is often a reaction to claims that genetic engineering can enrich certain animals – or even make them almost human – and may well dehumanize people. A common argument against the recombination of genetic information from plants and animals is that this science threatens to erase the barriers between species. Aside from the question of whether this formulation is correct, I would maintain that recombinant DNA technology does not deal with the barriers between species. The introduction of a small segment of DNA does not create a new species. Admittedly, new characteristics are added to a given species. But this kind of "species modification" has, quite rightly, long been standard practice in agriculture and in animal husbandry.

b. *Applications.* Recombinant DNA technology has led to large-scale applications, many of which have been very constructive, the ethical questions notwithstanding. A recent committee report dealing specifically with this biotechnology shows how the recombination of DNA has [119] led to new developments in pharmacology, agriculture, animal husbandry, medicine, and various industrial processes.

The insulin now used to treat diabetes is a breakthrough resulting from biotechnology. It is possible to produce various types of interferon which are claimed to have some effect on cancer. Similar combinations have led to the discovery of drugs to combat foot-and-mouth disease. It is no longer necessary to isolate these drugs from harmful viruses. They are produced by bacteria that have been infused with segments of alien DNA.

To a certain extent this new technology can remove the problems and dangers associated with producing synthetic chemicals. Recombinant DNA technology is also thought capable of producing bacteria that can absorb oil and degrade heavy metals. This would aid those whose responsibility it is to clean up the environment they have polluted. Science is currently working to "construct" plants that can absorb nitrogen directly from the air. This could make fertilizer more economical and more plentiful, thus helping to relieve hunger in the world.

Many of these proposals are still only daydreams. Much research remains to be done, especially on the potential negative effects of the industrial production of these bacteria. Another relevant ethical question is whether it is responsible to introduce newly created bacteria on a large scale into the environment. Will they not upset the delicate balance in nature and, if so, to what end? Some people consider this to be such a critical question that they strenuously resist the planned bioreactors and the general introduction of recombined bacteria into the environment, with the same energy that others protest nuclear reactors.

Indeed, we must be cautious. Current recombinant DNA research is justifiable only if it conforms to the safety regulations. Large-scale, indiscriminate application of these techniques will unquestionably have some harmful side effects. It may lead to the production of new biological weapons. But even with peaceful intent, we cannot ignore the fact [120] that we have no experience at all with many of these plans. It is impossible to anticipate what will happen. Current industrial technology teaches that there will always be negative results that could not have been foreseen. We must heed the lesson. It is too optimistic to expect that any new technology will automatically cure the problems created by other technologies.

c. *The Scientist's Responsibility.* Because of the problems and dangers associated with experimentation and its industrial application, and also in view of the possible wrong use of what may otherwise be sound scientific knowledge, the responsibility of the scientist has once again become our question. Such questions have been asked before. For example, was the nuclear physicist Otto Hann responsible for Hiroshima? Are nuclear physicists responsible for the development of nuclear weapons systems which, because they are becoming less manageable from a technical point of view, could lead to demonic catastrophes?

Obviously, such questions must be asked of scientific research leading to any new technological breakthroughs.

As pointed out earlier, recombining genetic material can have both positive and negative effects. Are researchers responsible for the effects? Some say they are responsible only for their own research. But how much greater would their responsibility be if the results of their research in DNA technology were to cause irreversible harm to humans, to animals, or to the environment!

Others rightly object to this kind of reasoning by noting that the harmful effects would be much greater if we halted all biotechnology altogether. World hunger, for example, would be much worse.

In view of this ambiguous situation, the question of the researcher's responsibility is not an easy one to answer. We are becoming aware that scientists must be very careful when introducing new technologies. Certainly, scientists and their support organizations must accept more responsibility. Yet the form of this responsibility, the way in which it [121] can be discharged, is one of the most crucial problems facing the scientific community and, to a large extent, the political community as well.

28. The Genetic Engineering of Human Beings

We turn to ethical questions concerning the potential of genetic engineering as it applies to man. Either directly or indirectly, recombinant DNA technology also enables the scientist to engineer human genetic information. Are there ethical limits to this power?

Charles Weissmann, Swiss molecular biologist, once said that in the near future every person will be able to apprise himself of the nature, structure, and function of his own genes. A gene is the smallest unit of hereditary information. The recombinant DNA enables the scientist to map our hereditary information. That information consists of roughly one hundred thousand genes, ten thousand of which have thus far been researched. Before long, it is thought, scientists will completely understand our total genetic being. Aided by the computer, such "gene mapping" will become a manageable research procedure.

If such gene profiles are not protected, they can easily be abused, as, for example, through discrimination in the workplace and by prejudicial conditions made by insurance policies if the gene profile shows that the insured poses a high risk to the company. Naturally, such ethical questions become even more crucial when a reading of parents' gene profiles in combination with the prenatal investigation of the fetus's gene profile shows that the child has an "unfavorable" gene structure. Such research could easily lead to a higher rate of abortions. Going even further, Weissmann explains: "Deep down I know that at some point in the future – I speak of decades – it will be possible to bring about complex genetic changes in man. Then we will have to confront head-on the associated ethical problems." [**122**]

We may not be in such a situation yet, but we are on the threshold of an age in which we will be able to manipulate procreation outside the mother's womb. Parents availing themselves of homologous insemination, can now be helped in the conception of their own child. Through heterologous insemination, it is possible to conceive a child whose father remains anonymous. The possibilities of the test-tube technique will increase to the point that fertilized ova of preselected parents can be implanted in other women. With the aid of test-tube technology, it will be possible not only to signal hereditary defects in advance, but also to "construct" a super-baby with the most favorable genetic structure. Thus the recombinant DNA technology will be applied, by way of the test tube, to the human embryo. Clearly we are here presented with significant ethical questions. Some people rightly warn that these techniques have already claimed the lives of many fetuses that were not considered suitable for implantation or that were simply superfluous. Others maintain that the sole criterion for application of this technology should be the quality, not the viability, of individual life.

The available literature that advocates this so-called high quality of life also promotes the "creation" of large numbers of such high-quality individuals. This is known as cloning. The technique of cloning was proven successful in mice and frogs, but has not yet been applied to human beings. But even if it were possible to apply it to human beings, would it be permissible?

I wish to discuss yet another experiment. In a certain area of research, scientists have tried to fertilize the ovum of a hamster with the sperm of a male human being. The result was a hybrid, a life-form somewhere between that of a human being and a hamster, that could develop to the stage of a few undifferentiated cells. Generally, experimentation with life forms between human beings and animals is not tolerated. Nonchristians appeal to the norm of human worth or humanity. Many nonchristians use this norm to argue for the prohibition of experiments [**123**] that could lead to further experiments involving the different life forms. In borderline cases, however, objections are less strenuous. In November 1982, the British medical research council established a number of guidelines for such experiments. One such guideline would halt fertility tests that could lead to "the human-hamster hybrid system" before cell division started.

Whereas these tests already signal the beginning of hybridization, they, too, should be rejected. This is true especially since these tests could signal the beginning of genetic engineering of those varied cells which include the hereditary information of the human. By way of such tests, the technology of recombining genes could easily be deflected into new, and in my estimation, illegitimate avenues. In the meantime, scientists should look for more ethical types of experiments to test the fertility of the ovum.

I do not expect the resistance to such experiments to stiffen. Rather, because of the positive potential of test-tube techniques, the experiments are likely to increase. One reason for this laissez-faire attitude can be found in the widely accepted legalization of abortion. To a large degree, the legalization of abortion has promoted experimentation with the human genetic structure. The American ban on federal funding for the transplantation of tissue from aborted fetuses into humans has recently been lifted. Hard and fast rules governing these experiments can no longer be established. Similarly, nothing is done about experimenting with aborted fetuses kept alive for research purposes. Recent research journals claim that experiments with aborted fetuses are necessary to test new medicines that can be of benefit to humankind, for example, in the struggle against Parkinson's disease.

a. *Objections.* My objections against possible genetic engineering of fertilized ova in vitro and against manipulating forms of life between human beings and animals, including life-forms in their initial stages, are based on my rejecting the current view of humans that has long been [124] subliminally accepted, but that has now found its way into the public domain through labs and hospitals.

Increasingly, people today are accepting and promoting a view of humanity that is thoroughly materialistic, a view inspired by the natural sciences and modern technology. It is said that human beings are machines, albeit very complicated ones, or a complex physical-chemical substance or, more recently still, an information programming system of complex DNA molecules. When one's view of human beings is tied to a specific dominant area of science, that particular view of human beings becomes the dominant one. In this way, specific, abstract views of human beings are absolutized. Their totality and unity becomes subservient to the theories of science and technology. The end result is a reductionist view of the human race. Having fallen this far, people can easily fall prey to the manipulation of other theories.

This reduced view of human beings was originally limited to scientific and philosophic circles. However, when the materialistic world-and-life view gained ground and the Christian world-and-life view went into retreat, this reductionistic view became commonplace. This secularizing of culture contributed to the fact that this thinking about human beings as purely material beings could be applied as theory more and more. Humans were dehumanized and compared to a thing. Fortunately, quite a few humanists also protested this dehumanization, since it challenges our humanity and human worth.

The materialistic view did not first become evident in genetic engineering. The modern way of using human resources in the labor process, as well as attempts to legalize abortion, have been clear evidence of this reduced view of human beings. Through the potentials promised by genetic engineering and the test-tube technique, the materialistic view has reached its zenith.

The Christian sees the individual, though corrupted and defaced, as the created image of God. God's love and care continually reach out to [125] the individual. Therefore, we must love every one of our fellow human beings. Out of this love we can use scientific-technological possibilities to combat sickness and relieve suffering, as is done in medical science. But the Christian's view of human beings is based on the inviolability and integrity of every individual, as well as on the realization that we can only moderate human suffering and can never entirely repair the effects of his corruption. Viewed from the perspective of creation and fall, the origin of this corruption lies in human beings willfully breaking their communion with God. This religious failure cannot be remedied by science and technology.

As indicated, the materialistic view of human beings does not accept the inviolability of human life. Instead, it focuses on the *quality* of life. The purpose of technology is to obtain the highest possible quality. In this view, our broken life may in theory appear to have been repaired by science and technology. In a real sense, genetic engineering then also has to guarantee the quality of its product. Needless to say, a poor product is not acceptable and must therefore be destroyed.

b. *The Test-Tube Baby.* Rejection of the test-tube technology invariably brings the counterargument that the method is ideal for helping childless parents who have found the more natural avenues of conception closed to them. This is a noteworthy argument, provided the method of test-tube technology is not used unless absolutely necessary. Yet closer inspection and a thorough acquaintance with the literature available on the subject will show that more is involved here than may at first appear.

The many questions and problems of test tube technology can be avoided by helping prospective parents in other ways. I have in mind the Craft method which involves the removal of ova from blocked fallopian tubes and their implantation into the womb, together with male sperm, *before fertilization takes place.* This method can be applied on a large scale and is socially more acceptable because its costs are much [126] smaller and thus its opportunities accrue not only to a small, privileged elite. From an ethical point of view, I believe this method to be acceptable, for the method does not allow for unjustified experiments and, secondly, the womb is where conception takes place and the embryo grows.

By contrast, if fertilization *takes place in the lab,* it is not possible to control the number and types of experiments conducted on ova. As well, it may mark the beginning of the recombination or manipulation of genes. As far as I know, no efforts have been made thus far to try to change human genetic information. However, the laboratory method of artificial insemination does require multiple ova, only one of which will ultimately be reimplanted. In the event of failure, any one of the remaining embryos will take the reject's place. Moreover, the criteria of artificial insemination demand that the ultimate product meets certain standards of excellence. Thus the test-tube technology itself must be able to guarantee the quality of life. This in turn means that selections must be made between products that are likely to be successful and those that are not, or perhaps less so. Is it up to us, however, to make such judgments and take decisive action? This can happen only if the individual is no longer seen as unique in the eyes of God and when human beings arbitrarily decide which human life meets current standards and which does not.

There is another equally important argument. From a number of publications, it appears that this technology is used to conduct uncontrolled scientific experiments on test-tube embryos. Through these experiments, researchers are encouraged to explore all kinds of new avenues. Any technology that offers such unlimited and uncontrolled opportunities to researchers must surely be rejected.

I previously mentioned that the test-tube technology has thus far not been used to bring about planned changes in human genetic information. However, this has been done with animals. Rat genes pertaining to **[127]** the rat's growth hormones have been implanted in the embryos of mice. The mice so manipulated turned out to be super-mice, whose offspring showed evidence of the earlier gene transplant.

The successful experiments conducted on animals opens the way to similar experiments with the genetic information of human beings. The very possibility requires us to deal with it at this time. Repairing genetic defects, infusing into people specific characteristics, such as greater intelligence, no longer belongs to the area of science fiction. However, before this can be put into practice, many experiments will have to be conducted, and this necessary experimenting with human beings cannot be justified. In a recent article, Professor H. J. J. Leenen of the University of Amsterdam raised a number of questions to which there are no unambiguous answers: What possible dislocations can the change of genetic information have on an organism? What unknown characteristics will be brought on by the new combination of genes? What is the effect of reinforcing a particular characteristic in terms of the overall balance in an organism? What happens when a genetic experiment fails?

My conclusion is that test-tube technology will surrender the total human genetic structure to scientistic and technicistic models. For that reason, it is not at all certain whether the high expectations will materialize. Quite possibly, the opposite could be true.

As a matter of fact, the opposite is certain. Materialistic human beings now think they can offset the ill effects of what was a religious failure, the fall into sin, by means of science and technology. Since human beings, without God, are no longer capable of bearing or absorbing the effects of sin, such as suffering and evil, they seek to redeem themselves, but in the process invite more suffering and evil upon themselves.

Does this mean that all genetic engineering of human beings must be rejected? No, not if the supremacy of science and technology is brought low in order to be turned to service. Only then will there be an end to **[128]** the scientistic and technicistic approach to humanity. Only then can we enlarge on the perspective of a normative development for science and technology. I can appreciate genetic engineering if it is applied to remedying certain hereditary deficiencies and diseases, but only if the approach zeroes in on limited genetic therapy on the *organic level* rather than on a general treatment of the *total genetic structure.* In any case, all the ethical rules of medical experiments must be strictly observed.

c. *Conclusion.* Objections have been raised against the use of the term test-tube baby, and rightly so, because the baby does not come from a test tube. But I hope that the term "test tube" continues to serve as a reminder that conception takes place in a laboratory. Temporarily, the pregnancy is interrupted. That brief time is filled with human technologies, for which the recipients will want some form of guarantee ensuring the quality of the "product." Right from the beginning, the materialistic process has degraded human beings to things to be manipulated.

Even the intent of this technology – although there may be no plans to experiment or keep back some spare embryos – must be rejected because of its many unforeseeable consequences. This is all the more so since research into the Craft method has not come to a dead-end.

Craft's alternative method, however, does not get the attention it deserves, primarily because it precludes any tinkering with the embryo. Generally, support for experimentation comes from secularized religion, the atheistic religion of Western man, which was developed from the Renaissance, through rationalism and the Enlightenment, to our increasingly scientistic and technicistic culture. The technology of the test-tube baby and the many possible experiments associated with it show clearly the post-humanity of our technicistic age.

Someone might possibly object that, because ethics is always undergoing change, I should be more careful with my dismissal. That is often true, especially in a secularized culture where Christian norms and values, long cast in stone, have suddenly begun to shift like sand. However, **[129]** I would hope that our ethics will not change when it comes to conducting experiments on children or the mentally impaired. Why should our ethical position change with respect to embryos? My position makes clear where the boundaries are; and even then, much reflection is needed to prevent responsible technologies from degenerating into acts of unspeakable evil.

Without a doubt, the problems surrounding modern science and technology are becoming more menacing. The responsibility of scientists and researchers, together with all those people involved in various ways in technological development, is becoming steadily heavier and more difficult. But, as I pointed out earlier, that responsibility must also have a different orientation than the current one.

We cannot continue to dwell on symptoms, but must search out the causes of the various critical traits. We cannot limit ourselves to partial problems, but must attack the many challenges with an integrated strategy. The combination of science, technology, and secularization encourages solutions that may provide some short-term gains but will bring agony in the long ran. If we elect to solve the problems caused by science and technology by applying more scientific and technological solutions, the process of escalating technicism will only be reinforced. If, furthermore, science and technology become political instruments, the result will be a technocracy in which we, though still making our contributions, will become the slave of anonymous, pseudo-independent forces.

I have pointed out in various ways the need to approach the question of responsibility differently from the way that is currently fashionable. In many respects, we will once again have to pose authentic questions, such as ones about the origin and meaning of things, about the coherence and diversity of everything, and especially questions about normativity for all of creation. [130] Christianity must accept this challenge. This means that the materialistic inclination, which in its diverse forms has reduced and sealed up a view of the world as a world without God, must make way for a profound, resolute, spiritual conviction. This conviction will attest to our dependence on Christ, our Redeemer from sin, guilt, and evil. It will also confess to our relationship to God, the Creator of the whole of reality. This renewal will offer a new direction to scientific and technological opportunities as it rejects autonomous, atheistic, and materialistic world-and-life views. In the latter, man remains the measure of all things, while in the Christian view, the law of God, as manifestation of God's will and summed up in the two love commandments, continues to provide direction to all things. Through it we can be liberated from the menacing super-force of science and technology.

5. Technological Culture Between the Times

A Christian Philosophic Assessment of Contemporary Society

29. Religious Groundmotives

[131] When evaluating the development of technology in human history, Christians must look carefully at its spiritual-historical background. A number of Christians working in the tradition of Reformational Philosophy have focused on this religious-historical dimension. Herman Dooyeweerd, in particular, with his philosophy of the cosmonomic idea and his view of religious driving forces, or *groundmotives,* provides a good vantage point for assessing what is happening in our culture. This theory of religious motives needs to be elaborated in a more comprehensive reading of our culture.

Dooyeweerd originally introduced the notion of religious groundmotive in 1941 to expose and clarify the relation between religion and theoretical thought. In doing so, he provided valuable insight into the depth, diversity, and turmoil within Western philosophic thought. From the start he realized, but never really expanded on the fact, that every groundmotive, given its religious character, is related to culture. Religion, under the sway of one or another religious groundmotive, conditions not only philosophic thinking and science, it also influences culture in general and our scientific-technological culture in particular.

Technology is today the mass, the motor, and the mark of nearly every cultural activity or field. By focussing on how religion relates to culture, we can broaden and deepen our insight into what is going on in [132] our culture, and into what we must do about it. Such a liberating perspective will help us to understand better the current crisis of our culture, also with respect to postmodernism.

Dooyeweerd identified three nonchristian groundmotives as moving three basic periods of history, namely: the pagan period; the period of synthesis between pagan thought patterns and biblical themes; and the humanistic period. The humanistic groundmotive, according to Dooyeweerd, originated in the pretension of human autonomy and is formulated as a dialectical relation between freedom and nature, whereby *freedom* is an ideal of the human personality and *nature* is the nonhuman world, scientifically conceived. Dooyeweerd originally spoke of science as a *religion;* later on, he speaks only of the science *ideal* as an expression of human freedom. The dialectic lies in that freedom when absolutized creates an absolutized science that threatens the freedom that gave it credibility. He claims that this humanistic religious groundmotive has influenced the entire development of postmedieval Western culture.

Dooyeweerd understands religious groundmotives to be the central driving force in developing culture (1979), but surprisingly restricted his discussion of that cultural influence to the field of philosophic and scientific thought. Although these motives are radical and integral in character, he seldom examines what they imply for culture in general, perhaps because he deals with religious groundmotives, especially in his later work (e.g., 1953), in the context of his transcendental critique of theoretical thought. In that critique he argues that every philosophic position is grounded in a religious motive. This motive is a driving force arising from the heart as the religious center of the human being. But he says little, for example, about the structural consequences of the humanistic groundmotive in the development of culture. He does stress losing or reducing freedom in the dialectical process between an absolutized human personality and an absolutized science, but he ignored other cul- [133] tural tendencies and problems that we are well acquainted with in our time. And so it is for us to answer the important question: Hew does the humanistic groundmotive influence the health and wholeness of our culture, a culture that philosophy and science influence greatly through technology?

Analyzing our technological culture in the light of the humanistic groundmotive will aid Christians and others who are seeking to find their bearings in modern culture. In this context we must also ask whether we should revise Dooyeweerd's formulation or description of that motive in light of the typically technological culture that has now evolved.

30. A Groundmotive Revised

Dooyeweerd spoke as early as 1939 about the groundmotive of "the control of nature" (see also 1960). "Control" here is not so much scientific as technological. There are advantages to speaking about a cultural qualification, but Dooyeweerd seems to apply this notion only to the field of philosophic and scientific thought. He wrote a great deal about the relation between Christian faith and science that is of interest to university scholars, but he did not broaden and deepen his analysis to include the influence of science on culture outside the university. For example, he says little about the influence of science on technology, about the influence of technology on science, or about cultural fields in which technology is a decisive factor, such as industry, economics, agriculture, health care, education, and politics. Certainly a groundmotive of technological control would leave its mark on developing science as well as, as is so often said, the other way around?

Historically, Dooyeweerd is right: the rise of modern science preceded the surprising rise of modern technology. But is the religious spirit of technological control not active earlier in both history and science? Dooyeweerd (1966) himself implicitly alludes to that spirit when he [134] describes the religious groundmotive as a creation power attributed to theoretic thought which, after it has methodically broken down the God-given creation order, reconstructs an order according to the ideas of human autonomy. Although Dooyeweerd states that the humanistic motive of creative freedom is concentrated in theoretical thought, the substance of what he holds can also be interpreted to mean that the motive of creative freedom is concentrated in scientific-technological control.

Other examples of the same shift in accent can be mentioned: In the *New Critique* (1953) Dooyeweerd refers more than once to the process of rationalization in modern society. Yet when he speaks about the relation between science and technological progress, he clearly holds that modern technology is a consequence of science or scientific rationality. He does not seem to perceive that under the influence of the humanistic groundmotive, the relation as such between science and technology is distorted to the extent that the reverse is more plausible. Dooyeweerd's later formulations in his *Roots of Western Culture* (1979) indicate that he indeed was aware of that effect. He speaks there about the ideal of science as an ideal of control, as a technological ideal: "Proudly conscious of his autonomy and freedom, modern man saw 'nature' as an expansive arena for the explorations of his free personality, as a field of infinite possibilities in which the sovereignty of human personality must be revealed by a complete mastery of the phenomena of nature" (150). What Dooyeweerd is describing here is technology, or better: a technicistic spirit. In this light it seems clear that Dooyeweerd's original science-ideal can better be understood as the ideal of scientific-technological control. Such an ideal actualizes itself first in science and modern technology and subsequently in many fields of culture.

The content of the humanistic groundmotive can therefore better be formulated as the motive of the creative freedom of human beings – the ideal of personality! – concentrated in scientific-technological control or power. [135]

If this is accurate, the consequences for understanding our culture are far-reaching. Given the guidance of this re-formulated humanistic groundmotive it would appear, for instance, that science in our culture is more often developed as applied technology than is technology as applied science. As early as 1936 Dooyeweerd in a certain sense already recognized that science is usually used as an instrument. Reality is brought under control with the help of scientific thinking. Instrumentalism here means subjecting creationally irreducible activities to absolutized technological control.

Henk Van Riessen expresses a similar view when he says in his *Society of the Future* (1957) that the crisis in our culture is caused by the spirit of absolutized technological power.

31. Faith, Science, and Technological Culture

In light of the above it is clear that our understanding of the relation between faith – as an expression of religion – and science deepens and broadens when extended to the relation between faith and technological culture – a process that may exceed the university context. In any case, we should always connect thinking about how faith relates to science with how faith relates to culture. Because we have not always done so in the past, Reformational Philosophy has contributed less than it might have to developing a normative perspective for modern culture. Now that our culture is in a profound crisis there is an opportunity to speak about this relation more pertinently than ever before.

Nowadays it is undeniable that the ideal of control has manifested itself in the history of culture as a disruptive force. The final result is, according to Dooyeweerd (1979), "God's judgment in world history." Today we see this historical judgment wherever people dehumanize fellow humans, destroy nature, pollute the environment, endanger lives with atomic energy, threaten nuclear war, and destabilize our highly developed technological culture. [136]

The instrumental use of science leads to a reality shaped according to science, including its functionalism and universalism. Large-scale and unrelenting instrumental use of science's abstractions, with their inherent reductions, warps and will eventually destroy reality and its meaning. This loss of meaning is already evident in the fragmentation of nature and society. In bioindustries, for example, reproductive and production technologies often reduce the integrity or essence of animals to a mere economic utility. Furthermore, as a consequence of the fragmentation of global society a sense of harmony and social justice between the over- and underdeveloped countries is absent. And these are just some of the problems (see §7).

Efforts are made to solve problems by uncritically introducing new forms of science and high technology, such as the systems approach, information technology (Laszlo, 1974), biotechnology, and even genetic manipulation. Under the influence of the humanistic groundmotive, every area of our culture has been technologized (Schuurman, 1987). It is seldom asked – and this is the critical question – whether the technology is suitable to all our problems, and especially to the problems technology itself has created, such as polluting the environment, deficiencies in agriculture, and so on.

32. The Dialectic of Culture and Nature

Reformulating one of the poles of the humanistic groundmotive – that of scientific-technological control – sheds light on the actual structural coherence and unity of many of our cultural problems. Insight into these problems gives us reason to reformulate the other pole of the humanistic groundmotive as well. Today the dialectic is oriented around (technological) culture and nature. *Nature* has come to mean *reality as organism,* and that is naturalism. As a result a dialectic rages between the anthropocentrism of the technological culture and the ecocentrism of a *counter-culture* committed to alternative technologies, alternative agriculture, and alternative medical care. [137]

Thus the real content of the humanistic groundmotive of our time is that of *culture* and *nature.* Absolutized *technological control* enjoys primacy over absolutized *nature as organism* because no way can be found from the existing technological culture to a future ecological culture. Culture depends upon this control.

This dialectic sheds a light on our time that reveals it to be postmodern and at the same time neo-pagan. The orientation of many people to the pole of *nature* demonstrates the influence of neo-paganism in our secularized culture. This essentially pre-Christian motive, which is connected with the religions of culture and nature associated with the Greek groundmotive of form and matter, has tended in the neo-pagan period of our time to deify either scientific-technological control (and often of the material welfare associated with it, as we shall see) or nature, "mother earth." The religion of nature, which is represented in several streams of the New Age movement, where the earth is adored as goddess Gaia, stands opposed to the religion of control, of technology. This dialectic characterizes the development of our culture. Such is the mind of our times (see §36).

Perhaps it is unnecessary to note that philosophers who are influenced by this groundmotive and who orient themselves to one of its poles dialectically manifest in their thinking the other pole as well. For instance, consistent environmentalists often speak of the "spaceship Earth." Some philosophers endeavor to achieve an impossible synthesis between the two poles (Bookchin, 1980 and 1982).

33. Technicism

So far I have tried to make clear that the crisis of our culture brought about by the humanistic religious groundmotive is not related first of all to science but rather to technology. Dooyeweerd concluded that scientism or rationalism is the dominant characteristic of our culture. Other representatives of Reformational Philosophy stress that economism is [138] the main characteristic of the crisis of our culture. Bob Goudzwaard, for instance, in his book *Capitalism and Progress* (1980) locates the main characteristic of our culture in the *economism* of capitalism. Such an analysis is very fruitful. Ironically, however, in elucidating the reduction in modern economics, Goudzwaard more than once speaks about capitalism in categories of technology. Thus we have economic *mechanisms,* the *tunnel*-economy, the *spaceship*-economy, and so on. It appears that economism, too, is reductive and in a certain sense insufficiently broad and deep to make perfectly clear what is going on in our culture, and perhaps especially in our economy.

The Belgian philosopher Vermeersch speaks of the complex of Science-Technology-Capitalism as he seeks to understand our culture, but focuses his critique on science and especially on capitalism. Such a critique of culture is, of course, generally well accepted, at least among philosophers. Yet I want to stress that neither scientism nor economism but technicism (see §21) is the primary point of contact between the religious humanistic groundmotive and human activities. This is true because technology is historically, in the sense of classical technology, and ontologically prior to science (Ihde, 1985). It is technicism, or one can even speak of the implicit or hidden ideology of technology, that influences science and economy, because there seems to be no room for critical distance in their relation to technology (Pippin, 1994). *Technological push* has priority over *economical pull.* Science and the economy as such are usually interpreted technicistically, whereupon via positive feedback, they reinforce technicism. Then taken together they feed a greedy society.

There are other arguments too for giving priority to technicism. For example, technicism is the best formulation for the secularized motive of creation and redemption. The utopias of Robert Bacon in the thirteenth century and of Francis Bacon in the seventeenth century are already splendidly technological. Francis Bacon in his *Nova Atlantis* [139] holds that we can overcome the consequences of the Fall by technology. Besides, history shows that economism. as materialistic economy, is not always all-determining, for instance during wars. Military technology or defense technology may require great financial sacrifices that have no positive effect on economic welfare. That happened, for instance, in the former Soviet Union. The technology of space travel cost a great deal to develop and reflected the conflict between the two superpowers. The competition between them was especially technologically qualified.

That technicism is much more important than scientism or economism becomes even more clear when we consider the matter of worldview. To speak of a "technological world picture or image" is more satisfying than to speak of a "scientific or economical world picture." Since the appearance of the theory of relativity and quantum physics, a scientific world picture offers no certainty and is fraught with many questions. The technological world picture seems to be stronger, broader, and reach deeper than an approach from scientism or economism (Dijksterhuis, 1986).

In short, the implicit ideology of technology, technicism, is the dominant expression of the humanistic groundmotive. Technicism entails the pretension of human autonomy to control the whole of reality: man as master seeks victory over the future; he is to have everything his way; he is to solve all the problems, including new problems caused by technicism; and to guarantee, as possible consequence, material progress. Technicism also always implies an obstacle or enemy; it may be God, nature, another country or state, or competitor. Isn't it amazing that technological development was the strongest during the Cold War?

Technicism reduces science to its instrumental use. The economy, as is obvious in Western culture today is also interpreted technicistically. with utilitarian economics as a complement. Moreover, the influence of [140] technicism on technology itself is also negative. Technological development becomes a destructive cultural power.

What I have said thus far entails many cultural consequences. Technicism has left its mark on many cultural activities, namely, reducing, disturbing, and fragmenting such activities. The symptoms can be found not only in science and the economy but also in agriculture, in health care, in the form of instrumental justice, and even in ethics, where people today are talking about *ethical engineering* (Caplan, 1980). Even the Christian religion of many Americans has, according to Wuthnow (1988), been increasingly influenced by technicism. It is the spiritual background not only of large cultural problems but also of micro-ethical problems, including abortion, euthanasia, and procreation technology.

To make the picture more complete, I want to add that a good deal of theology, although often unknown to the theologians themselves, is also influenced by technicism, when, for instance, one speaks about God as a human fabrication or construct. Some philosophers as well, such as Marvin Minsky (1986) with his ideas about artificial intelligence, express technicism very clearly when they conceive of both society and human beings as expressions of a very complicated information machine or system.

Generally speaking, one can argue convincingly from a technicistic standpoint, that the main trend of Western philosophic thought is best characterized as *thinking through technology*. This means that science and rationality in general are distorted because they have been used as technical instruments in the service of technological power.

I believe that a great variety of cultural problems can be understood more satisfactorily from the standpoint of a critique of technicism than through other more common approaches and that we also get a better grasp of several irrationalistic streams by considering them as reactions [141] to technicism rather than rationalism. Think, for instance, of existentialism, neo-Marxism, counterculture philosophy, New Age thinking, and postmodernism. They all express the increasingly shared sense that we live in a ruined world of our own making.

34. The (Hidden) Ideology of Technology and Postmodernism

Postmodernism is a form of technological pessimism. That helps explain the controversy between postmodernism and the mainstream of the Enlightenment and its idea of progress. Postmodernism demonstrates the failure of Western technicistic culture and philosophy.

Although postmodernism proclaims the end of ideology, the ideology of technology is still at work in it implicitly. Therein lies the continuity of postmodernism with modernism. Leo Marx (1994) says of postmodernism:

> A common feature ... of the umbrella concept of postmodernism, is the decisive role accorded to the new electronic communications technologies. The information or knowledge these technologies are able to generate and to disseminate is said to constitute a distinctively postmodern, increasingly dominant, form of capital, a "force of production," and in effect a new, dematerialized kind of power. This allegedly is the age of knowledge-based economies. (24)

Postmodernism is, in a way, the spirit and philosophy of the post-industrial society. Traditionally, power was thought of as entrenched. It could be attacked, removed, or replaced. But postmodernists like Jean-Francois Lyotard (1988) and Michel Foucault (Marx, 1994) envisage forms of power that have no central, single, fixed, distinctive, controllable locus. For the first time in history, concentrations of power and social hierarchies will disappear. An overabundance of information can result in incoherence, fragmentation, and disorientation. Thus it seems as if technicism is evolving from a central technocracy into an anarchic [142] technocracy. Technological power is present everywhere but concentrated nowhere. Hence postmodernism acknowledges no normative direction for technology. It is *comfortable with change.* Micro-electronics, information technology, telecommunications, and systems technology seem to hold sway over history without a controlling subject and to alter the meaning of time and place. Everything is technologically possible and everything is technologically allowed.

This postmodernist outlook, when combined with the operation of multimedia, tends to validate the idea that life is dominated by large technological systems. Enormous, unmanageable stores of information appear to function autonomously in information systems. Computer programs become incomprehensible. As a result, the postmodernist attitude towards technology is one mainly of melancholy resignation or fatalism. Fatalistic pessimism is an ambivalent tribute, however, to the decisive, all-determining power of technology.

Even so, when postmodernists become active as technological anarchists, they manifest a senseless optimism about modern technology. The (hidden) ideology of technology in the postmodern era gives priority to the individualistic instead of to the collectivistic version of it. As such it expresses a new, postmodern form of dialectics in relation to technology.

35. Philosophers Today

The analysis of the absolutized culturally formative power of technology presented in this chapter is confirmed by several present-day philosophers from different backgrounds, each in his own way. Some promote the current development positively or optimistically while others are negative or pessimistic. One sees no way out, another tries to find a liberating or saving perspective. I have learned a good deal from several of them. While remaining faithful to my biblical, reformational [143] basis and perspective, I have welcomed their insights with appreciation. Consider briefly Heidegger's exposition. He holds that Western philosophy is already technicistic at its core as a result of its Greek origin. Western thinking, he believes, is a controlling, ruling kind of thinking. Plato in constructing his world of ideas becomes, according to Heidegger, the first technicistic philosopher. Cybernetics and information technology to Heidegger fulfill and at the same time negate Western philosophy (Heidegger, 1977; Schuurman, 1980).

In this light it is interesting to notice that Jacques Ellul interprets the history of philosophy differently, yet comes to the same conclusion. He says that philosophers have thought about the future as something positive and glorious, but that in the meantime technicistically inspired scientists, engineers, economists, and politicians, all with good intentions, have really gotten this wrong and prepared a negative and disappointing future. We have been betrayed by technology (Ellul, 1964, 1978, 1980), but this betrayal has been hidden through technological bluff (Ellul, 1990).

However that may be, some postmodernists such as Toulmin (1990) interpret the leading edge of multimedia, things like the digital city, as helping to legitimate a positive evaluation of our cybernetic age and our information society. For that reason the individualistic, not the universalistic, approach should get a chance. He thinks that the individualizing tendencies in the postmodern era are a sign of hope. He doesn't recognize, however, that our society as a postmodern society depends on the aberrations and problems of the technological culture, from which norms and values are disappearing. The same wishful thinking is found in the view of the theologian George Vahanian. The philosophers Rafael Capurro and Heiner Hastedt also see positive connections between the information society and individuals with their personal interests. Similarly, the American philosopher Murray Bookchin (1980, 1982; Ferkiss, 1993) tries to synthesize postmodernism and naturalism. [144]

Dewey, Staudinger, Horkheimer, Sachsse, Ihde, Tillich, Lyotard, Postman, and Rivers, as philosophers of culture, all emphasize the decisive stamp that technology has made on culture, and they all, in one way or another, identify technicism in one way or another as the main cause of our cultural problems.

36. Christian Faith and Culture

I come now to assessing the position of Christians in a technological culture. Generally speaking, Christians have accepted the ongoing technological development uncritically. The main reason for this is that they have been spellbound by the positive effects of technology, such as enriching material life, enhancing the duration of life, turning the tide in the struggle against poverty and illness, and so on. Are those things not all signs of the kingdom of God? Abraham Kuyper, for instance, stressed the need for technology. But he did not recognize the danger of technicism, the ideology of technology. Perhaps he was already a victim of it himself, when he wrote in his *Pro Rege* (1911) that the wonders of technology are greater than the miracles of Jesus. Although Simon Ridderbos has made the point in his study of Kuyper's theological view of culture that Kuyper's view of technology was complex, it cannot be denied that Kuyper warns more against dance, theater, and card playing, than he does against a technicistic trend in our culture.

The error of Kuyper, as a child of his time, was that he accepted technology as being free of problems. In a certain sense he accepted technology as neutral with respect to values. Kuyper, and most Christians with him, forgot the biblical warning that since the Fall, technology in the sense of the technology of craftsmen has been a power that leads astray, the more so as the ages roll by. Technology has more than once been a means to rival God, to make one's name on the earth and to build a culture without God – a Babel-culture (Schuurman, 1987). The origin is the Fall into sin: human autonomy connected to **[145]** classical technicism. In its core, technicism wants to save human life without God. The Bible teaches us about Cain, about Lamech, about the building of the Tower of Babel, about Nebuchadnezzar, and so on. The Renaissance, Descartes, and the Enlightenment gave a new impetus to the modern form of technicism, namely absolutized scientific-technological control. The possibilities of modern technology have enhanced the ideology of technology. This ideology is mainly implicit. The gravity of this situation is often concealed by the numbing effects that emanate from a sweeping material prosperity.

Most of Kuyper's followers, too, have failed to evaluate technicism. Is it not astonishing that the Second Social Conference, held under Kuyper's leadership in the Netherlands in 1891, did not evaluate how technicism was guiding the development of technology? And can't the same be said of Christianity in general, when during the Conciliar Process in the churches the same theme is forgotten? Can it be that the biblical prophecy is not taken earnestly enough?

The last book of the Bible (Revelation chapters 17 and 18) speaks prophetically about the New Babel. That prophecy is a test for what I have said about the ideology of technology or technicism and its countless problems. When Christians accept technology only positively and uncritically – and, for instance, reduce the cultural mandate to a technological mandate – they are blind to what is really going on. They do not see that following the collapse of the collectivist materialism of Marxism experienced today in the countries of the former Soviet Union, the individualistic, liberal Western form of materialism will also collapse. Not seeing the religious background of our times, they are perhaps trying to Christianise the secularized ideas of the Enlightenment. In reality they have closed their eyes to the central fact that human autonomy is concentrated in the scientific-technological control of everything. In any case, they seem unable to resist the unrestricted scientific-technological manipulation of reality. **[146]**

People-without-God are trying to master God's creation. They want to realize a worldly paradise, a technological paradise. In the meantime the perspective of eternity is lost. Heaven is closed. Given the split between the divine world and earthly reality, they have set their hearts on technology and its results. This technicistic mind is the religious driving-force in our technological society and is enormously stimulated by welfare politics. If Christians are not aware of this religious background, they really lack the capacity to offer an adequate critique of our technological culture and to indicate the right, reforming normative direction to go. It is a pity that in the circles of Reformational Philosophy, we have spoken much about reforming philosophy and science but not enough so far about reforming culture. I think therefore that it will take a long time to analyze technological development critically and adequately. We are lagging far behind and as Christians have failed to develop a lifestyle suitable to a healthy culture.

In the technological culture the Christian worldview seems to have been reduced to a private belief lacking the capacity to reform culture. I am afraid that Christian thinking has indeed contributed little to reforming our technological culture. The reformation of scholarship and thought has not been placed in a broad cultural context but has instead been reduced and restricted to the inner circle of the university. Although such an approach has its place, it ought to be broadened and deepened so as to include the whole of culture. That means fighting against technicism, scientism, and economism, and at the same time looking for the authentic meaning of all cultural activities and developing, in both practice and thought, a normative approach to each cultural field. This perspective or new direction will be possible with the help of the Holy Spirit under the guidance of the biblical groundmotive.

But before considering that, we need to see again more clearly what is going on in our culture. There has never been a time as technological as ours, but there has also never been an age so spiritually empty. [147] Materialism, consumerism, greediness, redundancy, and "killing time" are expressions of the withering of culture and spiritual death. Technicism is the deepest background of secularization today and of the loss of faith and the experience of the remoteness of God (Tillich, 1986). Our spiritual situation hinders us from seeing the depth and the impossibility of solving our countless cultural problems, including exhausting our raw materials, polluting the environment, fragmenting society in the form of individualism, increasing the gap between the rich and poor countries, structural unemployment, and so on.

Meanwhile, our obedience to the main norms of technological development such as that of technological perfection, the technological imperative (what can be made, ought to be made), and efficiency has brought our culture into a gigantic, dynamic process, which is at the same time destructive. If the Second and Third Worlds are brought up to the same dynamic level of material welfare as the West – and do they not have the right to it? – tomorrow's problems will grow enormously. The future seems not to end in a technological paradise but in a colossal chaos. The future of the earth appears to be at stake.

Andrew Feenberg (1991) wants to democratize technological development such that there will be a balance between efficiency, justice, and humanity. But when we look at the results of such a democratization we can better speak about *relegitimation* than about *reformation*. The same can be said about the ideas of Habermas, when he tries to disclose rationality in a social and democratic sense (1970). He wants to bring rationality under the guiding norms of information and communication.

It is understandable that new philosophic movements are opting for the other pole of our reformulated humanistic groundmotive, the adoration of nature. Christians ought also to be aware that these reactions do not offer a way out. The Norwegian philosopher Arne Naess (1991), **[148]** with his concept of "deep ecology," is fully ecocentric. This movement desires in a certain sense to see a rebirth of animism: nature is divine in character and hence cannot be violated without punishment. A militant group within the movement of *deep ecology* is the "Earth First!" movement. They try to destroy the modern machinery used to control and gather what the earth produces.

Another reactionary group is guided by James Lovelock (1980) and Lynn Margulis (Ferkiss, 1993). The earth is seen and accepted as an organismic whole: Gaia. Gaia is involved or incorporated as a messiah. She can restore all that has been destroyed on earth, under the condition that humans accept themselves as a part of nature without a normative cultural calling to be stewards, without believing reality to be the creation of God the Creator, and without confessing the Fall into sin and Christ as Redeemer and Re-creator. Over against the technological paradise, these naturalistic thinkers are reclaiming the original paradise (McCormick, 1989).

37. The Biblical Groundmotive and Culture

The dialectic groundmotive of technological culture and nature stands in opposition to the biblical groundmotive of creation, fall, redemption through Christ in communion with the Holy Spirit and re-creation. This confrontation, which is founded neither in anthropocentrism nor ecocentrism (or naturalism), but in theocentrism or better Christocentrism, will dominate the next age. This conflict is concentrated in the struggle between the two groundmotives. The one is related to the *civitas terrena,* the terrestrial or technological paradise (dialectically related to "Mother Earth"), and the other one to the *civitas Dei,* the kingdom of God. Nevertheless, Babel (Rev. 17 and 18) parasitizes the New Jerusalem. The New Jerusalem is a gift of God, the fulfillment of creation as recreation, in which human cultural activities are interwoven in a way that is not fully understandable (Rev. 21). **[149]**

This gives a liberating and hopeful perspective. The dialectic in technological culture leaves room for the possibilities of a normative direction, even when the dialectic intensifies and assumes world-scale dimensions. Only when the whole of reality, including the relation between ourselves and nature or between us and technology, is seen as a unity founded in the Creation, disturbed in the Fall, restored in the Redemption, and soon to be fulfilled in the Re-creation as God's intervention, will a meaningful development of culture be possible.

The right, normative direction or meaningful perspective for cultural development stands over against both the accelerating technological culture with its problems and crisis, on the one side, and the naturalistic or ecocentric alternatives, on the other. Every field of cultural activity ought to be influenced by the biblical groundmotive. That groundmotive implies acknowledging that we are living between the Times, between the first and second coming of Christ. We are looking back in history and ought to be radically and integrally motivated by our calling for the future. Although the effects of the Fall will persist and be felt between the Times, we have to live in hope and expectation because the coming kingdom of God is the most decisive. Our position between the Times is fundamentally to struggle against every idol and to acknowledge God is the Creator, Redeemer, and Fulfiller. We are living in his creation, the meaning of which will be revealed, together with the divine mystery in history, in the kingdom of God.

Seen in that light, we must today primarily struggle against the power of technology and science, the power of absolutized scientific-technological control, and the influence of technological imperialism, in nearly every field in our culture. We ought to separate ourselves from the motives of secularized technological culture and turn to God in love, who loved us first in Christ, so that we can then return in love to our technological culture and our neighbors. Living in faith and hope and acting in love means that we are free both from technicism and its dialectic pole of naturalism. [150]

Amidst centralizing, large-scale technological power devoted only to effectiveness and efficiency, we ought to differentiate and to welcome diversity. Effectiveness and efficiency are not the most important, but in the first place there ought to be love, righteousness, justice, service, mercy, thankfulness, and a readiness to make sacrifices (see also Schumacher, 1978).

It is an enduring consolation to know that on our own and by ourselves we cannot negate this meaning of creation, the kingdom of God. On the contrary, the fact that the kingdom of God is already on the way means that at any moment people may be converted and led once again to seek the kingdom, even in a technological society.

A radical and integral change of heart, a *metanoia* to God as followers of Christ – which is too often left entirely unmentioned in philosophic discussions – implies a different position in culture. That is the position of a "cross bearer" in the struggle against technicism and naturalism. That does not mean separation, alienation, or distance from technology. We could not live without it and we need it, but our hearts ought not be set on it. Technology should be no more than an individual and societal prosthesis; that is the authentic meaning of technology. Technology would then contribute meaningfully to all cultural sectors. In such a wholesome culture one could speak about the blessing of technology. Although not all problems would be solved, they would be bearable, and life in society would be liveable.

In conclusion, the philosophy of the cosmonomic idea, not as controlling but as faithful thinking, gives us the integral framework of norms, resting in God's law as an expression of his will (see §24). God's ordinances, which are revealed through the Scriptures and the creational order, provide the guide which we need (Monsma, 1986; Schuurman, 1993). That guide expresses the relativity of all cultural work, but at the same time the individual and institutional responsibility (Unger, 1982) [151] to control culture and nature harmoniously. This struggle ought to be fought in the full armour of faith in the One Who holds all the power in heaven and on earth. When we fulfill Christ's power, the hidden ideology of technology will come to an end. Re-creation will overcome every distorting of creation. This final victory gives meaning to our cultural calling.

We all should orient our cultural task to this perspective of eternity at the end of the twentieth century.

> In the world you have tribulation; but be of good cheer, I have overcome the world (John 16:33).

Bibliography

Agassi, Joseph, *Technology: Philosophical and Social Aspects* (Hingham. MA: Reidel) 1986.

Altner, Günther, *Die Überlebenskrise in der Gegenwart* (Darmstadt: Wissenschaftliche Buchgesellschaft) 1988.

Beck, Heinrich, *Kulturphilosophie der Technik: Perspektiven zu Technik. Menschheit, Zukunft* (Trier: Spee-Verlag) 1979.

Bohring, Gunther, *Technik im Kampf der Weltanschauungen* (Berlin: Deutscher Verlag d. Wissenschaften) 1976.

Bookchin, Murray, *The Ecology of Freedom: The Emergence and Dissolution of Hierarchy* (Palo Alto, CA: Cheshire Books) 1982.

-----, *Toward an Ecological Society* (Montreal: Black Rose Books) 1980.

Bos, Abraham P., *Wetenschap en zin-ervaring* (Amsterdam: VU Uitgeverij) 1985.

Brinkmann, Donald, *Mensch und Technik: Grundzüge einer Philosophie der Technik* (Bem: A. Franke) 1946.

Caplan, Alan L., "Ethical Engineers," 33/6 *Science, Technology and Human Values* (1980) 24–32.

Capurro, Rafael, "Zur Computerethik—Ethische Fragen der Informationsgesellschaft", in H. Lenk and G. Ropohl, eds., *Technik und Ethik* (Stuttgart: Philip Reclam) 1987: 259–273.

Delschen, Karl-Heinz and Jochem Gieraths, *Philosophie der Technik: die Technikdiskussion in der Philosophie des 20 Jahrhunderts* (Frankfurt/Main: M. Diesterweg) 1982.

Dessauer, Friedrich, *Streit um die Technik* (Frankfurt/Main: Knecht) 1956.

Diels, Hermann, *Antike Technik,* 7th ed. (Leipzig: B.G. Teubner) 1924.

Dooyeweerd, Herman, "De gevaren van de geestelijke ontwapening der Christenheid op het gebied van de wetenschap," in J.H. de Goede, ed., *Geestelijk weerloos of weerbaar* (Amsterdam: Uitgevers-mij. Holland) 1936: 153–211.

-----, "De transcendentale kritiek van het wijsgerig denken," 4 *Synthese* (1939)314–339.

-----, "De vier religieuze grondthema's in den ontwikkelingsgang van het wijsgerig denken van het Avondland," 6 *Philosophia Reformata* (1941) 161–179.

-----, *Reformatie en Scholastiek in de Wijsbegeerte* (Franeker: Wever) 1949.

-----, *A New Critique of Theoretical Thought,* trans. by David H. Freeman and William S. Young (Philadelphia: Presbyterian and Reformed) 1953.

-----, *In the Twilight of Western Thought* (Philadelphia: Presbyterian and Reformed) 1960.

-----, "The Secularization of Science," *International Reformed Bulletin* (1966, no. 9) 2–17.

-----, *Roots of Western Culture: Pagan, Secular and Christian Options* (Toronto: Wedge) 1979.

Dijksterhuis, Eduard Jan, *The Mechanization of the World Picture: Pythagoras to Newton,* trans. by C. Dikshoorn (Princeton, NJ: Princeton University Press) 1986.

Durbin, Paul T., ed. *Research in Philosophy and Technology* (Greenwich, CT: Jai Press) annually since 1978

-----, ed. *Technology and Responsibility* (Hingham, MA: Reidel) 1987.

Ellul, Jacques, *The Technological Society* (New York: Knopf) 1964.

-----, *Betrayal of the West* (New York: Seabury) 1978.

-----, *The Technological System,* trans. by Joachim Neugroschel (New York: Continuum) 1980.

-----, *The Technological Bluff* (Grand Rapids: Eerdmans) 1990.

Feenberg, Andrew, *Critical Theory of Technology* (New York: Oxford University Press) 1991.

Ferkiss, Victor, *Nature. Technology and Society: Cultural Roots of The Current Environmental Crisis* (London: Adamatine Press) 1993.

Ferré, Frederick, *Philosophy of Technology* (Englewood Cliffs, NJ: Prentice Hall) 1988.

-----, *Hellfire and Lightning Rods* (New York: Maryknoll) 1993.

Forbes, R. J., *History of Science and Technology* (New York: Viking) 1963.

Goudzwaard, Bob, *Capitalism and Progress: A Diagnosis of Western Society.* Trans. Josina Van Nuis Zylstra (Grand Rapids, MI: Eerdmans) 1980.

Grant, George Parkin, *Technology and Empire: Perspectives on North America* (Don Mills, Ontario: Stoddart Pub.) 1969.

Habermas, Jürgen, *Toward a Rational Society: Student Protest, Science, and Politics,* trans. by Jeremy J. Shapiro (Boston: Beacon Press) 1970.

Hastedt, Heiner, *Aufklärung und Technik: Grundprobleme einer Ethik der Technik* (Frankfurt an Main: Suhrkamp) 1991.

Heidegger, Martin, *The Question Concerning Technology, and Other Essays* (New York: Harper and Row) 1977.

Hickman, Larry A. *John Dewey's Pragmatic Technology* (Bloomington. IN: Indiana University Press) 1990.

Hommes, Jakob, *Der Technische Eros: Das Wesen der materialistischen Geschichtsauffassung* (Freiburg: Herder) 1955.

Horkheimer, Max, *Critique of instrumental Reason: Lectures and essays since the end of WWII,* trans. by Matthew J. O'Connell (Minneapolis, MN: Seabury Press) 1976.

Howe, Gunter, *Gott und die Technik: die Verantwortung der Christenheit für die wissenschaftlich-technische Welt* (Hamburg: Fürche-Verlag) 1971.

Hüning, Alais, *Das Schaffen des Ingenieurs: Beitrage zu einer Philosophie der Technik* (Dusseldorf: VDI Verlag) 1987.

Ihde, Don, *Technics and Praxis,* Boston Studies in the Philosophy of Science, Vol. 24 (Kluwer: Norwell, MA) 1979.

-----, *Existential Technics* (Albany: State University of New York Press) 1983.

-----, "The Historical-ontological Priority of Technology over Science," in Larry Hickman, ed. *Philosophy, Technology and Human Affairs* (1985): 196–211.

Jonas, Hans, *The Imperative of Responsibility: In Search of an Ethics for the Technological Age* (Chicago: University of Chicago Press) 1984.

-----, *Technik, Medizin und Ethik: zur Praxis des Prinzips Verantwortungen* (Frankfurt/Main: Insel Verlag) 1985.

Jungk, Robert, *The Nuclear State,* trans. by Eric Mosbacher (New York: Riverrun Press) 1984.

Kalsbeek, L., *Contours of a Christian Philosophy: An Introduction to Herman Dooyeweerd's Thought,* edited by Bernard and Josina Zylstra (Jordan Station, Ontario: Wedge) 1975.

Kapp, Ernest, *Grundlinien einer Philosophie der Technik* (Düsseldorf: Braunschweig) 1877.

Kranzberg, Melvin, ed. *Ethics in an Age of Pervasive Technology* (Colorado: Westview Press) 1980.

Kristensen, Brede, *Het verraad van de techniek: over de sociologie van Jacques Ellul* (Amsterdam: VU Uitgeverij) 1986.

Kuyper, Abraham, *Pro Rege,* I (Kampen: Kok) 1911.

Laszlo, Ervin, *The Systems View of the World: The Natural Philosophy of The New Developments in The Sciences* (New York: G. Braziller) 1972.

-----, *Introduction to Systems Philosophy: Toward a New Paradigm of Contemporary Thought* (New York: Harper and Row) 1972.

-----, *A Strategy for the Future: The System Approach to World Order* (New York: G. Braziller) 1974.

Lenk, Hans, *Technokratie als Ideologie: Sozialphilosophische Beitrage zu einem politischen Dilemma* (Stuttgart: W. Kohlhammer) 1973.

-----, *Zur Sozialphilosophie der Technik* (Frankfurt/Main: Suhrkamp) 1982.

Lenk, Hans, ed., *Techne, Technik, Technologie: Philos. Perspektiven* (Pullach: Verlag Dokumentation) 1973.

Lovelock, James E., *Gaia: A New Look at Life on Earth* (New York: Oxford University Press) 1980

Lyotard, Jean-F., *The Inhuman: Reflections on Time* (Stanford, CA: Stanford University Press) 1992.

Maes, A.J.M.M., ed., *Vooruitkijken naar vooruitgaan: Technologie in de toekomst (Den* Haag: Directie Algemeen Technologiebeleid) 1993.

Marx, Leo, "The Idea of 'Technology' and Postmodern Pessimism," in Y. Ezrahi *et al.* eds., *Technology, Pessimism, and Postmodernism* (Dordrecht: Kluwer Academic Publishers) 1994: 11–29.

McCormick, John, *Reclaiming Paradise: The Global Environmental Movement* (Bloomington: Indiana University Press) 1989.

Meyer, Herman *J.*, *Die Technisierung der Welt: Herkunft, Wesen und Gefahren* (Tübingen: Niemeyer) 1961.

Minsky, Marvin, *The Society of Mind* (New York: Simon and Schuster) 1988.

Mitcham, Carl and Jim Grote, eds. *Theology and Technology: Essays in Christian Analysis and Exegesis* (Lanham, Md.: University Press of America) 1984.

Monsma, Stephen V., ed. *Responsible Technology: A Christian Perspective* (Grand Rapids: Eerdmans) 1986.

Müller-Schwefe, Hans Rudolf, *Technik und Glaube: eine permanente Herausforderung* (Mainz: Matthias-Grunewald-Verlag) 1971.

Naess, Arne, *Ecology, Community and Lifestyle: Outline of an Ecosophy* (Cambridge: Cambridge University Press) 1991.

Pippin, R.B., "On the Notion of Technology as Ideology," in Y. Ezrahi *et al.* eds., *Technology, Pessimism, and Postmodernism* (Dordrecht: Kluwer Academic Publishers) 1994: 93–114.

Postman, Neil, *Technopoly* (New York: Knopf) 1992.

Raby, W.H., *Über Sinn und Grenzen der Technik* (Düsseldorf) 1966.

Rapp, Friedrich, ed. *Contributions to a Philosophy of Technology: studies in the structure of thinking in the technological sciences* (Hingham, MA: Reidel) 1974.

Rapp, Friedrich, *Analytical Philosophy of Technology,* trans. by Stanley R. Carpenter and Theodor Langenbruch (Hingham, MA: Reidel) 1981.

Rapp, Friedrich and Paul T. Durbin, eds. *Philosophy and Technology* (Hingham, MA: Reidel) 1983.

Ridderbos, Simon J., *De theologische cultuurbeschouwing van Abraham Kuyper* (Kampen: Kok) 1947.

Rivers, Theodore J., *Contra Technologiam—The Crisis of Value in a Technological Age* (New York: University Press of America) 1993.

Rogers, Gordon E C., *The Nature of Engineering: A Philosophy of Technology* (London: Macmillan Press) 1983.

Sachsse, Hans, *Technik und Verantwortung: Probleme der Ethik im Technischen Zeitalter* (Freiburg: Rombach) 1972.

-----, *Anthropologie der Technik: Ein Beitrag zur Stellung des Menschen in der Welt* (Braunschweig: Vieweg) 1978.

Schumacher, Ernest E, *Small is Beautiful: Economics as if People Mattered* (New York: Harper and Row) 1973.

-----, *A Guide for the Perplexed* (New York: Harper and Row) 1978.

Schuurman, Egbert, *Techniek: middel of Moloch?* 2nd ed. (Kampen: Kok) 1980.

-----, *Technology and the Future: A Philosophical Challenge,* trans. by Herbert D. Morton (Toronto: Wedge Publishing Foundation) 1980.

-----, *Reflections on the Technological Society* 2nd ed. (Toronto: Wedge Pub. Foundation) 1982.

-----, "A Christian Philosophical Perspective on Technology," in Carl Mitcham and Jim Grote, eds. *Theology and Technology* (New York: University Press of America) 1984: 107–123.

-----, *Tussen technische overmacht en menselijke onmacht: Verantwoordelijkheid in een technische maatschappij* (Kampen: Kok) 1985.

-----, *Christians in Babel* (Jordan Station, Ontario: Paideia Press) 1987.

-----, "The Modern Babylonian Culture," in P. Durbin, ed., *Technology and Responsibility (Dordrecht:* Reidel) 1987: 229–243.

-----, "The Agricultural Crisis in Context: A reformational philosophical perspective," XVII *Pro Rege* (September, 1989) 2–14.

-----, *Het "Technische Paradijs": Om de gebrokenheid van heel de schepping* (Kampen: Kok/Voorhoeve) 1989.

-----, *The Future: Our choice or God's gift?* (New Zealand: Exile Publications) 1991.

-----, "Technicism and the Dynamics of Creation," 58 *Philosophia Reformata (W3)* 185–191.

Sizoo, A., *Techniek in de oudheid* (Kampen: Kok) 1961.

Staudinger, Hugo and Wolfgang Behler, eds.. *Chance und Risiko der Gegenwart: Eine kritische Analyse der wissenschaftlich-technischen Welt* (Paderborn: Schoningh) 1976.

Stork, H., *Einführung in die Philosophie der Technik* (Darmstadt: Wissenschaftliche Buchgesellschaft) 1977.

Strijbos, Sytze, *Het technische wereldbeeld: Een wijsgerig onderzoek van het systeem denken* (Amsterdam: Buijten & Schipperheijn) 1988.

Tillich, Paul, *The Spiritual Situation in Our Technical Society* (New York: Scribner) 1986.

Toulmin, Stephen, *Cosmopolis—The Hidden Agenda of Modernity* (New York: Free Press) 1990.

Unger, Stephen H., *Controlling Technology: Ethics and the Responsible Engineer (New* York: Holt, Rinehart, and Winston) 1982.

Vahanian, George, "Christliche Religion und Kultur," in A. Hertz *et al.* eds., *Handbuch der christlichen Ethik* (Freiburg: Herder) 1984: 439–453.

Van Griethuysen, A. J., ed. *Grenzen aan techniek* (Alphen aan den Rijn, The Netherlands: Samsom) 1989.

Van Meisen, Andrew G. M., *Science and Technology* (Pittsburg: Duquesne University Press) 1961.

-----, *Science and Responsibility,* trans. by Henry J. Koren (Pittsburg: Duquesne University Press) 1970.

Van Peursen, Cornelius A., *et al.* eds. *Informatie: een interdisciplinaire studie* (Utrecht: Spectrum) 1968.

-----, *Cultuur in Stroomversnelling* (Amsterdam: Elsevier) 1985.

-----, *Filosofie van de wetenschappen* (Leiden: Nijhoff) 1987.

Van Riessen, Hendrik, *Filosofie en techniek* (Kampen: Kok) 1949.

-----, *The Society of the Future,* trans. by David H. Freeman (Philadelphia: Presbyterian and Reformed) 1957.

-----, *Wijsbegeerte* (Kampen: Kok) 1970.

Van Woudenberg, René, *Gelovend denken: Inleiding tot een christelijke filosofie* (Amsterdam: Buijten en Schipperheijn) 1992.

Vermeersch, Emile, *De ogen van de panda—Een milieu-filosofisch essay* (Brugge: Van der Wiele) 1980.

Von Schilling, Kurt, *Philosophie der Technik: die geistige Entwicklung der Menschheit von den Anfängen bis zur Gegenwart* (Herford: Maximilian-Verlag Kurt Schober) 1968.

Von Weizsäcker, Carl F., *The Relevance of Science* (London: Collins) 1964.

Weizenbaum, Joseph, *Computer Power and Human Reason: From Judgment to Calculation* (San Francisco: W.H. Freeman) 1976.

Winner, Langdon, *Autonomous Technology: Technics-out-of-Control as a Theme in Political Thought* (Cambridge, MA: MIT Press) 1977.

Wuthnow, Robert, *The Restructuring of American Religion: Society and Faith since World War II* (Princeton: Princeton University Press) 1988.

Index of Persons and Subject-Matter

(page numbers refer to the original pagination)

BEYOND THE EMPIRICAL TURN: RESPONSIBLE TECHNOLOGY (2000)

1. Relation between Philosophy of Technology and General Philosophy

When the philosophy of technology first came on the scene at the beginning of the twentieth century, its practitioners did not devote much effort to the structural analysis of modern technology. Their aim at that time was primarily to defend technology as an independent segment of culture. They wanted to break the domination of science and economics over technology, and they rejected the idea that technology is simply neutral. Furthermore, because of the stormy development of modern technology, it was not until later that the philosophy of technology began to concentrate on the significance of technology for culture as a whole.

Engineers share the responsibility for the absence of any structural analysis of modern technology during the period in which the philosophy of technology was emerging. The engineers, deeply involved in the practice of technology, had so little interest in philosophy that they made almost no positive contribution to such an analysis. Besides, the lack of philosophical knowledge among engineers led all too easily to an overestimation of the role of technology on their part. More often than not, technological progress filled people with grand expectations for the future development of culture.

We must not forget that general philosophy – as an ontology or cosmology – in turn, paid very little attention to technology. It underestimated the significance of technology at first, and for the sake of convenience reduced it to a science or regarded it as a neutral tool in people's hands. Philosophers were unfamiliar with technology; they lacked a basic empirical knowledge of it.

Eventually, because of its enormous influence on all of culture, technology could no longer be disregarded. Only then did general philosophy begin to take notice of it. Nevertheless, because of the current widespread lack of a thorough knowledge of technology in our society, technology is still being disparaged as a dangerous power threatening human welfare. Technology is all too quickly blamed for the cultural crisis. Inherent in this negative appraisal is an idea which is also shared by the optimists who view progress in positive terms – the idea that technology is all-embracing and all-dominating. The optimistic and pessimistic views both lack an adequate perspective on technology. While the one view overestimates the cultural influence of technology, the other fails to appreciate the possibilities it offers.

A knowledge of technology and philosophy in their mutual interaction is essential to any effort to arrive at a structural analysis of modern technology. In the past, unfortunately, there was a great lack of communication between engineers and philosophers. "Their respective terminologies and the origins and orientations of their thought are so disparate that a consensus on problems of mutual interests becomes possible only through extraordinary effort and with much good will."[43]

When weighing the beginnings of the philosophy of technology, then, we must take into account the serious difficulties it originally faced. General philosophy virtually ignored the impressive phenomenon of technology and scarcely reserved a place for it. Philosophy of technology lacked a general framework. As a result, it developed almost completely independently of general philosophy. Engineers who had a philosophical interest in technology and its development but had an uneven background in philosophy (if at all) arrived at philosophies of technology most of which can only be regarded as pseudo-philosophical.

The proper task of general philosophy is to express the unity-in-diversity of total reality. By neglecting or overestimating technology as a disastrous power, general philosophy failed to do justice to the diversity of reality and thereby damaged its own insight into the whole of reality. The philosophy of technology, in turn, must take full account of the fact that it cannot be developed correctly outside an appropriate general framework. It deals with only one segment of culture. Its field of investigation is limited. Therefore it must be able to call upon general philosophy to account for the coherence of technology and reality as a whole. General philosophy, for example, should account for the significance of technology for culture. The fact that dynamic development has made technology a prominent cultural force renders it all the more important for the philosophy of technology to be able to appeal to general philosophy.

---

[43] Klaus Tuchel, *Die Philosophie der Technik bei Friedrich Dessauer: Ihre Entwicklung, Motive und Grenzen* [The Philosophy of Technology on the Part of Friedrich Dessauer: Its Development, Motive, and Boundaries], Frankfurt: Knecht, 1964, p. 15.

A philosophy of technology set within the framework of a general philosophy should provide insight into the rich diversity of technological objects, the designing process, the technological products and means of production, and of technological activities, all of which are increasing in variety. I agree fully with Mitcham and Kroes that philosophers should pay more attention "to what really goes on in engineering and technology," and that a philosophy of technology ought to be "an empirically informed philosophy of technology."[44]

2. Philosophical Analysis of the Structure of Modern Technology

A structural analysis of technology ought to define clearly both the potentialities and the limitations of technology, thereby obviating and guarding against misconceptions and false expectations concerning it. It ought to be related to the empirical practice of technology. All technological problems and all norms for technological development to promote responsible technology should be analyzed. In particular, a clear view is needed of both the relation between technology and science and the many cultural changes which their modern alliance has brought about.

Only with the aid of such a structural analysis does it become possible to deal philosophically with the cultural influence of technology. By giving prior attention to such an analysis, we can greatly reduce the danger of becoming too speculative about both the positive and the negative influence of technology on culture. We ought to go from a philosophy of technology to a cultural philosophy of technology, instead of the other way around as usually is done.

A structural analysis of modern technology within the framework of a general philosophy has been conducted within a special tradition of Dutch philosophy. This philosophy has not received that attention it deserves. The main reason of it is that it is developed as a Christian philosophy and that generally speaking philosophers are not interested in such an approach. But it is

---

[44] See Carl Mitcham, *Thinking through Technology*, Chicago: University of Chicago Press, 1994, p. ix.

worthwhile to give attention to the empirical and comprehensive approach of technology within that philosophy.

The Amsterdam school of reformational philosophy, also called the philosophy of the cosmonomic idea,[45] has developed a general philosophy within which technology may be treated. Because it provides a suitable place for the philosophy of technology, it can be used as a framework for analyzing the diversity of technology as well as exploring the significance of technology within the whole of reality – especially the significance of technology for special sectors of culture.

Hendrik van Riessen – an electrical engineer – in 1947 had already began to develop a philosophy of technology within the framework of the reformational philosophy. Hereby he applied what other philosophers had noted about the structure of technology. In giving his analysis of technology Van Riessen refers to a great number of examples from practical technology. In that sense he can be seen as the first philosopher of the empirical turn.[46] But more can be said.

A requisite of a structural analysis of technology is a good definition which refers to the empiricism or practice. Mostly philosophers who are claiming an empirical turn do not provide a strict definition of technology.[47] The same omission can be found in the circles of the so-called classical philosophy of technology. For instance, Heidegger and Ellul do not provide a strict definition. I will

---

[45] I refer here to a school of philosophy developed at the Free University in Amsterdam in the 1930s by Professors D. H. T. Vollenhoven and H. Dooyeweerd. See H. Dooyeweerd, *A New Critique of Theoretical Thought,* 4 volumes, Amsterdam/Philadelphia, 1953–1958.

[46] See H. van Riessen, *Filosofie en Techniek* [Philosophy and Technology], Kok, Kampen, 1949. In English see Schuurman, *Technology and the Future*, pp. 1-50, and chs. 2 and 3 of "Perspectives on Technology and Culture" (in this volume, pp. 203–258 below). New attention has recently been paid in the Netherlands to Van Riessen's philosophy of technology: see Hans Haaksma, *Van Riessen: Filosoof van de techniek* [Van Riessen: Philosopher of Technology], DAMON, 1999.

[47] A. Rip in *Technologie en samenleving* [Technology and Society], Leuven-Apeldoorn, Garant, 1995, says for instance that technology is so complex that precise distinctions need not be made (pp. 10, 15). I would stress that, given the complexity of technology, such is needed to define what technology is all about.

define technology as the activity by which people give form to nature for human ends, with the aid of tools. In modern technology the tool is developed in computers, robots and even automated factories.

Of course, the structural analysis of technology gives an insight in the development of tools, in technological form-giving, and in the designing process. In paying attention to the basic structure of modern technology we get a general overview of it.

3. Basic Structure of Modern Technology

The grand quest in modern technology has been to develop technological objects – tools – that can operate independently. To this end, human proficiency in forming is projected into and transferred to the technological object. A transfer of the decision-making capacity pertaining to the sequence of the activities of forming also occurs. By means of automatic switches, people make provision for the technological forming process to undergo discontinuous changes with the passage of time.

Taken together, the projection of proficiency, the transfer of "decisions," and the use of formed energy constitute the foundation of the independent operation of modern panoply of tools and instruments. This panoply is composed of what sometimes is called technological operators. Apart from design and installation, people need only give a command to set the technological operator going.

The "proficiency" of the technological operator surpasses that of human beings in speed, reliability, and accuracy. Even mechanical "decisions' are realized more quickly and faultlessly than "decisions" made and implemented by people. And the power of formed energy far exceeds human power. Now, what all this means is that people equipped with technological operators can accomplish a great deal more than people without them. Moreover, there are more recently created technological operators that work in ways that bear little or no resemblance to human activity. Electrotechnology and chemical technology offer examples of such a development.

The development of technological objects sketched here, namely, the realization of independent operation, has been made possible only by the scientific foundation and scientific method of designing.

Through the scientific approach of modern technology, a distinction has arisen between preparation (designing) and execution (technological forming).

Human responsibility and decision-making have been transferred to the phase of preparation, and the human activity of designing has thereby come to occupy a higher place. There has in fact been an intellectualization of technological labour. Preparation along the lines of the technological-scientific method leads to the accomplishment of a design for execution; this design is complemented by the independent operation of the technological panoply. As a result, human power in the phase of execution is enormously increased.

I summarize the basic structure of modern technology as follows: it is characterized by the technological operator, the scientific foundation, and the technological-scientific method.[48]

At this stage it is necessary to note that even in modern times, technological activity is and remains a human activity. This fact is decisive for proper insight into the limitations and possibilities of technology. One must never lose sight of the intimate connection between people and technology, especially when evaluating such a phenomenon as automation, which may sometimes appear to be rather autonomous with regard to people. In fact, the intimate connection between people and technology is crucial if we are to resist arguments to the effect that technology as a whole is autonomous.[49] This means that technological development is deterministic. This "autonomy of technology" has recently instigated "the empirical turn" in philosophical circles to make clear that the design process is not deterministic.[50] In my view we have to go beyond that turn to a responsible technology.[51] To make that clear we are looking more precisely to the design process and to responsibility in designing.

---

[48] See E. Schuurman, *Technology and the Future*, pp. 8–24.

[49] See J. Ellul, *The Technological Society*, London, 1965, pp. 14, 128, 134, and *The Technological System*, New York, 1980, pp .125ff.

[50] See recently: Andrew Feenberg, *Questioning Technology*, London/ New York: Routledge press, 1999, p. 76.

[51] See Stephen V. Monsma (ed.), *Responsible Technology*, Eerdmans, Grand Rapids, 1986, pp. 165ff.

4. Philosophy of Design

We have already seen that technology involves two stages of design and fabrication, closely linked as two interacting parts of the production process. They are also linked by the fact that the fabricating facilities and procedures are themselves the result of a design activity. It is necessary to give more attention to this.

Design is that structured, innovative activity whereby people creatively use theoretical and practical knowledge and available energy and materials in order to specify the size, shape, function, and material content of a technological object. So design not only depends upon theoretical and practical knowledge but also on the availability of energy and materials. It assumes the existence of the physical reality, and utilizes both knowledge and available energy and materials in its activity. Design results in a blueprint or set of detailed instructions for the physical characteristics of a technological object – either a product or a tool.

It should be noted that although the design process is a structured activity, it is rarely if ever carried out in a rigid, linear way. One step does not follow another in lockstep fashion until the tool or product "automatically" emerges. Feedback plays an important part in the process. The results of one step – for example, certain design specifications – may lead to changes in prior steps, such as redefining an identified need. Furthermore, this kind of structured thinking about design may enhance rather than inhibit creativity in that it more clearly focuses people's attention on the broad range of task that must be accomplished.

Although designing is dominated by the scientific design method (see next paragraph), which guarantees the continuity in technological development, we must attend to breaks in this continuity. We are talking about inventions. Technology continues to develop in the context of scientific consultation and formulation. In ways that are often unexpected and cannot be simulated, productive imagination can put people on the track of technological innovation. These innovations can interrupt the developmental process of technology and are clear evidence that technology is not autonomous.

In modern technology, invention takes place at a higher level than in classical technology. Today inventions build on mathematics, physics, information theory, and technology as science. This also means that when inventors are onto something new in technology, they try to develop it scientifically and to check it out with precision instruments to complete the invention. But at the decisive moment of inventing, theory plays no role. The invention and its effects will later belong to the field of technological science, which also served as its basis. This being so, the invention contributes an additional path to those presently available for new technological possibilities[52]. And such interventions give new possibilities for responsibility in design (see the last paragraph).

In recent discussions about "the empirical turn" it seems that attention is given to individuals doing design in a few specific industries and then to analyse their work in detail. This method would certainly lend itself to an explication of some of the details of design. The problem with this approach, however, is that one would tend to accumulate mountains of information about specific design activities, but only slowly – if at all – uncover the basic patterns present in design activity. Also, making design decisions and doing the actual work of designing – even within a specific design project – are tasks usually performed by many persons, and thus an enormous number of individual activities would need to be considered in detail before patterns useful for analysis would emerge. This method of analysis, which may have its place within a given industry, ends up being somewhat akin to studying a forest by studying the individual trees.

A better method – the one I like to follow – analyses general patterns of design. This broader approach is helpful both to general public, who need understand technology better, and to the decision makers within design activity, who need an overall perspective on their work. To achieve this broader approach, it is helpful to employ what Hans Lenk and Gunter Ropohl (following Max Weber) refer to as an "ideal type" analysis of design.[53] Instead of making a close scrutiny of a few individual efforts, this approach involves observing the design activity from some distance so that overall patterns can be seen more clearly. This approach takes a general perspective on the design activity based on an amalgamation of a wide range of experiences and analysis of design. It has to do with the concept of design, not with individual design activities. A method

---

[52] See Friedrich Dessauer, *Streit um die Technik* [Struggle over Technology], Frankfurt/Main, 1966, p. 167.

[53] See Hans Lenk and Gunther Ropohl, "Towards an Interdisciplinary and Pragmatic Philosophy of Technology," in *Research in Philosophy and Technology*, vol. 2, ed. Paul T. Durbin and Carl Mitcham, Greenwich, Jai Press, 1979, p. 32.

related to individual design belongs to the history of technology or sociology of technology, not to a philosophy of technology, specifically philosophy of design.

To a broad picture of design belongs a general analysis of the method of design, which is a scientific one. And this scientific method implies a holistic approach. Holism is the view that the integrated whole of a technological object has a reality independent of and greater than both the sum of its parts and its particular function. Every technological object or product is experienced as a unity.

The case for a holistic approach to design rests primarily on the fact that the technological tools and products produced by design and production enter everyday experience as unities. This being the case, design itself should be holistic. Indeed, people's lives are affected not by technological objects *per se*, but by how these objects function in all aspects of reality. A car not only conforms to the laws of physics and chemistry, but is also an object of proud ownership, has a value relative to that of alternate purchases, and is a focus of trust, because the buyer believes that if it is used correctly, it will not be harmful. If an object is to function well in all of its aspects, these must be taken into account in its design. And that is what holistic design seeks to do. The design of technological objects must take the diversity of reality into account if those objects are to function well.

Holistic design of this nature involves making value judgements in all of the various aspects of reality. Taking these value judgements into account in designing is the creative challenge for the designer. But is this enough? Does holism provide sufficient content for a design philosophy? One only has to image a horrific product – a gas chamber for the wholesale murder of innocent victims that is "well-designed" and "functional" in all aspect of reality (at least from the viewpoint of a Hitler) – to realize that a proper design philosophy requires something more than holism. One still must determine what normative principles should be followed in the practice of holistic design.

But before dealing with the value-ladenness of technology, I would like to describe in the next paragraph the general scientific method which is used in the process of design and which characterizes the technological objects for a great deal.

5. Scientific Design Method

In designing, the modern engineer uses the scientific method. To understand this method and the influence of this method in relation to technological operators or tools and products, it is necessary to give more attention to it.

The scientific method in designing is like that of the natural sciences, but is technological because it focuses on the design of technological things or processes. This scientific design method characterizes modern technology. As a result, the features or characteristics of science are projected in technology. Seeing this close intertwinement of science and technology also helps us see the characteristics of modern technology.[54]

Within the scientific method of the natural sciences we attempt to discover universal knowledge of reality. With the scientific design method we intend to formulate technological designs both for the product and for producing it. The ultimate aim is a temporally and spatially remote, scientific control of the individual technological process of forming or producing. But its result also belongs to the domain of technological science because the designs furnish us with knowledge regarding technology. Thus, physics is related to universal knowledge of the physical aspect of reality, technology as science is related to knowledge of technological objects as a whole[55]

Some of the characteristics of natural scientific knowledge and of its method are reflected in a technological kind of way in technological science. This is in large part due to the scientific basis of design. These characteristics can also be found in the technological production process and its results.

Analysis, abstraction, and logical synthesis constitute the method of science. That is to say, in science we consciously turn away from and seek to avoid the influence of the immediacies of the concrete individual, situation, or

---

[54] See E. Schuurman, *Filosofie van de Technische Wetenschappen* [Philosophy of the Technical Sciences], Martinus Nijhoff, Leiden, 1990, pp. 51–57; E. Schuurman, *Perspectives on Technology and Culture*, Potchefstroom, Zuid Afrika/ Dordt Press, Sioux Centre, IA, 1995, pp. 44–51 (in this volume, pp. 236ff.).

[55] This implies a problem. See "The Limits of the Scientific Design Method, and the Relation of Physics to Technological Science," in E. Schuurman, *Technology and the Future*, pp.25–31, 41.

environment. As a result, scientific knowledge is universal, enduring, logically coherent and functional, abstract knowledge.

The engineer, too, follows the same procedure. The technical question is broken down into parts which in turn are posed as universal (sub-) problems. Universal modular solutions are sought both for the components of the thing to be produced and for the steps of the production process. The division of functions continues until one has arrived at atomized, isolated, autonomous functions.

Analysis puts technological modular functions at one's disposal. Abstraction presents each of these functions on its own, bracketing out the others, so as to find a technological solution for that sub-function. The result consists of basic building blocks that have a standardized neutrality of purpose about them and, hence, can have a universal application in technology, as is the case for example with propellers, rivets, or dovetail joints. In this universal applicability, a result of analysis and abstraction, we find science's general or universal knowledge projected in technology. In other words, the universal suitability of the solution of a sub-function is an analogy of scientific knowledge in technology.

In addition to the neutralizing division of functions, the scientific design method, as was said, also aims at a spatially and temporally remote control of the process. Therefore, the tool or technical operator must be constant over the course of time. The control of technological formation at a distance and over times is an analogy of science as enduring abstract knowledge of reality.

This remote control, with respect to the whole, is further made possible by the integration of the neutral modular solutions. Depending on the modules that are merged, this integration of functions can represent many forms of individuality. As such, it is a technological analogy of the synthesis in science. The scientific design method aims than, on the one hand, to standardize how functions are divided and, on the other, to anticipate how these functions will be integrated for various specific ends. The production process that results is likewise an analogy of the coherence of scientific knowledge, with the product of mass production showing something akin to the universal character of scientific knowledge.

In like manner, an analogy of scientific abstraction can be noted in the integration of the solutions from each module. These problem sets were originally isolated by means of abstraction. After a standardized solution is found for each subset, the integration of these separate solutions requires that technological formation take place elsewhere, in isolation. Pertinent measures are taken with that in mind. Technological formation has an enduring character about it because influence from the surroundings, like temperature and humidity, are precluded as much as possible.

This scientific design method determines the basic structure of modern technology. Technological science and technology are virtually interwoven.

When developments in science lead to a new method, as was the case with systems analysis, then this method recurs analogously at the technological level. That is why general systems theory and computer technology are complementary.

Although the influence of science on technology is legitimate, when science is used in technology under the influence of power technology goes on the way of irresponsible technology, the design process becomes deterministic. We can also say that technology becomes autonomous technology. Then there is no place for normative principles in designing technology. And the normative principles ought to be at the heart of responsible design. A scientized tunnel vision in design and fabrication prevents one from asking questions of justice, stewardship, and cultural appropriateness. But when science and its methodologies are put in their proper place, holism can be practiced and normative principles can be given their due. Technology is then freed to be practiced responsibly.

6. Responsible Design

In Section 4 I have argued that technology is necessarily and intrinsically value-laden. The question is important which normative standards in technology ought to be observed.

As we have seen already, holistic design is important, but it needs to be undergirded by normative principles that take into account all to the aspects of reality within which the technological object is to function. I will restrict myself in dealing with headlines of the normative principles.

The basis of every technological development is that technology ought to be ecologically adaptable. This starting point prevents technology from causing ecological disruption. Next, the disclosure of the technological development should be guided by eight[56] normative principles. Although I will discuss these normative principles individually, they are to be followed simultaneously, not in isolation from each other.

Cultural Appropriateness. All those involved with design must make sure that their designing results in culturally appropriate technological objects. This means making appropriate decisions in relation to the five sets of opposites of the cultural-historical norm principle: continuity and discontinuity, differentiation and integration, centralization and decentralization, uniformity and pluriformity, and large scale and small scale. Those involved in design must not design technological objects that totally embrace one or the other of each of these five pairs of opposites, nor should they simply strike a balance between them.[57] This normative principle is not simply one of accommodation to anything that works in a given cultural setting. Important cultural manifestations should not be destroyed by an intrusive technological object. Indeed, the culturally appropriate tool or product is one that alleviates human burdens and preserves what is wholesome and good in a given culture. In this way it strikes an appropriate balance between continuity and discontinuity.

It is also important to balance the opposites of centralization and decentralization. Should our culture be more centralized and homogeneous or more decentralized and heterogeneous? Those involved in design should consider such matters of balance in following the normative principle of cultural appropriateness.

Openness and Communication. Openness means that we must clearly inform society about technological renovations. Only then can those working with the technology or purchasing its products, responsibly evaluate things and make decisions. Uniting the normative principles of openness and communication results in the principle of open communication. Without open communication, it is impossible for those who participate in technology to fulfil their communal and individual responsibilities. Those who develop technology have the

---

[56] See for more explanation: Stephen V. Monsma (ed.), *Responsible Technology*, Eerdmans, Grand Rapids, MI, Chapter V: "A guide to responsible technology," pp. 58–76.

[57] See Victor Papanek, *Design for the Real World: Human Ecology and Social Change*, New York, Bantam Books, 1973, p. 34.

responsibility to take the time to inform and communicate with the public. Quite obviously, people in the technological sciences in particular have to honour these norms of information and communication for these are the ones who stand, as it were, at the cradle of every new technological development.

Stewardship: efficiency and sustainability. The economic norm of efficiency must also be honoured, but not to the exclusion of the others. As it applies to technology we are talking here about the matter of efficiency. Its presently one-sided application is due especially to the prevailing influence of economic theory within industry. More than once industry has focussed fat too narrowly on a natural scientific sense of efficiency, in which only those goods that can be expressed in monetary terms determine what has value. We must integrate the economic norm into an integral framework of norms. We must apply it not only to production, but to the use of raw materials, energy, nature, the environment, landscapes, animals, and even of the people involved in developing technology. Problems arise when we limit the economic norm of efficiency exclusively to techniques of production. Technology and the economics of industry develop into a jumble. But when we apply the economic norms of efficiency and sustainability, in conjunction with the other norms, simultaneously across the board, we can prevent a kind over overdevelopment in producing surpluses and restore a kind of underdevelopment in being stewardly in dealing with nature. We will begin to attend more to nature and to the environment, to the scarcity of raw material and energy, and therefore to honour sustainability. Besides, we will come to see that people are much more than their ability to function economically within a technological context. We must acknowledge the responsibility of employer and employee.

Harmony. The normative development of technology is advanced when we abide by the norm of harmony. Non-essential surpluses, and the degradation of nature, make it abundantly clear that this norm is not being followed. If it were, given the other norms we have discussed, people would realize that technology ought to be developed in a balanced manner. This norm also requires that we introduce and incorporate new technology not in a revolutionary fashion, to prevent societal unrest and a loss of communal support. We must also consider this norm of harmony in the multifaceted interrelationships between nature, people, culture and technology. Technology ought to adapt to people, not people to technology. For example, it is not without reason that we appreciate user-friendly tools. That is the way all tools should function. When that is the case those who operate them will also have greater pleasure in doing so.

Justice. In honouring the normative principle of justice we oppose any and all injustice that the development of technology may bring about. Engineers, instructors, and employees must ask themselves whether their contribution to technology does justice to the plant and animal kingdom, to our sources for raw materials, to consumers, to society, to culture, to third world countries, and the like. This norm of justice is an intrinsic part of technology. When it is disregarded, the government must take specific measures to restore justice. It is worth emphasizing that the positive influence of technological developments that obey this norm will also be felt in the many sectors of society in which technology is playing an increasingly important role. That is especially true, for example, in modern agriculture and in health care.

Care and love. All the norms indicated above are opened up and deepened when people hold themselves to the ethical norm of care and love. We are called upon to nurture a concern and compassion for everything that has to do with technology. This, of course, includes caring for and loving our neighbours, far off and close by, but also protecting the great diversity of all other creatures. When love binds itself only to scientific-technological control, people become so obsessed with control that it is frightening. When this norm is not honoured on all sides, people become more and more alienated from their work, for example the farmer from the land, nature, and his animals.

Trust. The last norm to which designing is subject is the norm of trust. Those who use the tools of technology have to be able to trust that these tools work and are safe. When this norm is met, they will be. Or more broadly, when people accept and operate according to the framework of norm principles sketched above, technology will be safe. When all the normative principles will be followed, technological design will be truly responsible.[58]

---

[58] See also E. Schuurman, *Faith and Hope in Technology* (Toronto: Clements Publishing, 2003).

THE CHALLENGE OF THE ISLAMIC CRITICISM OF TECHNOLOGY (2022)

Introduction[59, 60]

Due to the rise of terrorism in the world, we have become very conscious that the Western world and the Islamic world, while having a common history, also have many differences. There is an increasing atmosphere of mutual confrontation. I would like to look at these two uneasy worlds as they face modern technology and the problems associated with it. This is not a common approach to the tensions between Western and Islamic culture, but – as we shall see – it is certainly a very significant and fascinating one.

When we place this subject in historical perspective, we cannot avoid the religious background of technology in the Islamic world and in the West. Here I find a connection with another burning issue; that of the vitality of religion in the world and the arguments regarding its influence upon culture (Haberman, 2005), in particular upon technological developments.

In relation to the concept of "religion," some clarification is called for. When, for example, the media give attention to religion, they treat it mainly as one of the many factors of human life, next to sports, politics, or science. However, if we look carefully at religious communities and at different social structures in the world, we can see that religion is not only one typical function or variable among others, but is the *root* from which the different branches of life sprout, grow, and are continually dependent. That will become evident in the course of this essay.

After I have given a brief sketch of the history of technology in the Islamic world, I will discuss the backgrounds of the current, growing tensions between Islam and the West. Different Islamic ideologies will be briefly looked at. Science and technology play an important role in their thought. The Islamic criticism of technology stems from two sources, on the one hand from the spiritual, peace-loving Muslims, and on the other hand from the radical, violent branch.

---

[59] Farewell address, September 20, 2007, Wageningen University.

[60] Various Muslim and Christian scholars have commented critically on this lecture: *Different Cultures, One World—Dialogue between Christians and Muslims about globalizing technology,* H. Jochemsen & J. van der Stoep (eds.), Amsterdam: Rozenberg, 2010.

I try to clarify the tensions which exist in this area for the West by examining the internal tensions which are present in the Western culture. It becomes evident that these especially go together with modern technology. These tensions have been present for a long time, but have been strengthened since Christian culture began to be secularized under the influence of the Enlightenment. The latter cultural movement does not want to be linked to religion, but has taken form as a radical movement in a definite and visible way which is more and more at odds with Christianity. When I describe this as a *religion* of a closed, material world – where people are blind to the non-material dimension of reality – I am doing so in order to portray more clearly the total picture of the area of tensions between Islam, Christianity, and the Enlightenment in relation to modern technological developments. This helps us to analyze the problems and formulate solutions to them.

In confrontation with both the Christian-philosophical and the Islamic criticism, Western culture is challenged to self-reflection. In my opinion, a real change to the dominating Western culture paradigm is needed because the ethical framework within which Western culture has developed and is developing. This is important because we have to do not just with an isolated Western culture, but with world-wide problems. The tensions, as well, with certain movements within the Islamic world could hereby possibly be able to be reduced. Islamic terrorists, however, will not be satisfied with this, for their attitude reflects – as they themselves say – a firm religious position which they will never give up. At the most it might be possible to dampen their hostility by seeking to overcome evil with good (Romans 12:21).

Technology and Islam

Let's start by looking at the place of science technology in the Islamic world.

Since the death of Mohammed in 632, history teaches us that early Islam was strongly influenced by the Greek-Hellenistic world. Thereby a favorable climate developed for carrying out and furthering science with a special Islamic tint (Stöklein und Daiber, 1990, 102). Science was seen as taking place in a universe created by Allah. That universe displays order and balance, and is therefore an aesthetic unity. The philosophy and science which developed in that perspective flourished in the Islamic world for more than 500 years. The climax of this development was the civilization of the Arabian world in the 9th and 10th centuries A.D. In this period, we also see much knowledge imported from foreign sources – for example from Persia, India, and even China. That was in accordance with the Islamic mandate that everyone increase one's knowledge one's whole life long. Scientific experimentation and posing technical questions were not at all strange for Islam at that time. Thereby, attention to Nature had to take place in a careful way, just like the care a husband and father ought to give his family. This development furthered trade (the economy), which then had a beneficial effect on science and technology. Historians say that at that time there was a symbiosis between Islamic religion and (practical) science. Building religious houses, mosques, schools, and carrying on water management in desert countries, in particular, were significant illustrations of this.

It is clear that the Islamic peoples were ahead of Western countries in the area of science and technology up till the Middle Ages. At the beginning of the Middle Ages, Islam even functioned as a bridge between the antique Western world of Greece and Rome and contemporary Europe. The West has a lot to owe to the Arab world as far as scientific development is concerned.

However, since the 11th century, the development of the sciences has been stagnant in the Arab states. All kinds of reasons are given – especially political and social-economic – for this decline. Since that time, traditionalism and isolation have been the chief characteristics of the Islamic world. This goes together with a change from a positive to a negative evaluation of science and technology (al-Hassan, 2001, Hoodbhoy, 2007).

Later, during the industrial and post-industrial ages, the Islamic countries contributed little to the development of science and technology, with the exception of the oil-rich Arab countries, which via oil production and industry have helped develop related technology, and also with the exception of weapon technology developed in Islamic countries. It is striking that at this moment in time there are Islamic scholars who are again – as I hope to show – seeking to advance modern science and technology in the light of Islam's own, original history and sources (Zayd, 2006, 31–35). Their criticism is not of science and technology as such, but of the "technological culture" of the West, thus the Western ethics bound up with this technology.

The Influence of the Enlightenment in the West

In the course of time, the West, with its dominating faith in progress – especially stimulated by the Enlightenment – began to feed the bias against Islam to the effect that Islam itself was opposed to science and technology. The contemplative character of Islam and the fatalistic attitude toward life of the Arab was blamed for this. Although contrary to its original attitude, this ethos has indeed had much influence in the world of Islam. The opposition to Western science and technology has been strengthened by it. Since the 12th century, the Islamic world has been looking more to the past than to the future.

In the 20th century, a change took place through the process of globalization. Universities in the Arab world were founded as a result. And much was taken over from the West (Huntington, 2001, 70, and Soroush, 2000). But it seems that modern technology has only been valued to the degree it could be used to serve Islamic religion. The idea is that science and technology must be brought under the Islamic flag. That cannot take place easily, however. Historically, Western ethics has accompanied the development of technology. Islamic resistance is being offered to such ethics just as resistance continues to be offered to the faith in progress present in the Western background of those ethics. Islamic acceptance of scientific and technological knowledge – modernization— continues to form a glaring contrast with the opposition to westernization, to secularism, to materialism, and to Western profanity (Soroush, 2000, XVII). The conviction is that modernization must be supplied with a moral compass via Islam (WRR, 2006, 38, 39).

Reactions to Islam

It is furthermore important to distinguish between different reactions from within the Islamic community. For more than one Muslim country, various reactions even go back to the colonial period. First, we have to do with the radical, violent, fundamentalist stream which rejects both the acceptance of science and technology and westernization (the ethics of the Enlightenment.) Then there is the stream which clearly takes over both elements from the West. This concerns especially the political and economic holders of power. But it is also true of some Muslim scholars (Hoodbhoy, 2007, 55). It is understandable that the first stream also opposes the latter phenomenon, and this is the reason why terrorist actions of radical Muslims take place just as often in Muslim countries as in Western ones.

We also have to do with Islamic reformers, as Huntington calls them (2001, 118 f.). Others call this group spiritual and peace-loving. They accept modern developments in science and technology, but without the dominating Western ethos. For them, it is a matter of accepting science and technology, whereby there is support for a thorough process of proper (non-secular) "rationalization" going together with spiritual conviction (Hassan, 2007, Soroush, 2004). In the same way they argue often for the approach to society of the Western democracies (Soroush, 200, WRR, 2006, 29–58).

As a result of substantial differences regarding convictions and the growing tensions between these three streams, political resistance and violent protest against the West will most likely grow, and within the Western world – where Islam is growing fast – cultural tensions will increase. The choice of the fanatical Islamic stream, the smallest group, is resulting in a violent threat to Western culture. Witnessing this passionate impulse to destruction leaves us with a somber perspective on the world situation.

Enemies of the West

The study of Ian Buruma and Avisah Margalit about Occidentalism (Buruma, 2004) provides us with good insights into this process. That which they call "Occidentalism" is the ugly image of the West which is painted by its enemies. According to this narrative, industrialists, capitalism, and economic laissez-faire policies have spread like diseases from the West, under the leadership of America, and infected the entire world. The fanatic Muslim sees in this "Americanism" a mechanical civilization which ruthlessly destroys cultures. Globalization strengthens the power of this evil civilization, which is cold, rationalistic, and soulless. Granted, the spirit of the West is capable of developing technology, bringing it to a higher level, and realizing great economic successes, but it cannot aspire to the higher things in life because it totally lacks spirituality. It is a hopeless situation since this spirituality is so important, indeed, the most important thing in life. The spirit of the West spreads atheistic scientism, a faith in science and technology as the only way to gather knowledge (Buruma, 2004, 76, 96). For Muslims, Western religion is identical to materialism and this religion is thus in conflict with the worship of the divine Spirit.

According to Buruma and Margalit, the roots of this antagonism to the West lie in the opposition to its "technological culture." The spirit of the West is brain-sick, arrogant, superficial, and hostile to piety. Instead, it is efficient, like a mechanical calculator. Western culture is therefore a non-spiritual, materialistic culture of technical, proud presumption that is power hungry, covetous, brutal, and decadent, which deserves to be destroyed. Suicide terrorism, on this basis, forms the pinnacle of hostility to Western technology. The suicide terrorists, as worshippers of the divine Spirit, carry out the slaughter of the worshippers of earthly matter, under the ambiguous motto (I quote): "Death for the sake of Allah is our highest ambition" (Buruma, 2004, 73; Al-Ansari, 2007). Their struggle against the West is a matter of holy dedication to the cause.

Islamic Terrorism and the Dialectic of Western Culture

In their analysis of Occidentalism, Buruma and Margalit do their best to understand the enemies of the West. They say that without understanding why these Islamic movements hate the West so much, we cannot prevent them from being destructive in the extreme (Buruma, 2004, 17).

It is quite easy to see parallels between Buruma and Margalit s analysis of Western culture and their search for reasons for the hostility to this culture, with what we call in the Reformed Philosophy movement the dialectic of Western culture. It is striking how often Buruma and Margalit look for the causes of terrorism in the internal tensions of Western "technological culture" itself. These tensions have been felt world-wide since the beginning of globalization. While the reactions up till a short while ago remained limited to Western culture, recently there are counter movements to be seen across the world, including the Far East. Islamic jihadistic terrorism is the strongest and most dangerous expression of this. Thereby, repeated use is made of Western cultural criticism. Thus Heidegger's criticism of "technological culture" is popular with radical Islamists (WRR, 2006, 45; Zayd, 2006).

What is the Dialectic of Culture?

What do we mean with the expression "dialectic of culture"?

My inaugural lecture treated the cultural tension between technocracy and revolution (Schuurman, 1973). Since that time, the subject of our cultural dialectic – tension or conflict – has been a recurring theme in my lectures in various different ways. This cultural dialectic reveals to us what is happening in our culture at the deepest level, what the problems are, how serious they are, and also – when we pay attention to the origin and the historical developments of it – how it should and can be resisted.

Herman Dooyeweerd (1959, 10f.) saw the origin of the Western dialectic in the pretentions of a humanity which imagines itself to be self-sufficient and autonomous, a humanity without God. As a consequence of this, the world is accepted as an anthropocentric, closed world, and history is regarded as a purely human history. Because in our culture the openness to a transcendent God has been shut down, human beings – in whatever variations conceivable – see themselves as only involved with a "diesseitige," earthly reality.

People of the West attempt to make a reality of the idea of a self-glorifying autonomy in science, and later, as confirmation, in technology. The thought that human beings and the world can be brought to perfection by modern technology has taken over. This development has set powers loose which have raised the tensions in the world to a fever pitch. The ideal of unimaginable material prosperity may partially have been realized but at the same time, it has become clear that this has taken place at the expense of human freedom and our environment, and that we with our prosperity are living on a volcano which is about to explode. Western culture is a culture which is deeply divided against itself. An absolutized freedom becomes antagonistic to the absolutizing of scientific-technical control and management, and vice versa. This tension works itself out in history.

The Development of the Cultural Dialectic

At the beginning this dialectic – which at its base has thus a religious character – was above all a philosophical, theoretical matter, but since the infiltration and eventual hegemony of the Enlightenment has become dominantly cultural in the broadest sense. It is certainly in the spirit of the Enlightenment to not only know reality but also to organize and give form to it rationally. The idea is to develop a society with the instrument of reason, a society in which human freedom would be able to express itself and become what it wants to be. The actual situation, however, is that when imagined, objective structures developed by autonomous reason are erected, they turn into independent powers in their own right, and as such become enemies of cultural freedom. This threat becomes greater in proportion to the dynamic and complex development of these powers, so that human beings do not have oversight anymore, let alone being able to change this development.

In my lectures on Modern Philosophical Streams, I have always tried to show how the powers of science, technology, and economy are recommended and strengthened by dominant philosophical streams such as positivism, pragmatism, and system thinking. This is so particularly because these streams consider new developments in technology to be necessary as solutions to the cultural problems which have been brought into being by obsolete technology.

Over against this are those philosophical streams which represent the opposite pole of the dialectic. Thus, we see existentialists demonstrating that human freedom is under threat in technical society, where human beings themselves have been degraded into technically manipulable objects. The neo-Marxists have called us to recognize how the economic and political powers exercise a consolidating influence, strengthening the development of science and technical applications, with the result being a threat to human beings as bearers of culture, a threat to human beings in their political roles. The thinkers of the counter-culture call us to pay attention to the suppression of Nature, and therefore argue for a reduction of technical applications which destroy Nature and pollute the environment. New Age thinkers oppose materialism and aspire to a more spiritual way of life, and philosophers of the Green movement emphasize above all the significance of Nature as a totality, over against a massive, artificial, and abstract technology (Schuurman, 2003, 135–161).

Characteristic of our age is how *all* people feel the tension in our technological culture; you could say that they experience this in both body and soul because the tension is growing greater by the minute between an infinite technical drive to expand itself and the finiteness of the creation and its hidden possibilities.

The Primacy of the Scientific-Technical Ideal of Control

That the scientific-technical ideal of control and management continues to win the battle against the other pole of the cultured dialectic— that is, the ideal of personal freedom – can be attributed to the fact that the former ideal makes use of the *objective* cultural powers, which manifest themselves in new scientific and technical possibilities such as system theory, computer science, computer technical applications, and genetic manipulation techniques. Moreover, economic powers strengthen that process. However much the criticism of that process is increasing, a far-reaching cultural revolution is actually almost impossible. The reason for this is to be found in the existence of economic powers which have unbounded ambitions and in the life of the masses of consumers who continue to support the present primary direction of our culture, because they believe in and hope for still more material benefits from science and technical applications.

The Threat of the Current Dialectic

It is necessary to emphasize that in this historical process the cultural dynamic is more and more assuming a malicious, life-threatening character. Modern technology and the use of its possibilities are reaching unheard-of heights and are taking on a despotic character. By means of scientific-technical control and management of the entire world, not only are human beings' freedoms being curtailed, but raw materials across the whole earth are being exhausted, Nature is being rapidly destroyed, and the environment is being irreversibly polluted.

Recently a lot of attention has been given to the issue of climate change. An unbridled scientific-technical dynamic is making havoc of natural, ecological, energy, and social borders, whereby the outbreak of wars between nations is a real possibility (Van der Wal et al., 2006, 223).

In developing countries, because of the influence of globalizing technical and economic developments, strong feelings of political powerlessness are growing, exacerbated by continuing economic malaise and stagnation. This is experienced as a direct humiliation. In other words, the scientific-technical culture of the West puts other cultures under pressure, via globalization. The dialectic manifests itself as a conflict between cultures, peoples, and nations. Through it, cultural tensions can discharge explosively and violent political conflicts can erupt.

The new dimension of the present-day configuration of the cultural dialectic has two components. Until now, resistance remained – as we have seen – limited to *subjective* resistance. Because people didn't have objective, cultural power at their disposal, that resistance could not be realized in terms of a change of the "technological culture," at most only adaptation was possible. The first thing which is genuinely new is that the resistance to the foreign "technological culture" – coming from Islam – is now a reality. It nestles in Western culture, and, at the same time – and that is the second thing which is new – makes use of *objective* cultural power. Terrorism is a true threat (Gray, 2007).

A Western philosopher such as Waskow – a revolutionary utopian – could call us to topple our technical culture by violence in the 1960s (Waskow, 1968). But he could reach no more than a verbal revolution. The present-day terrorists have considerable cultural power and extensive technical possibilities at their disposal and form a world-wide network by means of modern technology – think of the Internet – against which they are protesting, ironically enough. The destruction of the Twin Towers in New York in 2001 makes it clear that they are capable of destroying one kind of technology by another kind. It is this development which is so disturbing.

How do Muslim ideologues react to the current cultural situation?

Criticism from Islamic Intellectuals

Sayyid Qutb (Buruma, 2004, 36, 116 f., 124 f., 131), one of the most influential Islamic, Egyptian thinkers of the 20th century, defended pure Islamic community against a growing Americanism that is conceived as the empty, idolatrous materialism of the West. In the course of his life, he became more and more embittered toward the behavior of the West, and therefore, opposed every possible assimilation. Just as all dreams of total purity are, his ideal of spiritual community was a fantasy which contained the seeds of violence and destruction. He is the founder of the Islamist ideology which carried on a confrontation with the most important ideologies of the West. He responded to Western arrogance with Islamic intolerance (Huntington, 2001, 333). The radical purity of Islam and the destruction of the West are his aims. Thereby, Qutb has become a representative of a radical Islam which doesn't shrink back from violence in its resistance to the West. Instead, he even recommends it explicitly (Qutb, 1990). With him, the cultural dialectic becomes the fuming motor of destruction.

Thankfully, there are also Islamic reformists who seek a harmonious society, such as Mohammed Iqbal. This Pakistani thinker is not enamored of the Occident and expresses criticism of the West from his Muslim perspective. He focuses, in particular, on the visible, unbridled scientific-technical developments in the world, on the financial power of Western capitalism, on the economic exploitations inherent to capitalism, and on the secularism which accompanies it, or is even caused by it. He criticizes the influence of the West, because, in his view, humanity is being pulled away from Allah by the European Enlightenment. Now the West serves idols in His place. Iqbal therefore criticizes Western arrogance, imperialism, and public morality. But in doing so, he does not take a distance from science and technology (Buruma, 2004, 122, 152). On the contrary, he takes the Allah's Unity – so well-known to Muslims – as the basis for his thinking about science and technology. According to Iqbal, that unity must be reflected in human society by harmony that is expressed as justice, equality, solidarity, and care for Nature and the environment. With this he shows his support – in accordance with the spirit of early Islam – for reforms of science and technology (Iqbal, 1971; see also Foltz, 2003). He would like to help reduce the extant cultural tensions.

Points of Contact in Islam

The Pakistani Muslim and Nobel Prize winner for physics, Mohammed Abdus Salam, argues in the same spirit for an acceptance of technology. In a lecture entitled *Science and Technology in the Islamic World* (Salam, 1983), he says that Allah has given everything in heaven and earth to humanity in order to be made use of.

In Salam's view, the Muslim scientist must seek to gain insight into the world, and by doing so, into Allah's plan. Science must be an integral of the human community in order to further material welfare. That's why Salam aims at the universality of science and technology. In order to achieve success in this realm, humans must be grateful to Allah and, as a consequence, obey Allah's will.

Salam wants to return to the first age of Islam, the period in which the torch of scientific and technical development was passed from generation to generation in order to reappreciate the proper motives for the blossoming and development of science and technology. For Salam, Islam is thus *necessary* in order to have the proper motivation and ethics for science and technology. By expressing himself in this way, this Muslim scholar has spoken about the relation between or interplay of religion and technology in a way which is new in the present-day Muslim world, and which in Western Enlightenment thinking occurs too little or even not at all.

Criticism of Technology from Christian Philosophy

We observe that Western culture is judged quite one-sidedly by reformist representatives of Islam. It is historically justifiable to maintain that the Enlightenment has Christian roots. But this cultural movement has pulled itself more and more away from Christianity and now opposes Christianity repeatedly. That's why it is incorrect for Islam not to make much of a distinction between the influence of Christianity and that of the Enlightenment, as if both necessarily would lead to the same ethics for technology (Buruma, 2004). This, while it is orthodox Christianity that makes a trenchant critique of the Enlightenment dialectic, as I have shown.

In the course of the 20th century, both ideals of the Enlightenment – the ideal of personal freedom and the scientific-technical ideal of control and management – have landed us in crisis situations which have had disastrous effects on global culture. This internal Western cultural struggle between freedom and scientific control is increasing. And now a radical and violent Islam is becoming stronger and stronger in its opposition to the West. In other words, Western culture is being more and more internally undermined and externally threatened.

No one less than Habermas (2005) – originally a thoroughly Enlightenment philosopher – has recently shown that a "wrecked Enlightenment" needs religion. Huntington says in his book about the clash of civilizations that the clashes between Islamic and Western culture are above all a result of the weakening of Christianity as the central component of Western culture (Huntington, 2001, 335). The question immediately arises; can a culture, with the loss of its religious roots, survive (Hittinger, 1995)? A renewal of Western culture would mean that people return to the religious source of Christianity and that Christianity understand its cultural calling and follow it.

A renewed Christianity would, on the basis of a powerful faith conviction, call for a modern transformation of Western culture. This call would be responded to from many sources. Already we hear voices calling for change. I think of the work of the Roman Catholic theologian Hans Küng, who writes of coming to a world ethos in science and politics, "a global ethics" (Küng, 1997). As well, world-wide church organizations have already produced reports with criticism of the development of Western culture (Opschoor, 2007).

There is something very valuable in these appeals. Yet, in my opinion, in this analysis our social and cultural problems are attributed too much to dysfunctional, exploitive, *economic* relationships. There is not enough insight into the fact that the ideals of the Enlightenment, which concentrate on freedom, science, and technology, form the basis of the great cultural problems and tensions of our day. The Enlightenment ideals are, in fact, contradictory to each other. How can this contradiction be resolved? Instead of the idea of an autonomous freedom, free from all norms, we must discover and promote a freedom which corresponds to values such as order, discipline, authority, respect, trust, mutual assistance, and human solidarity That is, a freedom which is intimately bound up with responsibility.

A renewal of the motives for science and technology is also needed. The quest to exercise dominating power should be replaced by an attitude which seeks to *serve* in the perspective of *world-wide justice.* The values and norms for technology must no longer be borrowed from the technical worldview, which finally leads nowhere. This insight is needed precisely because technology is foundational for so many cultural activities.

It is clear that it is now quite automatic to seek technological solutions to the knotty problems which technology itself has caused, which in turn, predictably, become new problems and threats. That's why another vision of technology is needed in order to diminish and even find solutions to these kind of problems. The erratic, high flight of technology needs a firm, transcendent anchoring. How can this be achieved?

Acknowledging God as the Origin of creation and humanity as the responsible image of God that has a divine calling to disclose and unfold reality as God's creation, makes the purpose of science and technology subordinate to the divine purpose of the history of the Kingdom of God (Sweeringen, 2007, 271f.). Instead of the primacy of the technical worldview leading us forward, as the Enlightenment posited, the disclosure and unfolding of creation as God's Garden that is destined to become His glorious Garden city, should become primary and beckon us onward (Schuurman, 2005). In short, with respect to the transformation of "technological culture" which is so longed for, Christianity opposes the "religion of the material" just as much as reformist Islam.

It is evident that, thankfully, there are also many outside of Christianity and Islam who realize that fundamental changes in Western culture are needed. They are needed not only because of the inward threats coming from within Western culture itself but also due to the threats which come from the outside, such as radical Islam. Thereby extra support can be expected for a necessary paradigm shift within "technological culture" coming from the ethics of a reformist Islam (Hassan, 2007, Soroush, 2007), in its concern to care for Nature and the environment, and for social justice – in spite of the remaining big differences with Christianity.

Kuhn's Paradigm Theory

In order to have a better idea of what such a paradigm shift might mean, I would like to use Thomas Kuhn's paradigm theory regarding the development of science. Kuhn has made clear that the rise of scientific theories can be explained in sociological, psychological, economic, and even religious terms. By doing so, not only is the continual growth of scientific knowledge explained, but the unexpected "leaps" within science are clarified. The continual development of science demonstrates stability and agreement between scientists. In the case of there being a crisis in a field of scientific development, this leads to adopting a new framework – the paradigm – within which that scientific field can develop further. Only when the new paradigm emerges and triumphs can the rupture around the crisis be healed and a new period for that scientific field proceed. This means that, in the light of such a paradigm shift, the absolute truth claims of science are significantly relativized (Kuhn, 1962).

Kuhn teaches us that, in the midst of crises of scientific theory forming, suddenly big, fundamental questions are asked. The old scientific faith in certain theories shakes on its foundations. What were once commonly accepted ideas are no longer so. The once close-knit community of scientists in a particular field begins to dissolve into factions. Unanimity crumbles away. The "tacit knowledge" of the like-minded begins to waver. In short, the old paradigm begins to be seen as outmoded. A new theory emerges and begins to hold sway (Koningsveld, 2006, 110f.).

Could Kuhn's vision of the shift of paradigms in science be able to function as a model for a necessary change in our paradigm for culture? An analogous model teaches us something useful but has its limitations. For example, science is merely a branch or part of culture. Culture contains so much more than science. But precisely because our culture is more and more seen as "technological culture," or "scientific-technical culture," we are stimulated to nevertheless be inspired by Kuhn.

The Transformation of "Technological Culture"

Isn't something like this – a relativizing of the existing cultural paradigm and its transformation – possible in our conception of the current cultural developments? Within the dominating cultural paradigm of the West, we are faced with many problems. At the moment we're trying to solve these problems with the same means and methods which have actually caused these problems in the first place. Solutions are evidently part of the problems of our culture, particularly when support is sought from the economy and from politics. Slowly we are starting to realize that this is no longer possible. Is there a chance that we, in the midst of this crisis, might be able to find a way forward towards a new phase of culture wherein the problems of "technological culture" can truly be addressed and begun to be solved?

A cultural revolution or cultural about-face, analogous to a scientific revolution, will be necessarily accompanied by tense discussions, which finally must lead us back to assess what people believe and see as being true. Here the role of religion comes in view. Different critiques of culture and technology are being expressed from the standpoint of religion or religions, as we see coming from Christianity and a reforming Islam. The challenge is to see if another cultural paradigm can be formulated. This could limit or even begin to solve the existing problems and threats. That is not easy, for representatives of the old cultural model do not give up easily. They hold on to it, sometimes with the persistence of pit-bull terriers. These opposing forces have an economic, political, and cultural character. But at the same time, as the current developments continue to increase, we see more and more the weaknesses of the old paradigm. Isn't it true that we see an increase of world-threatening events as a result of the current scientific, technical, economic way of thinking?

The Conflict Between Industrial and Biological Agriculture

I believe that there are possibilities for change. One concrete, strikingly relevant example of cultural change, both in the West and in the Muslim world, is the fight which organic agriculture – with or without success, and with or without sufficient arguments – is carrying on against industrial agriculture (Foltz, 2003, 3f.; Petruccioli, 2003, 499f.; Schuurman, 2005, 49f.). Industrial agriculture is the cause of more and more problems. From the side of industrial agriculture, a growing number of opponents of organic farming are making their voices heard, while the supporters of organic agriculture are becoming more vocal as well.

As we see the vague contours of a new paradigm begin to take shape, more successes are being achieved. Conversely, the defenders of industrial agriculture are calling for an ecological approach to agriculture. Both developments show that the existing problems are being taken seriously, and that people are looking for new, sustainable paths to take (Simons, 2007, 240f., 340f.).

A Cultural About-Face

A comparable about-face ought to take place in the whole of "technological culture." The political and economic worlds are opening their eyes to cultural alternatives, sustainable development, and socially responsible enterprises. Our social-economic climate is becoming more favorable to drastic changes. For example, we see how recent reports coming from the business world and aimed at the political arena, are calling for more attention to be given to the environment and climate change (Willems, et al., 2007). As well, a recent United Nations report about climate, coming from a world-wide scientific cooperative project of some 2,500 scientists, labeled human beings, and their technical applications, economies, and consumption as the most important cause of the huge emissions of greenhouse gases, with all the risks that entails (Report of the IPCC, UN Climate Committee, 2007).

As attention is being given to climate change, rising ocean levels, the shift of climate zones, the disturbance of ecological systems, the loss of biodiversity, and new tropical sicknesses, an appeal is being made for a change in our cultural ethos. Bill Clinton and Al Gore have also been calling for changes in these areas. We must not underestimate the influence, in the course of the years, of the Greenpeace organization. More and more people are opening their eyes and discovering a new cultural paradigm. Increasingly, people are seeing that our modern society, with its patterns of producing, controlling, and consuming, is inherently, intentionally, non-sustainable (Van der Wal, et al., 2006, 8 f.).

The existing, old patterns of culture are being undermined by all of this. Political parties in The Netherlands are looking seriously at the idea of sustainability. At the same time, the dominating cultural elites are eliciting more and more doubts about their own character as an establishment. Political forces may eventually work in a positive way to change the cultural disposition of many.

When, in addition, the consumer also begins to get insight into what sustainability concretely means and that the quality of our life is to be benefited by new measures, the conditions are favorable for real cultural change. I do believe that a very necessary cultural about-face is about to happen. And this will hopefully include more attention given to world-wide justice, over against the injustice of the current globalizing developments.

It is thus very important that our post-industrial Western culture attempt to address the threats of industrial culture. It will be a necessary learning process. I think that the increased interest in religion has everything to do with this scenario. Fundamental questions are being brought forward by religious spokesmen after a long period of being disregarded. What is the essence, the purpose and meaning of human life, of culture, of technology, and of the economy? Starting with these fundamental questions – stemming from the religious roots of cultures – lines are being drawn to touch all the different sectors of life which shape culture.

Analogously to Kuhn's terminology, we can speak here of a "gestalt switch," a "turnabout," or "revolution." This means a sudden "leap." Correctly so, for it is "time to turn." We can conclude that, in terms of our Western cultural history, the cultural experiment of the Enlightenment has turned out to be a large-scale failure – however much we may give thanks for its benefits. A radical turnabout of our culture is called for. A new orientation, a new, meta-historical compass, is needed!

The Content of a New Cultural Paradigm

What does the new cultural paradigm look like? What is its essence? It must be substantially different from the previous paradigm, while still absorbing and preserving that which is of value in the old paradigm.

In the old cultural paradigm, Nature is seen as lifeless and mechanical, and within that framework, exploited by means of boundless manipulation. Nature, humanity, the environment, plants, and animals were, in accordance with the technical paradigm, looked at from a technical perspective – the so-called "machine model."

Change is now the order of the day. Now the protection of life itself must become the all-determining objective in the formation of culture. Science, technology, and economy must not be allowed to destroy life in all its variety and richness, but rather they must find their reason for existence in serving life.

However much Christianity and reformist Islam differ from each other – and I must stress how impossible it is to finally combine them with each other – at the same time they have much in common. This enables them to recognize each other as partners, participating in the new cultural about-face (Rohrmoser, 2006). For both Christianity and Islam, the model of the Garden is appropriate to this end (Petruccioli, 2003, 499f.; Schuurman, 2005, 37f.). Together they can agree with the following statement, "We love the whole creation for the sake of the Creator" (Foltz, 2003, 29). The Christian and the Islamic religions can, each in their own way, contribute to a globalizing culture in which life is no longer threatened, but enriched, and in which more genuine justice is put into practice, so that tensions are reduced. In spite of all the major mutual differences, there will be more social cohesion and mutual peace.

If that is not successful – for example, because of a lack of faith vitality in Christianity and/or lack of support from reformist Islam – the struggle between the ambitions of the Enlightenment and radical Islam will intensify and Islamic violence increase. Then there will indeed be reason for pessimism with respect to the future (Bawer, 2006).

Summarizing Conclusion

In industrial society, technical thinking is still dominant. Virtually everything is viewed in the light of the technical model or the machine model. In this model, life itself is missing as a fundamental, decisive factor. The power of technology is worked out in a despotic and destructive way on the basis of this model.

Granted, in the new cultural phase we cannot and will not do away with much of technology itself. But it is clear that technology must function as a servant to life and to society. Reality can no longer be regarded as merely an object of technical manipulation, but must be seen as a creation and a gift of God, which we ought to accept with love and gratitude. This demands from us awe for the Owner, receptivity, modesty, humility, a sense of wonder, reverence, and prudence.

This new approach to technology will be characterized not by destructiveness as a result of seeking domination, but by seeking to carefully disclose and unfold reality, however much they may differ. Christianity and reformist Islam cherish the perspective of the living Garden city, shown us in Genesis 2. In this perspective there is appropriate care for the environment and Nature. The preservation and well-being of life on earth become thus more important than material prosperity.

This new framework for culture, by which life and love become the grounding categories, also seeks to strengthen justice as a mandate, and seeks to honor the supra-subjective, normative boundaries in creation that are established by God. Hereby it becomes possible to reduce cultural tensions and begin to realize a balanced, sustainable, peaceful, and richly varied development. An attitude of moderation will also help to reduce cultural tensions and threats, within both Western culture, and also in relation to Islamic culture. Reformist Islam, when we see its ethical priorities, must be able to be won for such a cultural about-face.

One big consequence is that to the degree radical, violent Muslims refuse to go along with this scenario, they must be isolated politically, and prevented from obtaining the cultural powers of science, technology, and economy, including financial funds, subsidies, and weapons.

I believe that, in this concrete way of following this new paradigm, a more sustainable, just, healthy, and peaceful globalizing development is possible.

Literature

Al-Ansari, Abd Al-Hamid, 2007, *The Root of Terrorism is the Culture of Hate,* http://www.memri.org/bin/opener_latest,cgi?ID=SD162507.

Barbour, Ian, 1990, *Religion in an Age of Science,* San Francisco: Harper.

Bawer, Bruce, 2007, *While Europe Slept,* New York: Broadway Books.

van Bommel, Abdulwahid, 2002, "Islamitisch wijsgerig denken," in *Cultuurfilosofie,* p. 295–340, edited by Edith Brugmans, Open Universiteit Nederland, Budel: Damon.

Buruma, Ian and Margalit, Avishai, 2004, *Occidentalism,* New York: Penguin.

Daiber, Hans, "Die Technik im Islam," in: Stöklein, Ansgar und Rassem, in Mohammed, *Technik und Religion,* Düsseldorf: VDI Verlag, p. 102–117.

Dooyeweerd, Herman, 1959, *Vernieuwing en Bezinning: Om het Reformatorisch Grondmotief,* Zutphen: J.B. van den Brink & Co.

Foltz, Richard C., Denny, Frederick M. and Baharuddin, Azizan (eds.), 2003, *Islam and Ecology – A Bestowed Trust,* Cambridge, Massachusetts: Harvard University Press.

Gray, John, 2007, *Black Mass: Apocalyptic Religion and the Death of Utopia,* New York: Farrar, Straus & Giroux.

Habermas, Jürgen, 2005, *Zwischen Naturalismus und Religion,* Frankfurt am Main: Suhrkamp Verlag.

Al-Hassan, Ahmad Y., 2001, "Factors Behind the Decline of Islamic Science after the Sixteenth Century," Epilogue to *Science and Technology in Islam,* Part II, UNESCO.

Hittinger, Russel, 1995, "Christopher Dawsons Insights: Can a Culture Survive the Loss of Religious Roots?" in *Christianity and Western Civilization.* Ft. Collins, CO: Ignatius Press.

Hoodbhoy, Pervez, 2007, "Science and the Islamic World – The Quest for Rapprochement," p. 49–55, in *Physics Today,* August.

Huntington, Samuel, 1996, *The Clash of Civilizations and the Remaking of World Order,* New York: Simon & Schuster.

Iqbal, Mohammad, 1971, *The Reconstruction of Religious Thought in Islam,* Lahore: Shaikl Muhammad Ashraf.

Koningsveld, Herman, 2006, *Het verschijnsel wetenschap,* Boom: Amsterdam.

Küng, Hans, *Weltethos für Weltpolitik und Weltwirtschaft,* Wissenschaftliche Buchgesellschaft Darmstadt, München: Piper Verlag GmbH.

Kuhn, Thomas S., *The Structure of Scientific Revolutions,* Chicago: University of Chicago Press.

Newman, Jay, 1997, *Religion and Technology – A Study in the Philosophy of Culture,* London: Praeger.

Noble, David F., 1997, *The Religion of Technology – The Divinity of Man and the Spirit of Invention,* New York: Alfred A. Knopf.

Opschoor, Hans, J.B., 2007, "Wealth of Nations or a 'Common Future': Religion-based Responses to Unsustainability and Globalisation," p. 247–281, in: Klein Goldewijk, Berma, ed., *Religion: International Relations and Development Cooperation,* Wageningen: Academic Publishers.

Petruccioli, Attilio, 2003, "Nature in Islamic Urbanism: The Garden in Practice and in Metaphor," in Richard C. Foltz, Frederick M. Denny and Azizan Baharuddin (eds.), *Islam and Ecology: A Bestowed Trust,* Cambridge, Massachusetts: Harvard University Press, pp. 499–511.

Qutb, Sayyid, 1990, *Milestones,* Indianapolis: American Trust.

Rohrmoser, Günter, 2006, *Islam – die unverstandene Herausforderung – Kurzkommentar.* Bietigheim: Gesellschaft für Kulturwissenschaft.

Sadri, Mahmoud and Sadri, Ahmad (eds.), 2000, *Reason, Freedom & Democracy in Islam – Essential Writings of Abdolkarim Soroush,* Oxford University Press.

Salam, Mohammed Abdus, 1983, *Science and Technology in the Islamic World,* Keynote Address delivered at the Science and Technology Conference, Islamabad.

Schuurman, Egbert, 1973, 1980 (2nd ed.), "De spanning tussen technocratie en revolutie," 1973, in: *Techniek: Middel of Moloch,* Kampen: Kok.

----, 2003, (Vriend, John, translator) *Faith and Hope in Technology,* Toronto: Clements Publishing.

----, 2005, *The Technological World Picture and an Ethics of Responsibility,* Sioux Center: Dordt College Press.

Simons, Petrus, 2007, *Tilling the Good Earth – The Impact of Technicism and Economism on Agriculture,* Potchefstroom, South Africa: North-West University Press.

Soroush, Abdolkarim, 2000, *Reason, Freedom, and Democracy in Islam,* Oxford University Press.

----, 2004, *Ethics and Ethical Critiques,* http://www.drsoroush.com/en/category/interviews/.

----, 2007, *Dialogue of Cultures instead of Dialogue of Civilizations,* www.drsoroush.com/en/.

Stöcklein, Ansgar and Rassem, Mohammed (publishers), 1999, *Technik und Religion,* Düsseldorf: VDI Verlag.

Swearengen, Jack Clayton, 2007, *Beyond Paradise – Technology and the Kingdom of God,* Oregon: Wipf & Stock Publishers.

van der Wal, Koo and Goudzwaard, Bob, (eds.), 2006, *Van grenzen weten—Aanzetten tot een nieuw denken over duurzaamheid,* Budel: Uitgeverij Damon.

Waskow, Arthur, I., 1968, "Creating the Future in the Present," in: *Futurist,* Vol. 2, nr. 4, WWR (Wetenschappelijke Raad voor het Regeringsbeleid), 2006, *Dynamiek in islamitisch activisme – Aanknopingspunten voor democratisering en mensenrechten,* Amsterdam University Press.

Willems, Rein, et al., December, 2006. *Pleidooi voor een kabinet met een mondiale visie op natuur- en klimaatbehoud,* Open Brief aan de leiders van de politieke partijen in de Tweede Kamer der Staren Generaal, Den Haag.

Zayd, N. Abu, 2006, *Reformation of Islamic Thought: A Critical Historical Analysis,* WRR-Verkenningen, nr. 10.

THE TECHNOLOGICAL WORLD PICTURE AND AN ETHICS OF RESPONSIBILITY (2022)

Foreword[61]

Since my first appointment, more than thirty-five years ago, as special professor of reformational philosophy[62]' at eventually three technical universities in the Netherlands, the ethics of technology has proven to be an obvious focus for my teaching and writing. From the start, I felt called to address culture-related topics, especially issues related to the development of modern technology. The big questions concerning modern technology deserve a critical review from the standpoint of the Christian faith. Over the years, I have championed the need to evaluate the development of technological practice, especially its underlying motives, given its far-reaching influence.

In this little book, I want to consider the fluctuating state of ethical reflection on technological practice.[63] I have wrestled with this topic for decades, and it is a complicated subject. In this context, I will try to limit myself to the main lines of argument, I hope without doing injustice to the complexity of the matter.

Many, including those in academe, have raised and are raising ethical questions regarding technology. In the Netherlands, for example, the Minister of Education Jo Ritzen in 1986 urged that Dutch technical universities give more

---

[61] I want to thank John Kok for the definitive translation of this book. His precision and feel for both languages, as well as his desire to publish a reader-friendly rendition of my thoughts, improved the text significantly. I also want to acknowledge and thank Alida Sewell and Don Morton for the initial contribution they made to this translation.

[62] In the Dutch educational context, organizations outside any university may request and fund a "special chair" [bijzondere leerstoel] devoted to a framework or ideology or focusing on a topic that is not represented or highlighted within a particular institution of higher learning.

[63] In Dutch, there are two words relating to technology: technologie, meaning the technical science, and techniek, referring to the result of that science, which has been developed so impressively in modern society. Both of these words are commonly translated as "technology" in English, although "technique" can also be used. In this translation, where clarity demands it, I will make the distinction by referring to "the science of technology" or just "technology" for the first and "the practice of technology" or "technique" for the latter.

attention to the ethics of technology. In 1994, the Technical University of Delft published reports on the topic, and the Dutch Royal Academy of Sciences reaffirmed its role in the discussion of the ethical aspects of scientific research. In 1999, the University of Eindhoven likewise took a stand regarding issues related to teaching engineering and ethics. One result has been that for a number of years, ethics has been a compulsory course for all students in Delft. Eindhoven is also testing this approach with some of its students. Moreover, these universities have set up "Platforms for Ethics and Technology" to systematically address the ethical questions engineers have to deal with when doing their work. These initiatives are all very understandable, because engineers, not only as researchers, designers, and developers, but also as managers of technical systems, increasingly have to wrestle with ethical problems.

I approach the ethics of technology in the perspective of reformational philosophy – a normative ethics rooted in a Christian-philosophical standpoint. This perspective differs from other current approaches. Most of the time, people are content to analyze and evaluate case studies and practical examples. Within those parameters, a world-and-life view is touched on at most in a descriptive manner. By contrast, I have a broad, normative ethics in view that contains directives grounded in the convictions of a Reformed Christian world-and-life view. In assuming this stance, my intention is to contribute to the cause of critical discernment – something urgently needed when it comes to the phenomenon of modern technology.

Lest some be put off: Although there are clearly different approaches – different responses – to these questions, the pervasive problems related to technology do provide a common ground. Moreover, there is a general awareness that technology is a shared experience that we may appreciate. Most agree as well that we should strive to do good things with technology (see §13).

1. Introduction

Technology to date has fulfilled many promises. And it offers many more for the future. Unfortunately, people, often mesmerized by technology's phenomenal growth, fail to foresee potentially disastrous consequences. How is this possible? I suspect that our culture is predisposed to see only the positive side of technology. Filled with the spirit of modernity – and postmodernity – most see little need for a critical look at technology.

Is a quiet admiration of technology possibly the very heart of modernity? Technology is often fascinating and exciting. And yet, there is a downside. Many grant that technology contributes in important ways to the quality of life, but few suggest that the opposite could also be true. My sense is that it will become increasingly clear that the key issue in our culture is our attitude toward and our vision of technical development. Unfortunately, few people in our culture recognize this to be a problem because we allow ourselves to be guided by a technical view of reality. Everything – all of reality – is seen and evaluated in light of comprehensive technological control.

Technology helps us to become ever more familiar with the world. At the same time, however, we are increasingly alienated from it because the technical models that guide us when interpreting reality and when changing how things are also flatten and reduce – and sometimes even destroy – that reality. Repercussions of modern technology are manifest in nature and the environment but also in the coldly pragmatic and objective nature of a technological society with all of its inherent tensions.

Even though many reactionary movements have responded to these degradations, an uncritical embrace of the practice of technology persists, creating a great need for ethical reflection. Many advocate an "environmental ethics," an "ethics of development," or "business ethics," but few see the need for an ethics of technology – which to my mind is foundational to all of these issues.

Someone in the reformational-philosophical tradition may not sidestep an ethics of technology. Modern science and technology are part of the Judeo-Christian legacy. Does this tradition not have anything to say about a responsible ethics of technology? This question is all the more urgent because many view that same tradition as at least one of the historic roots of the dire consequences of technology.

In the following chapters, I wish to discuss briefly the need for an ethics of technology (§2), the question of what ethics is (§3), and the spiritual-historical background of modern technology (§4). The spirit of the Enlightenment has produced a scientific-technological picture of the world (§5) with an ethical recipe of intentions, values, and norms that is still current (§6). But there are, it seems, alternative approaches (§7). After drawing attention to the cosmological and the ethical deficiencies in contemporary ethics (§8), I make a case for a different approach that starts with the "enlightenment of the Enlightenment" (§9), that addresses both the cosmological and the ethical deficiencies (§10), and that implies a cultural paradigm different from the current one (§11). I pay a great deal of attention to ethics as an ethics of responsibility and the motives, values, and norms that such an ethic implies (§12). I will also highlight concrete possibilities inherent to this approach that could reorient our technological culture (§13) and that would lead to political changes as well (§14). This new perspective will undoubtedly involve struggle, but it will also afford hope (§15).

2. The Call for an Ethics of Technology

The New Situation: "Technological Culture"

The need for an ethics of technology is not evident to all. But things are changing.

Traditional technologies of the guild and, later, artisans were characterized by an interpersonal dimension. A sense of the whole was apparent. The effects of their technologies were short-term and their negative impact minute and predictable. Furthermore, these technologies did not put their stamp on culture but were a part of it. Handicraft technologies were embedded, as it were, in the natural order. In this rather static situation, there was no demand for an ethics of technology. It was something self-evident and not at all problematic.

Compared to even a century ago, however, we find ourselves in a completely new situation. Modern technology is dynamic and has expanded tremendously. It has left its mark on culture and has become a world-encompassing system. In modern technology, everything is connected to everything else. The result is a technological environment. Take away this technology, and our culture collapses. It has become an essential precondition for our whole life. In particular, the connections between technology, business, and industry have changed the lay of the land. Clearly, modern technology and the economy are tightly interwoven. One cannot do without the other.

Therefore, although I attend to the ethics of technology, an ethics of the economy cannot be separated from it. As we will see, they are connected by an extremely technical mindset, a technological spirit or frame of reference that necessarily generates problems for the ethics of both.

A good degree of uncertainty exists concerning the fast pace of changes in technology and the economy – of which we are a part, but with which we have little experience. To date, we have been able to draw few lessons from the past. Given the problems that we face, however, such lessons would be extremely valuable. Our lack of experience combined with our apparent inability or ignorance about ways to solve new problems makes an ethics of technology challenging. Some people do not sense this challenge because they see no reason to question the inevitability of these developments. But technology surely is not an autonomous process. Technology is a human endeavor for which we are personally and communally responsible – even though the increase in responsibility that technology brings may be difficult to realize and to bear.

Specialization in general, technology's complexity in particular, and the connections of both to still other developments add to the urgency of the situation. Technology has penetrated our individual lives to such an extent that we can scarcely create the distance we need to assess and evaluate it, let alone possibly change its direction. So let us first try to get a sense of the balance between the advantages and disadvantages of technology.

Advantages

Comparing our age with that of a few centuries back, one notices the great advantages of modern technology. The average life span has increased. Sanitary methods and water purification have led to a healthier environment. Mechanization, automation, and robots have relieved people of much hard manual and routine labor. Because of the connection with the economy, material wealth is greater than ever. We gratefully make use of medical techniques, many of which can help heal diseases. Simultaneously, the hunger of many has been abated. Modern means of communication supply us with unprecedented information. In short, the possibilities for shaping reality according to our wishes have increased enormously.

It is no wonder that technology's potential has received laud and honor for some time now. "The Wonders of Technology," "The Age of Technology," and "The Triumph of Technology" are titles of books or slogans from some thirty years ago that point to the abundant blessings of technology. Imagined worlds and the picture of reality many people held to was defined by the possibilities of technology. In other words, technological products increasingly directed the development of new technological projects. Possibility became the grounds for applied implementation.

The Downside: Problems and Threats

In today's culture, however, the darker side of changes in science and technology is evident as well. Influenced by the human desire to master everything, we can and do arrange things as we would have them. Modern technology is beginning to penetrate and direct all of culture. The eventual result will be a "technological culture" in which technology puts a stamp on most everything and most everything becomes dependent on technology. When this dependence is connected with the economy, culture tends to become one-dimensional. Environmental problems arise, and the degradation of nature ensues. Likewise, human development becomes one-sided, and society begins to fall apart.

People discuss the threat of nuclear weapons or radioactive waste from nuclear power plants, the depletion of raw materials, the extinction of many plant and animal species, deforestation, the loss of useable land to salinization and desertification, the depletion of the ozone layer and the increase of exhaust fumes with their far-reaching consequences for life and climate, and the scale of urban sprawl and reduction of arable land. Then there is the growing threat of overestimating genetic modification techniques and underestimating the repercussions of cloning and human gene therapy. Finally, the latest information and communication technologies promise ample information and communication – enough to drown in. Nevertheless, there is actually less face-to-face contact between people than ever before, leading to mutual alienation, loneliness, and social disintegration.

People are hoping to safeguard their culture through a development of technology that knows no bounds; all the while, what grounds their humanity seems threatened with destruction. The brutalities of current technological developments jeopardize the sustainability of the natural environment and of the biosphere. The values that hold here are being ignored. If – as seems to be the case – the disadvantages are going to exceed the advantages, we are going to be stuck with a major ethical problem.

Vulnerable Technology

There is something else. Technological development also faces threats from within. Large-scale technical developments regularly prove to be vulnerable and risky. Due to human error or poorly functioning technology, we are sometimes confronted with far-reaching, unexpected consequences. Think of Chernobyl, the chemical disaster in Bhopal, the "I love you" virus, or the Code Rep-worm. In a similar way, recent terrorist attacks have underscored how vulnerable modern culture can be because of its dependence on technology.

Power over Technical Power

Our conclusion can be that whereas people used to be mainly threatened by the forces of nature, they now also face the threat of a technological attempt to control everything. The pressing question today is whether we can contain and control technology itself.

Given the potential for negative consequences to the risks taken, human responsibility for these moves is becoming unbearable. Have we appropriated more technical power than we can manage? Can we prevent what Albert Einstein referred to as "the degradation of the scientific-technological culture"? These are ethical questions *par excellence.*

3. What Is Ethics?

Now we come to the question of what we mean by *ethics.* The term has many definitions. Fortunately, most overlap, at least in part, resulting in a large measure of agreement despite the differences. Most will agree that ethics is a theoretical discipline that reflects on the good or responsible actions of human beings. Ethics is not so much a specialized science as a multidisciplinary or interdisciplinary endeavor. Ethics has to do with human compliance to all normative aspects of reality.

An ethics of technology must therefore concern itself with people's good or responsible conduct in and with technology as well as with complying with the legitimate motives, sound values, and norms that hold for technology and its use.

Basic questions and answers regarding such issues do differ. What does it mean to be a human being? What are the criteria for good and responsible human actions? Where does technology fit in when it comes to nature and culture? Which norms and whose values hold here? Responses to these kinds of questions ground one's ethics. Differences in this regard make the task of ethics all the more arduous because there is no longer a unity of vision about humanity, history, the meaning of technology, culture, and the future. Differences in these matters have to do with one's philosophical orientation and the different convictions about life and the world in which these are rooted.

Consensus on values and norms may be what is called for, but a diverging pluralism seems to be the reality that confronts us. My position, however, is that in spite of this diversity, a dominant theme is evident in the spiritual background of technological development.

4. Spiritual-Historical Background

When discussing the problems and threats of the Technological Age, people often do not get beyond addressing the symptoms. This discussion needs to be more thorough and to include the deep-seated causes that developed a long time ago.

I would like to call the dominant way of thinking about things in Western culture "technical thinking." The legacy of this approach is made plain when one attends to the spiritual-historical background of the West. As things have changed, faith in the creation – and with it, belief in the Creator – has steadily disappeared. Technical thinking as a means of domination is rooted in the autonomy or acclaimed self-sufficiency of the thinker. It does not recognize the limits and limitations of human thinking.

Descartes is the father of modern thought. Descartes dealt with technical rationality in such a way that especially the natural sciences – and, in line with this tradition, later the technological sciences – were used as instruments with the pretense of putting everything under the control of human beings to solve human and cultural problems, both old and new. Descartes says that the laws of mechanics are the same as the laws that hold for nature. He sees nature as a set of automatons. In other words, nature is made up of material mechanisms. The mechanization of the world picture, to use a phrase of Dijksterhuis, is the result. "Nature is a machine, as easy to understand as clocks and automatons, as long as they are investigated precisely enough," says Descartes. Supposedly, once people know the power connections in nature, nature can be deciphered and directed.

Because, for Descartes, humankind is "*maître et possesseur de la nature*" – the master and owner of nature – humans can rule over nature and control it.

Descartes no longer acknowledges the integrity and intrinsic worth of plants and animals but sees them simply as manipulable things. He is sure that manipulation will put us in a position in one way or another to make advantageous use of these "things." Reality in its manipulability is viewed exclusively in terms of its use to us. The fullness of reality is reduced to the technical use that people make of it. This Cartesian mindset is evident today in bioindustries and genetic manipulation. Technical thinking seems insatiable and increasingly totalitarian and imperialistic. We encounter this spirit already in a somewhat older contemporary of Descartes, Francis Bacon (1561–1626). Bacon is sometimes called "the herald of the new age." With his slogans "Knowledge is power" and "To conquer nature we must obey her," he anticipated everything that was going to be possible in a technical sense. Nature must be forced to serve humanity and, in that way, be made into a slave. Bacon says that nature must "be penetrated to her most intimate core." Instead of seeing through the nature-nihilism of this position, he pleads for man as absolute ruler over nature.

Although Reijer Hooykaas designates this development as "Christian-religious," it cannot be denied that Bacon was driven by godless pride. In his utopia *New Atlantis,* Bacon describes an ideal society in which all power is in the hands of natural scientists and engineers who will make sure that "Progress" happens. He contends that the development of science and technology should be applauded as imitating the divine works of creation. So, too, biblical-eschatological perspectives are reinterpreted into the prospect of progress. Bacon was even of the opinion that science and technology could help humankind rise above the results of the fall into sin. He regarded his plans for the progress of science and technology as a restoration of the power that human beings possessed before the fall. His concern was not to ameliorate or prevent suffering with the help of technology – no, science and technology would be able to repair what the fall into sin had damaged. Themes of creation and redemption become tightly bound into one, something the twentieth-century philosopher Oswald Spengler captures in his pithy description: "Technology is eternal like God the Father, it redeems life like the Son, and sanctifies life like the Spirit."

Technical thinking, once dominant, is unstoppable – it refuses to acknowledge the impenetrable mysteries that most deeply characterize reality as creation. Once in gear, the tireless process of constructing and reconstructing all of reality ensues. What is there that cannot be measured, weighed, counted, and thus controlled? Reality is just one big machine or, to use more modern terminology, just one huge information-processing system.

The notion of technological control arose from the pretended autonomy of humankind, from the claim to absolute freedom— and the assumption that scientific-technical control will enhance this freedom. More and more, problems are presumed to be opportunities for scientific-technical resolution. In a certain sense, only those problems that are recognized can be solved through science and technology. Positivism later declared all questions relating to spiritual reflection and religious problems as nonsensical; they are, therefore, denied. It is not surprising that the technological culture that took shape is accompanied by secularization and a spiritual void on a scale previously unheard of. We could say that hidden behind the facade of modern technology and the mask of autonomous individual freedom is a spiritual vacuum. That people are not inclined to deny this makes the situation even worse. The result is that a technical way of thinking, a technological mindset, pervades the entire culture. Its influence is evident in many sectors of society. In turn, the interrelationships of science, technology, and the economy are likewise influenced by an overextended technical spirit.

The spirit of the Enlightenment, in particular, promoted the influence of the technical control mentality. This movement, which started in the eighteenth century, linked the spirit of the Renaissance – an unlimited confidence in humankind's ability to renew life – with the development of the natural sciences. The pretense of human autonomy, humanity as Prometheus, attached itself to a scientific engagement that knew no bounds. Inspired by the successful development of the natural sciences, heroic Enlightenment figures believed that they would be able to overcome all problems and to renew themselves and society by means of the natural sciences. Because no other norm except the standards of instrumentalistic science itself was recognized, the way lay open for the limitless scientific-technical manipulation of all of reality. This dominating role of scientific thinking meant that every nonscientific authority was dismissed. With that, the definitive breach from God as the Origin of all things was accomplished.

In the course of time, the power of science soon knew no peers. As Christian convictions were secularized and Enlightenment trends were uncritically adopted, the Christian faith was secularized and resistance to the absolutization of science gradually diminished, a thoroughly secular vision for the future gained sway. Given that spiritual climate, positivism and pragmatism easily undid any resistance to the unhindered scientific-technological control of reality. The greater the influence of secularized science and technology have become, the more all of reality is seen as matter-of-factly material and hence as controllable in a completely technical and rational manner.

In a certain way, this course of affairs confirms what Habermas called "the ideology of technology" – an ideology inspired by the Enlightenment, which, as do most, restricts or obstructs one's purview and rules. In this case, fundamental questions about what is behind the development of technology, its origin and meaning, and the values and norms that hold for technology are simply not asked. Modernity and expecting too much from modern technology go hand in hand. We will soon see that many people remain committed to unarticulated, deeply materialistic priorities, values, and norms.

This situation gives rise to problems. Allow me to give a current example. Some think that the greatest danger of the Internet is the polluting of "information" it makes available and that cleaning up the Internet will solve the ethical problems. Yet here, too, technical thinking promotes a certain pattern of behavior that ends up reducing the fullness of life. The patterns of thought built into the computer influence the uncritical user, who will increasingly follow that same pattern. When one's head and hands are busy day in and day out with technology, one's heart will soon be filled with the same as well.

Preoccupation with one thing inevitably decreases one's sensitivity to others. As data and information multiply, their cumulative meaning usually decreases. Contacts increase, but their significance is blurred. Information abounds while wisdom wanes and understanding hardly scratches the surface. Internet use may be on the rise, but spirituality is on the decline. Overestimating technology's hardware ossifies humans' spiritual software. That is why secularization grows as the technological mindset gains precedence.

David Noble's recent book, *The Religion of Technology,* incorporates probing examples of the expectation of technological salvation. He shows that since the Renaissance, many have claimed that technological practice puts us in a position to behave like gods. Technology is linked here, for the first time, with the idea of co-creation and co-redemption. Notwithstanding the continued effect of evil, people in philosophical and scientific circles thought they could restore the original paradise with the help of technology. Technical Man is the new Adam. With this as its starting point, the religion of technology focuses on the earth's future. Technological paradise comes to replace the Kingdom of God and the Christian's hope for the future.

According to Noble, the expectation of salvation through technology lives in all new areas of technological development. This is not Noble's conjecture. With the help of quotations from space scientists, representatives of Artificial Intelligence, developers of cyberspace and virtual reality, and representatives of genetic manipulation, he documents their religious adoration of technology. Limits of space and time are transcended by technology; people strive to achieve machinelike immortality and long to perfect a digital presence of mind that will be omnipresent in the cyberage. Genetic manipulation likewise assures them of a re-created, new humanity.

5. The Technological World Picture

Whatever does not fit into the technological model is usually disregarded or forgotten. As the purview of technical thought expands, the extent of reality shrinks. What remains is taken to be a conglomerate whole that is open for technological improvement. This overextended, technical way of looking at things translates into a technological world picture to which our culture has become enslaved.

This world picture, like the technological developments that it produced, is not static. In fact, the discoveries and innovations, and the technological advancements to which they give rise, render this world picture more dynamic and more easily adaptable. The technological world picture is therefore continually revised by new technological developments. It is, however, a human construct that functions as a cultural paradigm – a type of ethical framework within which people think and act. It sets the norm; priorities, values, and standards of excellence are derived from it.

Whatever science can analyze and explain, whatever it can manipulate, fits into this picture of the world – what science cannot analyze or manipulate does not fit. This picture of the world has with time come to define the development of Western culture, and it continues to characterize the current globalization.

There should be no misunderstanding, however: technology and the technological world picture are not the same. The problems do not lie with technology as such, but with the technological world picture.

This picture of the world, derived from technical developments, has a far-reaching influence throughout and beyond the scope and realm of technology. Not only has it put a stamp on the relationship to nature and the environment, the relationship to human society is colored by it as well. By using technology, it strives to dominate or control both nature and society. Technological-economic powers, in particular, are the driving forces behind this picture of the world, and yet we all breathe its air. We all compromise ourselves with the desire for power and control by being touched as we are by the greed of consumerism.

This picture of the world is actually a scientifically technical take on the world. The picture it presents reflects the image of abstract science, emphasizing functionality, rationality, and universality. As such, it tends to reduce and level out reality. Sometimes its destructive influence even affects nature, from ecosystems to the biosphere, as well as society and the social environment. The ecological crisis has been in the limelight of late, which cannot be said about parallel problems in society.

6. The Current Ethical Orientation: Its Priorities, Values, and Standards

The technological world picture generates a breadth of problems. That is not all. It is also usually decisive in defining desirable ethical solutions. Because of the technical way of thinking, there is a good deal of coherence in the most current approaches to these matters. In other words, the technological picture of the world defines contemporary ethics as well.

It is difficult to keep oneself from conforming to a technical systems approach to ethical questions. Current discussions in the ethics of technology are, generally speaking, limited to calculating precautions for behavior with an eye to reducing risk. I have sometimes called this "technical ethics." Ethics becomes a technique because people want to streamline and guide the technical development. In the ethics of technology, the "control-technical perspective" is then dominant. People restrict their attention to the adverse symptoms of an otherwise limitlessly developing scientific-technical control. In doing so, this ethics does bring relief to some of the problems technical developments create.

Changes in existing developments, the search for alternatives, and proposals to reject earlier decisions seldom occur. People have become rather entangled in technology. Many may wrestle with that fact but do not really know which way to turn. Information and communication technologies do not help matters. It is increasingly difficult to adopt another starting point, a different picture of reality with different priorities, values, and standards. In short, industrial and postindustrial societies are permeated by strong technical values, attitudes, and ways of thinking – few of which are being questioned critically.

Attaining power over reality is the implicit priority of this ethical stance. It follows closely on the heels of technological innovation. Values behind this project include economic self-interest (greed) and an across-the-board increase in consumption. The presumed outcome will find humankind front and center, in control, as lord and master of technical progress. What this will do to us as individuals or as a society, let alone how this will affect the environment, are questions few take time even to ask.

The norms that follow from the values of the technological world picture are effectiveness, standardization, efficiency, success, safety, reliability, and maximum profit, with little or no attention given to the cost to humanity, society, the environment, and nature. In summary, the first and great commandment of "technological culture" is, "Be as effective as is technically possible," and the second like unto it is, "Be as efficient as is economically possible." The breadth and depth of a technological-materialistic culture hang on these two commandments.

Material values and standards clearly have the upper hand in the technological world picture. Given recent degradations of nature and related environmental problems, however, some are convinced that these values and standards, which continue to control culture under the banner of "progress," need to be transformed with an eye to "survival." That said, adjustments to date have come after the fact and are seldom more than politically correct.

We continue to encounter more problems in which the technological world picture and the ethics that accompany it fail us. This is especially clear from problems related to preserving biodiversity and sustainability. Recent decreases in biodiversity are shocking. Within the time span of a single generation, the number of species has been halved. This must surely be the result of looking at living reality from a predominantly technical perspective.

Sustainability ought to satisfy the requirements of the present generation without jeopardizing the ability of future generations to provide for their needs. Why is sustainability under pressure?

The prevailing technological world picture and its model of control dominate today's economy as well. Lopsided growth is engrained in the process from the start. As a result, sustainable development is by definition out of the question. Environmental technology may make some steps in the right direction, but these are often subsequently undone or negated by the technical economy, which provides the infrastructure for these environmentally friendly technical innovations. The technological world picture also stands in the way of resolving growing concerns about climate change. Our way of dealing with creation prevents us from gaining a new perspective within which to revise the problematics of this impasse.

The current cultural picture continues to be fed by a technological expectation of salvation. Spiritually, the focus is on technology. Discussion of basic assumptions and questions about meaning are usually excluded, and reality is reduced to a reality that has to be controlled. The guiding principle is the picture of a technical construction that continually increases in strength. Reality has no essential value. The focus is always on its instrumental value. Plants and animals are prized primarily for their material use to us in science and technology. Even human beings are increasingly considered remake-able.

Werner Heisenberg has drawn an impressive picture of this situation: "In what appears to be its unlimited development of material powers, humanity finds itself in the position of a captain whose ship has been built so strongly of steel and iron that the magnetic needle of its compass no longer responds to anything but the iron structures of the ship; it no longer points north. The ship can no longer be steered to reach any goal, but will go round in circles, a victim of wind and currents."[64] We have abandoned our culture to just such a lack of orientation. Technical power has undoubtedly increased, but the threat of devastation has also increased. Technological advancement as such is turning

---

[64] Werner Heisenberg, *The Physicist's Conception of Nature*, trans. Arnold J. Pomerans (New York: Harcourt Brace, 1958), 30.

against man and his environment. These threats are frequently veiled by the vaunted superiority of technological effectiveness and economic efficiency. The ethical reduction these involve is scarcely discerned.

7. The "Empirical Turn" and Postmodernism: An Intermezzo

The recent, so-called "empirical turn" in the philosophy of technology[65] quite rightly challenges philosophers like Martin Heidegger and Jacques Ellul, who are of the opinion that the development of technology is an autonomous process; in this sense, people contribute to technological development, to be sure, but have, in fact, very little say over it. In underscoring this point, Heidegger and Ellul frame these developments as our inevitable fate. Thinkers of the empirical turn are correct in resisting this kind of fatalism. They concentrate their ethical reflections on specific problems, "case studies" in the practice of technology, which they abstract from the length and breadth of the history of technology. They pay very little attention to the structural development of technology or to a situating of technology within the whole of reality. That neglect is regrettable, because there is coherence both in the development of technology and in ethical problems, although this coherence takes on a special coloring in particular cases.

Attending to differences within technology is their strong suit. Not all ethical problems are equally urgent. In the reformational-philosophical conception of structures, due consideration is always given to that diversity. The idea of the autonomy of technology as a massive and impenetrable phenomenon, in which there is no place for responsibility, is also discounted. Reformational-philosophical teaching emphasizes, moreover, that technology takes place in a historical, cultural, social, and political context and that various groups within these contexts actively pursue various interests and goals. Each context or group influences the development of technology, but they cannot cut themselves off from the continuity of this development. We should consider whether the same spiritual-historical background that is an important cause of ethical problems conditions them all. Philosophers with an empirical bent too easily dismiss this issue.

---

[65] See Peter Kroes and Anthonie Meyers, eds., *The Empirical Turn in the Philosophy of Technology* (Amsterdam: Jai, 2001).

The current formula for dealing with ethical problems is to investigate cases in order to derive a rule of thumb for the future. People seldom go back to the roots of a problem or to the connection between problems. By combating symptoms, people merely displace problems and ignore their common roots. From an ethical point of view, they remain stuck in the "labyrinth of the technology."[66]

Only when we see technological development as a whole do the separate practical examples receive their specific place. In my opinion, it is important that we do justice to both the general basic structure of technology as well as the individuality of the separate technical phenomena. Concentrating on just one of the poles does an injustice to the other. By attending only to the individual problems (the trees), one loses sight of tendencies in technological development as a whole (the forest). Analysis that is restricted just to the players who have some influence on the technological development ignores common motives that are operative at a deeper level. Superficiality and a lack of coherence are the result.

It is possible to attain a deeper understanding of particular cases by seeing them in total and by not just connecting them with a description of what people have actually done. In a philosophical-ethical reflection, we are concerned in the first place not with the ethics of particular technical phenomena but with an all-encompassing ethical approach via cultural pictures, ethos, motives, values, and norms on the basis of which certain techniques can be evaluated later. We prefer to grasp at empirical examples so as not to have to pay attention to structural roots and backgrounds. The specialization prevalent in our time promotes this approach because it means that we know more and more about less and less. That is why we have become vulnerable in our appreciation of the whole situation. In other words, attention to individual technical phenomena can distract us from the main concern, meaning, the ethical quest for a reorientation such that technology will have a place and value in culture different than is currently the case. We can contribute to this reorientation on a small scale only if we know the priorities, aims, values, and standards that hold at the macrolevel. Or, to say it as a variation on a well-known slogan: "Think universally (globally), but act individually (locally)." It is not possible to evaluate the place of particular

---

[66] See Willem H. Vandenburg, *The Labyrinth of Technology* (Toronto: University of Toronto Press, 2000).

technical phenomena without a thorough understanding of technology as a universal phenomenon. One without the other leads to misguided appraisals. In other words, particular technical phenomena must be appraised in the light of the general phenomenon of technology, and vice versa.

There seems to be some connection between the so-called empirical turn and postmodernism. Postmodernism arose as a protest against the metanarratives of the Enlightenment and in reaction to the problems of modern technology. That is why it is sometimes said that postmodernism is technologically pessimistic. At the same time, it still draws from the roots of the Enlightenment and, in that regard, may be characterized as hypermodern. Postmodernism puts more emphasis on one pole of the dialectic of the Enlightenment. It chooses freedom over control, and individuality beats out universality. As a result, postmodernism shows more fragmentation and discontinuity than coherence and continuity. Both postmodernism and the philosophy of the empirical turn lack a coherent view of reality and promote an ethical relativism. The attention shifts, as the postmodernist Jean-Francois Lyotard has taught us, from the ends told of in grand narratives to means for action.

Remarkably enough, recent developments, like information technology and nanotechnology, encourage postmodernism because they provide more, very concrete ways in which to individualize the technology. Assuming that this happens in a sound ethical context, doing so is an improvement on how things used to be.

By attending to specific technological phenomena, the empirical turn in philosophy as well as postmodernism have made important contributions, but by failing to deal with modern technology as a whole they have left the dominant ethics of technology that I have been sketching pretty much as it was. There are many examples of this, but I will limit myself to one.

Recent environmental technology has helped to make hothouse horticultural production cleaner than before. At the local level or microlevel, compared to the past, that is an improvement. To realize this improvement, however, requires so much additional investment up front that it simultaneously leads to an expansion in production – such that ultimately the energy taken from the source and the concomitant pollution increases correspondingly. At the macrolevel, when one factors in all of the links in the energy chain, the situation could turn out to be worse than before. Being blind to that fact is the result of reducing the ethics of technology to an ethics of particular technological phenomena.

8. Cosmological and Ethical Deficiencies

I now return to the main line of my argument. Life as many understand it today was shaped and nourished by the spirit of the Enlightenment. Much good can be attributed to that spirit, but also much that is evil. In my estimation, current views about technology, generally speaking, suffer from a cosmological deficiency and from an ethical deficiency. Conceptions about the cosmos are often very limited because justice is not being done to the multifaceted depth and breadth of reality. Reality is often reduced to the world that science and technology aim to control – to a positivistic cosmology, a view of the cosmos to which technology is the key. This lopsided take on the world does not do justice to the many-sided dimensions and coherence of reality in its fullness and pays no attention to its dependence on and orientation with respect to its divine Origin, no heed to the transcendental direction of everything.

In addition to a cosmological shortfall, there is also an ethical deficit. The world around us is taken to consist of things to be manipulated. Scientific-technological thinking reduces everything to the status of useful object. The unique value and meaning of things is dissolved into the use or benefit that that reality has for humankind. This ethical deficiency can best be characterized as the lack of love, because justice is not done to the peculiar nature, individuality, or uniqueness of things. That is evident today in how the technological model dictates how we deal with animals. We see animals more and more as production units that supply the technologically defined functions we say we need. The ideas behind the therapeutic and reproductive cloning of humans also fit in with this technological world picture.

The German philosopher Peter Sloterdijk provides a second illustration. He maintains that the influence of the Enlightenment in shaping humankind has not gone far enough and actually cannot do so. Human formation needs to be augmented with technological innovation. Developments in the area of genetic manipulation make that possible, and Sloterdijk is convinced that we should move in that direction.

These things being the case, what constitutes a critical approach to the technological world picture?

9. Enlightening the Enlightenment

In general, because of their high expectations, people are oblivious to the deepest background of today's "technological culture" and the prevalent ethics of technology. The spirit of modernity seems to coincide with unrestrained technological development. Yet, as a result of the tangible problems and threats tied to that development, we are confronted in the meantime with all kinds of protest movements. Even so, the majority of people still support the Enlightenment project.

A thoroughgoing evaluation of "technological culture" cannot avoid dealing with the Enlightenment. To suggest that we are dealing with nature and society in too scientific and technological a fashion is actually to take issue with the Enlightenment's abstract postulate of autonomy. The crisis is imminent: It is increasingly clear that our culture cannot handle both absolute freedom and absolute controlling power.

The great philosopher Immanuel Kant answered the question as to what the Enlightenment is. People of the Enlightenment have come of age and do not accept any guidance from above: "Have the courage to avail yourself of your own intellect." Kant is concerned not only with the growth of knowledge or a spontaneous act of the will to liberate oneself but especially with the courageous decision to control praxis by means of scientific knowledge. Human reason is accepted as the controlling instrument: People set out to re-create the world as they wish it to be by means of science and technology. The spirit of the Enlightenment connects itself via technology and the economy with happiness and freedom, with optimism, progress, and utility (but closes its eyes to the possible ill effects of its striving).

Many current cultural-philosophical critiques highlight the shortcomings of the Enlightenment. People are increasingly convinced that its instrumental rationality has and will continue to have devastating results. Technology is no longer the liberator, but itself stands in the service of power over humans and nature and, as such, binds humanity, destroys nature, and threatens culture. It is no wonder that philosophers of culture continue to discuss the meaning of the Enlightenment, but few are ready to let go of its point of departure. To ward off criticism of the practice and science of technology, people try to tweak what the Enlightenment is. Some, like Theodor Adorno and Max Horkheimer, want to unpack and deepen the Enlightenment. Others, like Heiner Hastedt, champion a broad outworking of the Enlightenment project, one that will include, for example, a new ecological ethics and an ethics for the management of systems technology. Yet none of these adjustments involve abandoning the autonomy of the scientific-technological person. It is almost impossible for these thinkers to relinquish autonomy. Instead, they seek to expand the *ratio* to a fuller or broader, multisided rationality that covers more areas. The technological world picture is constantly being adapted. Even when people plead for a second Enlightenment in which attention is given to metaphysical or spiritual questions that have been neglected in the course of time, they remain loyal to the Enlightenment's point of departure.

Reformational philosophy stands in a tradition that grew out of a fundamental critique of the world-and-life view of the Enlightenment, particularly, of the pretense of human autonomy and the intellectual hubris and will to power that are connected with it. Although we cannot turn the clock back on the Enlightenment, we must acknowledge its devastating consequences and find a revised ethical framework for its positive results. Addressing the cosmological and ethical deficiencies will also require a different perspective. With Günter Rohrmoser, I advocate the "enlightenment of the Enlightenment"; to put it in terms similar to Psalm 36:9, "In God's light we see light." The divine light of Revelation must enlighten the "Enlightenment" itself. In the light of God's Word, a path can be found between technological paradise and technological apocalypse or, better said, a way that rises above that dilemma. To refer again to Heisenberg's metaphor of the ship: If the captain wants to set his ship on the right course, he will once again have to orient himself with respect to the starry heavens. So, too, a technological culture needs to be evaluated with reference to viewpoints outside technology.

The central point of the enlightenment of the Enlightenment is that we acknowledge that we live in a created reality, in the context of which a breach occurred between God and humankind; we acknowledge as well, in the perspective of the Kingdom of love and peace, that restoration has been made possible in Christ. His is a Kingdom in which nature and culture will be filled with the glory of God. This religious recognition cannot but throw new light on the ethics of technology.

10. Covering the Deficit

When we acknowledge that reality is a created reality, we know that the cosmological and ethical deficit resulting from a reductionistic-scientific approach to reality cannot be solved by more science or more technology. For example, however much systems analysis is presented as a holistic approach – and its merits should be appreciated – it remains an abstract scientific approach rooted in an anthropocentric outlook. Human beings are no longer seen as "lord and master," but when push comes to shove, it is still they who make the final decisions.

What we need is a vista and a vehicle that embraces more dimensions, something that provides a more comprehensive, holistic approach. People need to acknowledge that the breadth of reality is a given, bestowed – including for scientific analysis – long before science ever was, and that this reality does not depend on itself, but is in all respects dependent on and involved with God as its Origin. A cosmology grounded in such a vista clearly will lack the deficiencies referred to earlier.

The most intimate involvement of God with created reality is characterized by his love. Accepting this unity in love covers the ethical deficit of love. With good reason, the command to love God and one's neighbor is at the core of all of the directives, commandments, values, and norms in the Christian religion. All of the law and the prophets hinge on this love. This twofold love must be the starting point for an ethics of technology. This means that everything must be evaluated from the get-go according to its individual nature, including that which is weak and vulnerable. With love as our starting point, we must acknowledge that every created thing is characterized by earthly and – by even more than earthly – divine secrets. This affirmation means that besides attending to technology's values and norms, we will pay attention to ecological and social – contextual – values as well.

Our task must be to trace what the covering of the cosmological and ethical deficit means for a responsible development of technology.

11. A Renewed Cultural Picture

Is there a picture of reality more basic than science and technology (or the economy) provide that can help us understand how to reorient ourselves with respect to technological development? The cultural philosopher Hans Jonas can be helpful here. Imagine, he says, that we found ourselves on the Moon. We would be impressed by the vastness of the cosmos and, even more so, by how unique the Earth is compared to everything else there is to see. It is the only green planet in our solar system, filled to overflowing with a rich diversity of life. If we Moon travelers are to survive, we will have to return to Earth. But to our horror, says Jonas, we find that planet Earth is in danger, with so many of its life forms threatened by current technological-economic practices. Something is going to have to give. Technology and the economy may not threaten life. They must be used to serve life.

Responsible cultural development evokes an image reminiscent of the earth as a garden, tended by humans with the goal of creating a "communal home" within which nature, technology, and culture are in harmony and there is a meaningful place for everyone living and everything that lives. Foremost in that picture is an integral coherence in which everything participates, all the while retaining its own individual value or nature. Before getting involved in scientific-technological activities, one needs to respect this intrinsic value of things. Every human activity should begin with caring contact and respectful treatment. Creation and its creatures have to be dealt with according to their nature; otherwise, life will disappear. This is no idolization of nature; on the contrary, it is acknowledging the care of the Creator, a care we are to imitate. Technology and the economy ought to be directed to inhabiting the garden and to maintaining and strengthening every living thing.

The metaphor of a garden developing in the direction of a communal home also expresses the human connection with and dependence on the whole of the creation. Reality is given to us. We are not its lord or master, but creation's guardians and caregivers. We are allowed to unpack and unfold creation. Just as we carefully unwrap a precious gift, we should treat the gift of God's earth with a sense of awe and gratitude. A change in attitude and behavior is called for.

This image of the garden is also clearly linked to the original meaning of *oikonomos (oikos* house + *nemein* to manage). Caring, keeping, cherishing, and protecting go hand in hand with cultivating, harvesting, and producing. In the cultural paradigm of the managed garden, accelerated increases in scope and scale will be transposed to levels that will benefit the coexistence of human beings and the creation. The carrying capacity of nature will be respected and the goal of long-term occupancy set the direction for sustainable cultural development. Sustainability is possible within the metaphor of the garden. Technology, along with the economy, need not travel the road of manipulation, extortion, and pollution. As World Bank economist Herman Daly puts it, they should maintain the fruit-bearing capacity of the earth and, where possible, increase it. What we take from the earth should be offered to all human beings but limited to what we can use and enjoy, now and in the future. Responsible cultural development means living off the interest on the capital given to us, but it does not allow us to touch or deplete the capital sum itself. This notion of living off the interest fits well with human beings as stewards. (And even though some refuse to use the term, what it entails is nevertheless often very attractive to their enlightened self-interest.)

The cultural picture sketched here differs from the current one and calls for a fundamental reorientation of the technological-economic order. It allows room for growth, but growth that is more proportional and selective. Besides the technology and economy involved in building and producing, more attention needs to be given to maintaining, protecting, conserving, guarding, and caring for – in a word, preserving – the diversity of life forms in the plant and animal kingdoms. Ecology, technology, and the economy will be in equilibrium as long as the natural cycles are not broken and the natural resources do not dry up. The whole earth will come to be seen as one big garden city.

While we must hang on to the original image of the garden that is unfolding in the direction of a communal home, it is also true that the conditions under which human beings are allowed to work in the garden were severely altered by the fall into sin. Since the breach between God and those who were to image him, thorns, thistles, and death abound, and their effects are evident in technology, too. Through God's love in Christ, there is a new perspective for the sin-marred creation. The meaning of it all beckons: the Kingdom of God. Acknowledging this implies struggle. This struggle is inherent to the human position. To orientate oneself to that Kingdom differs enormously from the materialistic and hedonistic attitude of our age: "What good will it be for a man if he gains the whole world, yet forfeits his soul? Or what can a man give in exchange for his soul?" (Matthew 16:26).

12. An Ethics of Responsibility

Which ethical approach is best suited to cover the described ethical deficit of love in the current ethics of technology?

Deontological and teleological approaches to ethics are centuries old, but they no longer fit the dynamic and complex phenomenon of modern technology. Technology is no longer only characterized by the relationship between people and their tools. Due to the influence of science on technology, it has become a dynamic system with global ramifications. Modern technology is likewise intertwined with big business. This complex of scientific, technological, and corporate powers has become a dynamic force with so many members in the cast that an alternative to the older ethical approaches is called for.

Furthermore, deontology, which is an ethics of moral obligation, eventually has resulted in a more pragmatic or even pragmatistic ethics that has relativized what were formerly self-evident norms. Likewise, given technology's many unintended and deferred adverse results, the teleological approach, which evaluates actions according to their consequences, is also not up to the task: Not only must one attend to ascertaining the appropriate goals, but the many possible means to those ends need testing as well. As a result, some have turned to a "game theory" of ethics that allows even basic rules of the game to change and differs little from the ethics of pragmatism. In a certain sense, both approaches eventually culminate in what I have termed technological ethics (see §6).

An ethics of responsibility is, in my opinion, the most suitable approach for an ethics of technology, because it integrates ethos, intention, values, and norms in a coherent way.

Many assume that the ethics of responsibility has philosophical roots, but the theologians H. Richard Niebuhr and Karl Barth were writing about an ethics of responsibility long before Hans Jonas. Already in 1948, when the World Council of Churches was established, an ethics of responsibility formed the guiding principle in discussions about matters relevant to society.

The word *responsibility* – in the sense of accountable *for* and accountable *to* – is very apt because it also indicates that everyone involved in scientific-technological development must act as proxy or steward with reference to one another. In other words, every stakeholder must indicate the priorities, values, principles, and norms – the cultural picture – that constitute the ethos and grounds for one's actions and define one's contribution to the scientific-technological event. As a result, the ethics of responsibility nurtures a positive sense of vocation or calling. In current discussions about problematic developments, ethics is usually associated with "what ought not be allowed," whereas in an ethics of responsibility, one has to begin with an emphasis on the positive. For example, given new technological means to alleviate human needs and suffering, the ethical sense of possibly helping has become one of ethical obligation. In general, a good starting point for an ethics of responsibility seems to be that the participants are aware of the positive tenor of their actions in or with technology and give account of the same to the public.[67] In the image of the garden that is being developed into a communal home, the first concerns are to make and keep it habitable, to provide the basic necessities of life, and to alleviate the needs and suffering of all people.

I would now like to turn briefly to the implications of an ethics of responsibility for priorities, values, and norms.

Renewal of Priorities

We have seen from its historical-spiritual roots that the ethos pervading technology today takes power and control as absolutes. The same is true in those sectors, like agriculture, politics, and the economy, where technology holds sway. In science, this ethos is unpacked in terms of "knowledge is power." In technology, this becomes technology for the sake of technology, or technological perfection – what can be made must be made.

In industrialized agriculture, harvesting with unbridled scientific-technological power eventually leads to exploitation and land degradation. Materialistic politics and economics, in which people are only concerned with the power of money and material gain, weave cultural powers into a mutually dependent conglomerate. This convergence of powers proves to have a disruptive effect on both nature and culture. To think that these powers could serve other values than increased power, expansion, and intensity is a delusion.

[67] This is not to say that "responsibility" has the same content or profile for everyone. There are different kinds of responsibility: substantial, functional, individual, and professional responsibility. See also J. O. Kroesen, *Ethics and Technology* (Delft: TUD, 2001), 12–22.

When it comes to cultural activities undertaken within the perspective of developing the garden into a communal home, one should turn away from oneself out of love for God and one's neighbor. A healthy ethos like this affords a broad field of cultural activities a different sense of priority and even of content than is currently possible. Instead of encircling the ethos of power in self-interest around oneself, the ethos of love has a referential focus that draws one beyond oneself and into a variety of different cultural activities. In science, growth in wisdom becomes the objective; in technology, building and preservation; in agriculture, conservation and care as well as harvesting; in the economy, stewardship; and in politics, service and the promotion of law and public justice. This kind of differentiation helps culture flourish meaningfully and protects the variously qualified responsibilities in these diverse cultural activities.

I now turn to what these authentic priorities mean for science and technology.

Science: Growth in Wisdom

The desire for technological control conditions modern science. The results of its applications do not make science technological; science is technological because it sees reality only to the extent that it is quantifiable and predictable. It is only interested in organizing and controlling reality. Even though curiosity and awe drive the aspirations of many scientists to know reality better, my thesis still stands: Modern science became technological in core and character because of prevailing cultural powers and priorities.

For one's scientific endeavors to stand in right relation to the fullness of reality we experience daily, one must first acknowledge the Origin and meaning of reality and reject an instrumentalist view of science. Science needs to be integrated into that full experience of reality, thereby deepening experiential knowledge. When that happens, science will serve the cause of growing in wisdom.

When approached in this way, science promotes increasingly comprehensive insight. Reality will no longer be reduced to logically independent, causally related factors or subjected to a human definition of the meaning of reality, for example, the benefit that functional reality affords a materialistically oriented humanity. In brief, science will then contribute to a comprehensively wise insight and enhance human responsibility with a view to directing everything that is going on in the world into a garden that is developing in the direction of a communal home. Seen in this light, an interdisciplinary approach is also highly desirable! Too few recognize that technology needs a more comprehensive scientific basis. A broader basis will facilitate a growth in wisdom, lead to more creative and circumspect actions, and help modern technology better serve life. If a more interdisciplinary approach to technology were in place, biology and ecology would have been accepted as foundational sciences long ago.

Technology in the Service of Life: Technology as Prosthesis

What I have said of science in general also applies to technological science, to technology: It should not be allowed to serve as an instrument of scientific-technological control. When it does, technology is robbed of its distinct character or meaning. The practice of technology should not be (as oft is said) the result of an instrumental use or application of science. That route will more readily blind and derail than provide room for the responsible development of technology.

At the same time, instrumental views also diminish the place of invention in technological development. Invention is in many ways the heart of technology. Science ought not to put a limit on human creativity, but old and new scientific knowledge can nourish and even foster human creativity. When training engineers, more attention should be given to responsible creativity through invention and innovation. Doing so will render technology more serviceable to life.

Technology's number one priority should be serving life and society. Technology must function, so to speak, as prosthesis,[68] both individually and collectively. Then people will (continue to) have a say in the matter. Nor is small-scale technology the only option. For example, when it comes to transporting goods, building underground is the way to go wherever possible, especially if doing so will preserve the environment and cause less disruption to society. Just one example of thinking outside the box and doing the right thing with respect to safety, the environment, and nature, would be the causeway built across the East Scheldt (an estuary in the Dutch province of Zeeland). This half-open, always

---

[68] "Prosthesis" narrowly defined means an artificial device to replace a missing part of the body. A pair of eyeglasses is one example.

closable dam is designed to protect the coastline during storms, but in the meantime, it allows tidal sea life to flourish.

Yet what about the costs? I can hear people protest. They will indeed be higher. In general, we have been getting modern technology too cheaply. We are good at figuring out the economic – or rather, the production – costs, but we usually fail to calculate in the damage to nature and the environment. Alternatively, to keep costs down, we take risks, for example, with respect to safety, that are greater than they ought to be. These are some of the consequences of accepting the technological world picture as a guide.

Modern technology too often neglects the criterion of service. What it delivers does dazzle, but as the complexity of it all increases, overconfidence does as well. We need to be more wary when turning to technology and to work harder at resisting its temptations. Instead of being haughty, we need to become more humble. Timid awe befits us better as we involve ourselves with God's creation. Technological formation should help us cherish life.

Other Values: Ecological, Technological, and Social

Besides having the right posture (ethos) and sense of priorities, the preeminent ethical challenge for a responsible, well-directed technology is defining the values one embraces with respect to nature and the environment as well as to technology, the economy, and society.

Principal ecological values must certainly include preserving biodiversity and clean water and air, keeping and making the soil fertile, and improving the living environment. The biosphere must remain unharmed; therefore, a war must be waged against dangerous emissions. Technology must adapt to natural life environs, not crushing the diversity found there but maintaining it.

Technical (and economic) values include being fit for habitation; being safe and reliable; providing basic necessities for life, such as food and health; battling sickness and suffering; countering threats from nature; fostering healing and sustainability; lessening the physical burdens of work; and so on. Beyond our physical wants, real fulfillment is found in spiritual growth, personal relations, and communal life – where technical values also touch social values.

The social values are those of community spirit, sobriety, justice, civility, mutual care and respect – of improving information, communication, and therefore social welfare in general. Is it too daring to suggest that "rest," "having time," and "spiritual flourishing" should be mentioned here as the forgotten social values of technology?

An Integral Framework of Norms: Simultaneous and Multifaceted Application

In my discussion of technological development, I have been focusing on the cultural picture – the ethos, priorities, and values that have to be considered: Technology must be serviceable to a great diversity of life forms and befit a responsible garden development, always with an eye on the current situation.

In addition to connecting with nature and culture and being defined by the right ethos and appropriate priorities, one needs to work at keeping to the normative course. A good number of normative principles and related norms are necessary to test the correct direction. These normative principles concern not only technology, but also the multifaceted relationship that technology has with people, nature, and society.

The integral framework of normative principles derived from the philosophical cosmology (or theory of structures) articulated by Reformational philosophy constitutes a guide for responsible technological development. These norms include: cultural-historical norms; the norm of effectiveness; the norm of harmony between continuity and discontinuity, large-scale and small-scale projects, integration and differentiation, and universality and individuality; and the norms of clear information and open communication (among all participants); of harmony between people, technology, nature and society; of stewardship and efficiency; of always doing the right thing for all concerned, including nature and culture; of care and respect for everything and everyone involved in technological development; of service, trust, and faith. Within that framework, we distance ourselves from the suggestion that because "safety does not sell," irresponsible risks are an option.

Honoring such a normative framework[69] ensures that a one-sided technological development makes room for a responsible, richly varied disclosure of nature and society. The practice of technology may not oppress but must serve

---

[69] For more on this integral framework of norms, see my *Faith and Hope in Technology* (Toronto: Clements Publishing, 2003).

nature and society. Not a one-sided or one-dimensional technological culture but a rich flowering of culture should be our aim.

Following this extensive ethical approach implies a broad point of view and, at the same time, a steady course.

13. Consequences of a Cultural Reorientation

I now turn to some of the implications of this new approach to culture. To this point, I have set two different perspectives as ideal types over against each other. In reality, however, the differences are less stark. For example, when acknowledging that reality is a created reality, there is a clear relationship. The technological world picture is a parasite when it comes to created reality. It causes disturbances, but it can never distance itself from that reality. That is one reason why there is a good deal of versatility within the technological world picture, but it also explains why this take on the world often does not satisfy. Fortunately, there is an abundance of instructive inconsistencies. Conversely, those who acknowledge the world as created are often bound – in many ways and much more than one ought to be – to the technological world picture.

In this context, it is interesting to verify whether the cultural mandate of Genesis 1:28 ("fill the earth and subdue it") plays a role in discussions about the cause of the degradation of the earth. How one answers the question depends on the cultural picture one embraces. Appealing to that mandate from within the technological world picture leads to big problems; problems people are then party to, especially given that so little attention is given to protecting, preserving, and caring for creation. In my opinion, an appeal to the Genesis 1:28 mandate when seeing the garden as communal home leads to a much more harmonious development of technology. That is why it is better to speak of a *creation* mandate.

In any case, it must be clear that choosing the right direction remains an ongoing task, one that implies struggle and excludes laziness.

So, what are the practical consequences of the ethical-philosophical views I am advocating, and how do they differ from what is currently the case? We need to attend here to the meaning of this ethical perspective for technology today but also to indicate that precisely because of the creative responsibility of the technician when it comes to inventions and innovations, new technologies do provide opportunities for a change of direction. We need to stay away from the one-sided development of science, technology, and economy as a model for the whole culture. Such a one-sided "ladder-model" is very common, for example, in assessing so-called "developing countries."

If we want to express the development of culture in a model, it would be better to think of a folding wooden fan, just one segment of which would represent the development of science and technology in connection with the economy. This one segment would not characterize the whole culture, as is the case with the lowest rungs in the model of the "ladder."

Allow me to review a number of the implications of what I am suggesting.

An Ethical Assessment Framework as "Guide for the Perplexed"

I have endeavored to approach the phenomenon of technology from a different perspective than is currently in practice. What does this perspective mean for specific, individual technical phenomena, for what many refer to as "case studies"? I said earlier that when these cases are considered individually, at the microlevel, some ethical improvements are almost always possible. However, in relation to the phenomenon of technology at the macrolevel, ethical tweaking usually amounts to little more than temporary solutions.

The perspective I that have outlined, which highlights the importance of cultural projects, ethos, priorities, values, and norms, will have to be transcribed into a coherent ethical assessment framework. Such a framework could be used as a checklist for the analysis and evaluation of practical, individual technical matters. Responsible cases would then be evaluated relative to a responsible development of technology in general. In other words, those who are perplexed by the technological maze would be able to use this framework as a reliable guide and assessment tool. That does not always mean doing different things, but doing things differently. Nevertheless, choosing another direction can also mean choosing different technologies or doing different things.

First Things First

It is commonplace in science and technology to strive for fireworks, for the spectacular. When that happens, social justice is sometimes violated as less attention is paid to technologies that could help many people in their struggles against hunger and disease. It is painful to realize that attempts to address social injustice receive less funding and attention than, for example, prestigious, costly endeavors in space. I am referring not to communication satellites but to interplanetary space travel. This is an interesting project, but would it not be better to comply with our ethical obligations by setting different priorities? To mention another example of injustice, should we not say of the raw materials given to us that they require a just distribution so that the poor and needy who inhabit our communal home can also share in them?

By carefully setting priorities, we can ensure that there is enough for everyone; hunger results from a one-sided technological development: "There is enough for every need, but not for every greed."

Precaution with respect to future technical-economical development is crucial and needs to be emphasized in politics. Politics is the place to attend to the correct priorities beforehand, instead of what usually happens: reflection after the event.

Ecologically, Culturally, or Socially Applied Technology

We have to devote more attention, with the aforementioned principles in mind, to making modern technology a better fit for the unique situation of today, with regard to people, culture, nature, the environment, and even the landscape. Modern technology ought to be more ecologically and culturally responsible and accommodating. Where disruption has already occurred, restoration where possible should set the agenda.

The damage that has and continues to be done to nature and the landscape can be repaired or prevented by showing greater care. Care may cost more, but the price is worth it! This does not mean that I am advocating a return to the technology of the guild. Applied appropriately, technology needs to gain a breadth that it usually lacks today. Such a differentiation in technological development will obviously also need to have a cultural dimension.

Technological developments should not be at odds with the state of cultural development and the rich variety within it but should seek to fit in with the same. A technology that adapts to its culture will simultaneously enrich it. Unfortunately, we often see the opposite phenomenon in developing nations. In those countries, modern technology, rather than accommodating to the existing cultural diversity, often means a break with the existing culture. Atrocious cities filled with both myriads of poor people and, strangely enough, often venerable industries are the flip side of an exodus from the country, which also results in the destruction of cultures that are centuries old.

Industrial countries, however, also have to deal with serious problems, for example, between technology and nature. These are often due to overdevelopment. Sometimes industrial waste is very damaging to people and nature. Engineers and those in engineering technology need to realize that dangerous by-products have no place in a responsible technology. It is the duty of engineers to find solutions for processing such waste products. Allowing economic profit to define the bottom line will result in poorly developed technology.

It is a pity that the work of E. F. Schumacher, with his call for the wider application of intermediary or small technologies, no longer receives much attention. His was an influential voice during the energy crisis in the 1970s. When the crisis subsided, people made a caricature of him. Schumacher did not advocate a return to primitive and prescientific technology but to a technology that is in accord with nature and culture and thus a technology that fits the human scale. Such a technology should be adapted to external boundaries and human limitations.

We have mighty powers and forces at our disposal, but we remain dependent on fragile ecosystems. This is frequently not taken into consideration by the existing technological-economical powers, with well-known, serious consequences. Therefore, with a view to a healthy future, we must insist that what is needed is a creative technology, one with room for invention and innovation, a technology that is economically productive, accommodating to ecology and culture, socially just, and personally and communally fulfilling. Such technology could even be integrated with computers or the Internet because it would allow for a good degree of decentralization rather than the concentration of power. It is precisely when our technology provides us with more power that we must strengthen our commitment to use this technology with wisdom.

Culturally and Bioecologically Appropriate Agriculture

The power of technology is especially evident when it comes to industrialized agriculture. The problems here are huge. That some are switching to bioecologically and culturally adapted agriculture shows that the picture I have sketched may soon become reality. Unless agriculture wrongheadedly commits itself to an almost divine worship of nature, this alternative approach will increasingly find support as the problems of industrial agriculture increase.

What happens when we technologize agriculture? Modern technology usually does lead to larger yields and lower production costs; however, these increases parallel increased harm to farmers, animals, nature, and the environment. Overproduction precipitates uncertainties in agriculture and concerns about future possibilities: the loss of animal welfare, ground water pollution, land degradation, various new diseases of the soil, disturbance of the landscape, global toxification, loss of biodiversity, and impaired prospects for rural areas. Agriculture involves cooperating with a living reality, but we have been treating the land as though it were inorganic. When we subject agriculture to the grip of the ideal of scientific-technological control, we divorce it from its ecological, biotic, and cultural context.

Ecologically sensitive agriculture works to restore healthy biological relationships. Quality yields and environmental benefit are not mutually exclusive. It is not a matter of reverting to an earlier age, but qualitatively insightful biology, ecology, and soil science will help people handle soils, plants, and animals more wisely and ultimately preserve the fertility of the land.

Genetic Manipulation

The absence of a normative framework for new technologies is especially evident in the research and development of genetically manipulated organisms. In general, biotechnology usually treats life incorrectly from the start. The technical or machine model of organisms misjudges life. That happens, for example, when, in a popular way, one compares the genetic structure of all that lives to something made of interchangeable Lego blocks. Because of this reduction or even denial of life, it is no wonder that biotechnology (genetic manipulation) has to deal with so many hidden problems.

Genetic manipulation requires a more thorough critique and must be restricted ethically and legally. Current developments are unpredictable, risky, and possibly irreversible in their negative consequences. We need a model that is different from the technological one when approaching living organisms. Such a model should represent the organism as a living whole. Then life will be protected and not misunderstood, as happens in the technological model.

In general, regarding the possibilities of the genetic manipulation of plants, animals, and humans, I want to honor the ethical "No, unless ... " principle that many other people have already accepted. That "No" must keep us from causing maladies, natural disturbances, or a loss of biodiversity with genetically manipulated plants. Should the introduction of genetic manipulation be contemplated anyway, then people must make a reasoned appeal regarding the "unless." Likewise, the entrepreneurs who are going to make use of this technology should probably also be made responsible from the start for the risks involved. Even then, the government should be obliged to establish a framework to assess the outcomes.

When people consider the genetic manipulation of human beings, the "unless" will obviously only apply to the level of organs. Genetic manipulation via stem cells, which involve the whole human being, must remain prohibited. Fortunately, few people choose to pursue the therapeutic and reproductive cloning of humans, even though doing so could be taken as a logical consequence of the technological world picture. Where animals are involved, such as genetic manipulation for the purpose of producing medicines, the "unless" can have a more open interpretation.

Alternative and Sustainable Energy

In general, given looming problems, we cannot pay too much attention to alternatives. When we acknowledge the relative nature of science and technology and recognize the threats that do exist, we have every reason to become more creative in developing alternatives. Researching renewable energy sources is an attractive option. By extracting clean energy from garbage treatment plants, for example, process technology is making an important contribution. Process technology is also making use of new carriers of energy, such as hydrogen, photocells, biomass, wind, tidal currents, and the natural heat, hot water, and steam within the Earth. The switch to sustainable raw materials from the agricultural world, particularly from industrialized agriculture, becomes feasible by applying much more efficient biochemical processes. Separation technology, based on a much-improved fundamental process-science, plays a key role in these developments. The use of less or fewer materials ("dematerialization") is also an attractive alternative: solar energy, for example, can be directly converted to electricity or hydrogen. So as not to waste any materials, we need to promote recycling as much as possible. The life cycle of products should be controlled in such a way that almost 100 percent of the raw materials can be recycled.

Dilemmas

Even when technology is reoriented, acute dilemmas will remain. In other words, more needs to be done. Research must continue to seek ways to render harmless the radioactive waste from nuclear power generators. The aim should be a more responsible use of these facilities, which of late we are learning to better control. Only then may this source of energy be considered a resource more sustainable than the fossil fuels currently used. Nevertheless, until technology can render radioactive waste harmless, nuclear energy remains very risky.

14. Political Consequences

Before concluding, I want to highlight the milieu in which the development of technological science and modern technology takes place. The framework I have sketched needs to be followed in a "free-market economy"—assuming that "free" is understood as "freedom in the context of responsibility." Reality teaches us, however, that economic powers usually intensify technicism and technologizing tendencies. To ward that off, many have turned to the government. However, the materialistic politics that usually prevails tends itself to bend with the economy and encourage the process. That notwithstanding, because the arrangement of society does concern everyone and because limitations can be put on economic enterprises, the political realm is the place to bring missteps to light and to counter wrongheaded developments. The political discussion, of course, will need to be attuned to the normative framework outlined earlier and, in keeping with the nature of the political realm, focus on law and public justice.

In light of the present state of disruption, precaution demands that we check the drive to trust in technology. We can use the political process to help weigh our choices: concerning a different direction for technology; regarding a broadly normative and differently constituted technology; focusing on technological investments that are friendlier toward the environment, nature, animals, and culture. Then ethically correct actions will be legally mandated. Such a national agenda, of course, can only be politically effective if it is based on law and a sense of public justice, and – given our globalized world – is backed in the international political arena. The prophetic message of Amos is of current global interest: "But let justice roll on like a river, righteousness like a never-failing stream!" (Amos 5:24).

There are examples that indicate that politics can correct malformed developments. Over the course of time, legislation has been able to address all kinds of important issues: child labor, workers' safety, social security, guidelines for business and industry, wage and pricing policies, endangered species and environmental protection, quality control, and product safety, to name just a few. In doing so, the government has created a framework to protect responsible businesses and to encourage responsible technology. Technology's role as service provider needs to remain a priority.

Obliged Accountability

When new developments arise, people rightly appeal to "the precautionary principle." Rather than having the government act after the fact to correct matters, the intent is to prevent new processes or products that will not be safe for humans, society, nature, or the environment. An important question here is: Can the government foresee all possible implications and consequences of new technological-economical developments? I do not think so. The government is not sufficiently equipped for such a task, and making these calls is not its responsibility in the first place. And yet, the government must do something, because it can be assumed from the start that egoistic motives will have detrimental public consequences.

It would therefore be interesting to see if we can gain consensus about the possibility of general obligations articulated by the government for innovative companies such that, should they not follow through on these, they could still be called upon after the fact to take responsibility for these obligations to society. To date, the cost of dealing with the effects of this kind of corporate irresponsibility has usually been carried by the state and therefore by the taxpayer. This burden should be shifted. Moreover, the societal responsibility of all who are involved with technological-economic development would become more explicit. The rights of corporations to pursue research and development carry with them societal duties and responsibilities. It would be wonderful if this sense of societal responsibility could be worked out in the normative direction I have attempted to indicate!

15. Struggle and Hope

I do not mean to suggest that we are in a position to realize fully the perspective outlined here. But it will provide an enlightening framework in the middle of cultural problems and threats. Thorns and thistles will continue to accompany our work until, through God's intervention, the developed earth, now characterized by distortion, will be transformed into a godly, radiant garden-city (Revelation 21:9–22:5), where people will be revealed as liberated children of God, "brought into the glorious freedom of the children of God" (Romans 8:21). Then the work of science and technology will be seen in surprising ways, and in spite of people themselves, be involved in the new creation. That perspective gives hope and creates obligations. These, in turn, should inspire us to embrace an ethics in which people are expected to invest in their responsibility to seek the meaning of technology – not in isolation, but as woven into the full meaning of reality: the Kingdom of God. This is an ethical perspective that will continue to inspire us and that needs to be proclaimed and heard in our technological world.

ENLIGHTENING THE ENLIGHTENMENT (2020)

Enlightenment Culture

Reflection on where we are in our culture and where we are headed always remains relevant. Especially for a Christian and especially in a culture as rapidly changing as ours. The Gospel calls us to be imitators of Christ. What direction do we choose for this in our time? Christians at intersections of our culture and certainly pastors in their sermons will constantly be confronted with that question.

Our culture is historically and spiritually Enlightenment culture. The basic dogma of that culture is that man must triumph over the prejudices of the past; man must summon up the courage to divest himself of every form of authority in order to think autonomously. By the light of his own mind, man faces the future. This is the driving force of Western science with the enlightened university exerting especially great influence on man and culture through scientific thinking and scientific-technical mastery. A tremendous dynamic of innovation, change and improvement arises. Meanwhile, the thereby reinforced religion of matter – materialism – displaces the Christian religion. Historically, Christianity – through the Reformation – contributed much to the development of science and technology. But the spiritual background of Renaissance and Enlightenment eventually took over. A thoroughly secularized Western culture presents itself to the whole world as the pre-eminent path to the future of humanity. This new worldview presupposes, in addition to the autonomy of man, a radically different relationship between man and nature compared with the pre-Enlightenment era. Modern science with its rationality is the light within which reality – nature, human relations and man himself – is considered and technically controlled. Man cannot escape the dynamics of the scientific-technical world as the engine of the modern economy and must – it seems – adapt to the scientific-technical culture and its spirit. An artificial paradise emerges as a substitute for divine, transcendent reality.

It is this development that has greatly increased man's power, created a fully materialistic culture supported by the autonomous, absolute freedom of (individual) man. This Western culture also permeates non-Western cultures and there too, if it is to succeed, represents a turning of the spirits, unprecedented in the history of mankind.

Meanwhile, that Western culture is less harmonious than was promised. Its progress has a threatening flip side. Indeed, with his technology, man is more powerful than ever before. Computers, robots and even satellites are at his disposal. But at the same time, man also becomes the prisoner of that power. He becomes encapsulated in his own creation and can do nothing with his supposed absolute freedom. Instead of being in control of this growing power, he is mostly controlled by it. This can even become a threat to his future, as witness the threat posed by nuclear technology. Man's future is at stake if nuclear weapons are ever used in global wars. And we now know all too well how the changed relationship between man and nature is hurting us. Even with the pretence of being "lord and master" over nature, man(-kind) has been misled. The view of nature as a mechanism that man can perfect with his technical thinking avenges itself painfully. Currently, there is justified attention to the global climate problem as an exponent of the environmental and natural problem.

Again and again, the question of the cause of all these dangers, threats and problems is rightly raised. Quite commonly, the finger is pointed at Christianity. This accusation is echoed worldwide in non-Christian cultures when they refer to Western culture as a Christian culture. And many in these cultures believe they have the right to do so because they were previously acquainted with that culture and its colonial politics.

Christians should perhaps take this extremely critical accusation more seriously than they usually do. Is there truth in the accusation? Some theologians have seen secularization as the completion of Christianity, thereby legitimizing the decline in church membership. That secularization continues. Sometimes because of aversion to the Gospel of Christ, to the redemptive Cross. But perhaps also more than once because the church has indeed itself been secularized by conforming – mostly tacitly – to the prevailing culture and the power that goes with it. And that process, too, is ongoing today. Reflection is needed on the meaning of Christ and the church for today's culture.

Christ and Enlightenment Culture

It is by no means easy to criticize the Enlightenment culture as here outlined. For there is much to appreciate about what it has brought about in terms of education, health care, poverty and hunger alleviation, material prosperity and democracy. Much of it can be defended with the appeal to the Biblical message that at creation God gave man the assignment to subdue and manage the earth. That command stands regardless of – or precisely because of – criticism of the Enlightenment. Given the problems of Enlightenment culture, it is an urgent question: what responsible direction do we take?

To emerge from this problem, we must realize that our culture is part of history. How can we get a clear view through history of what is to be valued in our culture and what is not? How can Christians say meaningful things about that relationship?

This requires a spiritual in-depth study of history. In the light of God's Word, we recognize that Christ leads and directs history. According to John's Gospel, He was there at creation. Through Christ the Word, everything came into being. The cosmic significance of Christ – He directs history – is described especially in the letter to the Colossians. Christ – the Incarnate Word – also redeems creation history from sin and its consequences. And He brings a new perspective of the completion of everything in God's Kingdom. That perspective passes through the Cross – of deliverance from human apostasy and death. In the yearning of creation, the Kingdom of God is invincible. In other words: Christ is the propelling force – the driving force – of all creation, in which we are included. No one can escape this *dunamis* of history. All of reality as creation is in His hand. In short, *Christ the Meaning of History*, as the theologian Hendrikus Berkhof titled one of his books. From, through and to Christ everything exists. He has shown the way of love and justice for history, and is focused on life, justice and peace for all and for everything. That path must be sought in Enlightenment culture. Only beyond the horizon of earthly time does that Kingdom come in fullness.

So there is one dominant, all-controlling motif. That is that of Christ as Lord of history. All other motives, including those of the (alleged) autonomy of the Western Enlightenment, parasitize on it. Even the concentration and source of all evil, the Evil One (1 Cor. 4: 4), depends on it, and against this dependence he resists, in order to undo it. But he does not succeed.

In opposition to the dynamics of creation – grounded in opposition to Christ and following the law of sin – all kinds of disruption and disturbance, dialectics, struggles, conflicts and tensions arise, with which culture becomes like a safety net and people begin to lose their bearings. But even then one remains bound to the history of creation. One is even judged by it: therefore, neither the problems and tensions, the many crises of our time have the last word. The pretensions of overconfident and haughty man must – against all appearances to the contrary – give way to the supremacy of Christ's government. *Actions directed against the great future of Christ are rectified in due course.*

Meanwhile, in that development – no matter how grand the pretensions of enlightened man – suffering and evil in great variety can take on massive proportions. As the time of world history progresses this – according to Bible prophecy – will get worse. Nothing can be said exactly about when and how a crisis and accompanying realignment will occur. But we do know that God does not allow man in his self-conceit to disrupt everything all the way to the end. Therein lies something of the divine secret of history. Sometimes disasters and threats bring man back on track. We then euphemistically say that the shore turns the ship. We saw this writ large in the unexpected collapse of the Soviet Union in 1989 and in the rise and fall of the Arab Spring in our days. Usually, after revolutions, we see tensions in culture develop again – sometimes very quickly. The same happens around the tensions, threats and disasters of the scientific-technical and economic development of our time. But always in the history of constant and changing tensions, and sometimes even hopelessness, because of the dominance of Christ's government, there is again a perspective for human society and culture. Marked changes in cultural contexts do not change Christ's government, but confrontation with the spirit of the times and listening to God's Word do require us to test old ways and take new ones, in the power of Christ and His Spirit. And again and again, with the new cultural possibilities the direction of the Kingdom must be chosen. So, with the times but not of the times. This challenge faces not only the individual Christian but certainly and above all the church of Christ.

Enlightening the Enlightenment

The influence of the church on culture is actually decreasing significantly in our time. The difference between church and culture is widening. How should the church judge our secularized culture, and how should the church as the body of Christ behave in that culture?

Usually in culture we look – rightly so – to the past. Yet it is also good especially in our time as the crises increase and anti-Christian and demonic powers grow, to look with greater attention to the future. The great Future of the Kingdom of God-though its completion is beyond our time horizon-determines past and present. Past and present is absorbed and determined by that future. Such thinking from the future, can only be done by the Christian because he knows of the fulfillment of everything in the Kingdom of God. For where man closes the world, there can be no prospect. Thinking from the future of a Christian is also the best weapon against the Enlightenment thinking of many in which the future of man is absorbed into the future of technology and economics. And if one takes materialistic evolutionism as a guide in assessing that future, that future looks bleak. Ultimately, it means the demise of everything in death. The Christian vision, on the other hand, is very hopeful. As followers of Christ, even in today's culture with its many possibilities, Christians may point the way out of the swamp of Enlightenment to that of love and justice with a view to life, right and peace for all and everything. That means, on the one hand, rejecting the motives of the Enlightenment. But on the other hand pointing the results of that development through the Enlightenment to a different perspective. It is obvious that the church, especially in preaching, must emphasize both elements, rejection and appreciation of culture. History is decline and ascent at the same time.

In preaching, the church must enlighten the way of the Enlightenment with the word of God. That means, above all, not adapting to the culture. When that happens, I speak of secularization of the church. Recognizing and combating this secularization of the church and the spiritual poverty that accompanies it must be done primarily through the content of the proclamation of God's Word – preaching. More than ever, God's Word must resound in the godless Enlightenment culture. Much could be said about this. I will limit myself to what I think is the main issue. In our materially wealthy culture which is at the same time spiritually poverty-stricken, the church must be an oasis, a refuge in the materialistic wilderness. A place to catch one's breath. Preaching that enlightens the Enlightenment cannot (must not) be left unheard in our culture. After all, history and therefore culture is all about God reigning in Christ, redeeming people from sin in Christ and placing people led by the Holy Spirit on the path to eternity. That flight to God and that turn toward the real future is worlds apart from the anthropocentrism of Enlightenment culture and the building of modern towers of Babel, with death as the terminus of every life and culture. Over against Enlightenment culture, in the church it is about being "rich in God" in that culture.

We saw that the engine of Enlightenment culture is autonomous science. The center of man is seen in reason. Therefore, scientific knowledge is accepted as knowledge par excellence. Should the same be true in preaching? In sermons it should be about penetrating the human heart as the religious center of man. After all, from the heart are the issues of life. Preaching must be existential preaching that touches the heart: it is about being or not being before the face of God. It is about hearing and meeting God in our fathomless depth of apostasy from Him as well as hearing of His saving love in Christ. That is the unprecedented thing about preaching according to God's Word: there occurs a radical turning away from apostasy toward faith, hope and love through Christ in communion with the Holy Spirit. There comes room for the true knowledge of being created, of being fallen, and of being redeemed through Jesus Christ. This knowing reaches far and has many dimensions arising from the center of the knowledge of faith. This is a different knowing than the one-dimensional, flat, cold, rational knowing of science – however valuable that may be within its necessary limits. This knowing involves the whole fullness of life; it governs all expressions of life. This knowing lives from Christ the meaning of everything, and therefore extends to the depth and mystery of life. All our vision of man and world, man and nature, past, present and future is decisively governed by it. Christ is – as mentioned earlier – the driving force of history and therefore of our lives. This is the full content of Christ as the Way, the Truth and the Life. Abstract, impersonal scientific knowledge can never win out over this knowledge of faith. And sermons must keep away from that abstract, rational Enlightenment knowledge, for otherwise the church will come to favor secularization and become secularized. In opposition to the word of man, the Word of God stands in the church; in opposition to the individualism of culture, community in the church as the body of Christ. The church lives by the great secret of history and shows that secret in everything – listening, believing, praying, singing, gratitude, sacrificial readiness, prophetic speaking and priestly action. In humility, the church preaches that mystery in order to make many more and more curious, who have lost their way in the crisis of Enlightenment culture.

There is a common world shared by all – yes, the Christian shares in it – in guilt and participation in Enlightenment culture, but at the same time there is a deep chasm between the path of Enlightenment and the path of Light that the church preaches. Fortunately.

The church of Christ – the testimony of Christ is the spirit of prophecy!!! (Revelation 19: 10) – adheres to the prophetic word, and warns, but above all encourages.

TECHNICISM AND THE MEANING OF TECHNOLOGY (2023)

According to Herman Dooyeweerd[70] – the father of the influential Reformational "cosmonomic" philosophy – the humanistic groundmotive after the Middle Ages influences "the whole development of Western culture" (NC 1:169). It is surprising however that although Dooyeweerd speaks about the cultural influence of the religious groundmotive, he nevertheless restricts his discussion of that cultural influence mainly to the field of philosophical and scientific thought.

Still the important question is this: what influence does the humanistic groundmotive have on the health and wholeness of our culture, a culture which philosophy and science influence greatly, through technology and economy?

Stressing the *cultural* influence of the groundmotive, I would like to speak of the *scientific-technical ideal of control*, or *technicism* (Schuurman, 2003: 63–93).[71] This ideal of control puts the whole of reality under pressure. The ideal of technical control threatens not just man in his freedom but also nature and the social structures within which people function. As a result of the ideal of control, people reduce reality, and so culture, to scientific-technical categories and deny the distinctive character of things.

Perhaps the reader might wish to counter my interpretation by saying that not technicism but economism – as a reduced and at the same time absolutized economy – is the basic ailment of our culture. There is some merit to this view. But a little historical research reveals that in wartime and, for example, in the development of space travel in the era of the Cold War, the spirit of technicism, rather than bringing economic prosperity, demanded numerous economic sacrifices. Meanwhile we have learned that the population of the Soviet Union was starved in the interest of realizing the ideals of technicism. For the sake of the geopolitical power and name of the Soviet Union everything was made subservient to space travel technology, which therefore came at the expense of the population. Accordingly, when today the liberal market economy – i.e. capitalism – is singled out as the cause of much misery, that analysis is inadequate. Such an analysis, to which Bob Goudzwaard[72] in particular has devoted attention, is

---

[70] Dooyeweerd, *A New Critique of Theoretical Thought,* Vols. I and II, Amsterdam/Philadelphia, 1953, 1955.

[71] Schuurman, *Faith and Hope in Technology*, Toronto: Clements Publishing, 2003.

[72] Goudzwaard, *Capitalism and Progress: A Diagnosis of Western Society*, Toronto/Grand Rapids: Wedge/Eerdmans, 1979.

indeed fruitful. Yet I believe that the ideal of technical control works in the background of capitalism. Today's free market economy is possible only through the drive to achieve technical control. In other words, capitalism's lifeline is technicism.

Governments in particular are driven by technicism in their plans to control society. While the power of technical control may be pivotal for industrial enterprises, they still have to take account of what consumers want. The liberal market economy cannot thrive without the approval of consumers. For that reason, it can also be said that economism as capitalism in turn strengthens technicism. The two are, as it were, Siamese twins. Both suffer from a contraction of their outlook. As a result, in part of the dynamism evoked and the enormous scale of it – there is good reason why we speak of the globalization of technology and the economy – the concurrence of an overestimated technology and a contracted economy is dangerous to humanity, society, and the environment. *We all are confronted with problems like climate crisis and with the progress of digitalization.*

Technicism is a very ancient pretension. We already encounter its hubris in the building of the Tower of Babel: humans, viewing themselves as gods, want to storm the heavens: "nothing they propose to do will now be impossible for them" (Genesis 11:6). The same mentality arises after the Middle Ages when the development of the natural sciences gains momentum and is united with a philosophy from which God has disappeared and in which man is enthroned. Once this occurs, the original bond between humanity and nature is broken and nature is regarded as a mechanism to be controlled.

In the light of the fundamental attitude sketched above I can more accurately describe what technicism is. *Technicism is the pretension of humans, as self-declared lords and masters using the scientific-technical method of control, to bend all of reality to their will in order to solve all problems, old and new, and to guarantee increasing material prosperity and progress.* By means of their technology, humans wish control and safeguard the future. This technicism answers to two important norms as though they are the two great commandments: the norm of technical perfection or effectiveness, and the economic norm of efficiency. In other words, by means of the scientific-technical method of control, the stated goals must be reached as directly and efficiently as possible. The entire technical process, therefore, is clearly set within a narrow framework. Everything outside that narrow framework is denied recognition. This concerns the value of human freedom, nature and the distinctive character of plants and animals. The whole of reality is seen as a mechanism or machine which we have to control. The ***cultural paradigm*** under the influence of technicism is the ***machine model.*** Norms as that of appreciation, care, love, harmony, doing justice, and so forth are, accordingly, discounted.

Thus, when I speak of *"technicism"* I especially have in mind the overestimation of the scientific method of technical control. Hence technologies in which that method remains in the background (as in civil engineering) also produce fewer problems; but precisely those technologies which cannot exist without this method give rise to more uneasiness (such as chemical technology, electrotechnology, information technology, biotechnology, and so forth). In addition, the method of those technologies is also being introduced in non-technical areas. In particular, the method employed by modern technology in the designing phase inspires people to apply it outside the realm of technology as well. The economical praxis is more than once marked by excessive scientific control, being dominated by a scientific-technical model: we hear of the *mechanism* of the marketplace. I suspect besides that more organizations are also stamped by technicism and thereby causes huge problems for the world of labor relations and the role of responsible people in that world.

This "evil" is not inherent in the method itself, of course, but in its overestimation and imperialism. It is inherent in the conviction that with this scientific-technical method of control, humans can accomplish ever better and bigger results both in technology itself (where the method is appropriate) but outside of it.

The dark side of technicism is becoming ever clearer in our day. When in the post-World War II era the results in the form of growth in prosperity became truly impressive, consumers too began to venerate modern technology. For a long time, this veneration made people blind to its negative aspects. In the meantime, we have discovered that the world surrounding us – plants, animals, but also humanity itself – can easily become victims of modern technology. Present-day technical developments are in many ways very threatening. The very basis of our existence is being ruined.

In thinking of dangerous manifestations of technicism, I do not just have in mind the threat of nuclear weaponry or the radioactive waste generated by nuclear power plants, but equally certain phenomena of our technological culture such as deforestation, desertification of large parts of the earth, depletion of the ozone layer, the emission of greenhouse gases along with climate change, the destruction and pollution of nature, the overestimation of genetic manipulations techniques and of the latest information and communication technologies which produce increasingly less information and communication among people. An overvalued technology makes for mutual estrangement and social disintegration. Under the influence of technicism and economism, man becomes the prisoner and victim of his own culture. Present-day technical en economic culture is clearly paradoxical or dialectical. How can we overcome this deep crisis?[73]

The Meaning of Technology

When I write or speak critically about technicism mostly people think that I am a 'culture pessimist' who has a negative attitude over against technology itself. That is not the case. Nor can it be, which is why we need to deal with the meaning of technology.

In Reformational philosophy, the idea of the meaning-character of reality occupies a central place (Schuurman, 2009, 380). All things, technology included, are related to God as the Origin of all meaning.

What this meaning-character means with respect to scientific-technological development is that the direction of this development ought to be subjected to the meaning-*dunamis*. Human responsibility can help to disclose meaning, and scientific-technological thought can render meaningful service. When humanity fails to submit radically and integrally to the Radix of all meaning and to the meaning-*dunamis* of the creation, then the original dynamic *of* the creation turns into an original dialectic *within* the creation. The dialectic resists the disclosure of meaning; it dislocates and stifles development. And it is technicism that implies disruption or perversion of meaning of technology. But this dialectic process is a parasite on the meaning-*dunamis*.

When we develop technology, we ought to honor the *dunamis* of creation. What this means concretely for technology is that humanity submits to the

---

[73] Schuurman, *Technology and the Future – A Philosophical Challenge*, 2nd ed., Grand Rapids: Paideia Press, 2009.

given normative principles which obtain for the cultural and supra-cultural modalities, and that people positivize these principles. These principles motivate people and make them wise to the perspective of the fullness of the meaning-disclosure toward which all structures of creation point.

Technology is in the first place qualified by and related to the cultural/historical aspect of reality, with the meaning of free and responsible forming of technological objects and processes in obedience to the cultural norm of effectiveness.

Technological meaning-disclosure implies the deepening of meaning or enrichment of meaning whenever the post-cultural aspects lead technological meaning (see Dooyeweerd, NC-II: 285–305; Schuurman, 2005, 45–48,[74] and 2009, 382–433).

Technological meaning that is disclosed is in the first instance meaning that is expressed symbolically. A design is indicated in formulas and drawings, which serve as guidelines for technological forming. Everyone who is involved in the technological process must participate with his specific responsibility, which implies social meaning-disclosure. Re the economic disclosure of the technological process, man is a steward: waste of materials and damaging the environment have to be prevented. Profit-making should be incorporated into the enterprise and disclosed in connection with service to God, nature and men. The aesthetic meaning-disclosure of technology should appear in a harmonious development among all persons, and harmony should also exist between nature and technology. Technology may not be permitted to make a wasteland of nature. Technological meaning-disclosure should also be led by the juridical aspect. This has many dimensions (Schuurman, 2005). Too many to mention them all. But in any case, this disclosure should prevent the destruction of the environment.

After juridical disclosure, technological meaning-deepening ought to be carried forward into ethical or moral disclosure. Again, the protection of neighbor and of nature are here at stake in caring and loving. Finally, technological meaning-disclosure ought to be led by the belief that humanity is called to the task of technology and that people are obliged to accept this mission as a responsibility before God. Faith is always the boundary-function and horizon of meaning-disclosure.

---

[74] Schuurman, *The Technological World Picture and an Ethics of Responsibility*, Iowa: Dordt Press, 2005; above, pp. 406ff.

In its analysis of meaning-disclosure at the point where it involves faith, philosophical-scientific thought approaches a fundamental boundary. It thus becomes apparent that philosophical-scientific thought is not self-sufficient. Theoretical thought can serve only to point to this *transcending in faith;* it can only approach its substance *by describing an idea,* as it were – an *idea* that ought to lead people in their work in technology (Dooyeweerd, NC-II, 304, 305; Schuurman, 2009, 421).

We have seen so far that in an analysis of the meaning-disclosure of technological development, philosophical-scientific thought ultimately arrives at faith as the boundary-function of the process of meaning-disclosure. Humanity should faithfully transcend technological development in the direction of the fullness of meaning and thus the Origin of all meaning. Humanity does not dispose over the fullness of meaning as the ground of the disclosure of the meaning of technology. This ground should rather be presupposed in all technologically activity and in all reflection on it. Its meaning can only be approached – in the idea which ought to lead and inspire people in their work in technology, the idea through which people are filled with hope.

In approaching the technological development idea, we have come back again to the question of the ground for a liberating perspective for technological development. This ground appears to be religious in character. I reject as unsuitable for such a ground both an absolutized idea of control or power and an absolutized idea of freedom, since such ideas imply a closed view of the world and humankind. Nor do these ideas offer a meaningful perspective for the dynamic development of modern technology. On the contrary, they are the very source of the tremendous problems called forth by this technology and the real dislocations that are presently making themselves felt in our culture.

To get at the substance of the technological development idea, we shall have to *listen to and search for* the fullness of meaning. When people listen, they understand God's purpose with the creation more and more. Man received at his creation the command to be steward of God's completed work of creation and to disclose that work. Contained in this calling is the task of technology as the disclosure of the nature side of creation and as the realization of its technological side.

It is especially difficult – and perhaps impossible – to discern to what extent the pristine creation is present in nature as it now exists. Certainly, the original elements are dominant. The pretension of human autonomy did not and does not have the power to lay creation waste in an integral way, to destroy it as a whole.

A liberated technology – liberated from technicism – will be able to ease the difficult circumstances in which people live 'by nature'. All norms are concentrated in the paradigm of the **garden model;** all technology has to **serve life.** It will afford an enlargement of life's opportunities, relieve the aches and pains and difficulties of work, resist natural catastrophes, conquer disease, protect wealthy ecological conditions, improve social security, expand communication, multiply information, augment responsibility, vastly increase material prosperity in harmony with spiritual well-being, and abolish alienation from self, nature and culture. Technology frees man's time and fosters the developments of new possibilities. Given these possibilities, culture will advance to new disclosures. Technology also will create room for multifaceted work – for careful, creative, love-filled work. And not to forget, caring for and loving all kinds of plants, trees, animals, and humanity and their very varied community associations according to the garden model.

In all of this, humanity finds its share and portion of the *meaning of technology* in the disclosure of the meaning of the creation as a whole (Schuurman, 2003: 64–70, 160–168, 201–205; Schuurman, 2005: 41–59; Schuurman, 2009: 416–433).

INDEX OF SCRIPTURE REFERENCES

GENERAL INDEX

ABOUT THE AUTHOR

Egbert Schuurman (born in 1937 in Borger, The Netherlands) studied civil engineering at Delft University of Technology and philosophy at the Free University of Amsterdam. He received his doctorate in 1972 with a dissertation entitled *Techniek* and *Toekomst: Confrontatie met wijsgerige beschouwingen* (English translation: *Technology and the Future: A Philosophical Challenge* (2009)). From 1964 to 1966 he worked in the area of soil mechanics at Delft University of Technology, and from 1966 to 1984 in the area of cultural philosophy at the Free University of Amsterdam. In 1972, he became special Professor of Christian Philosophy at Eindhoven University of Technology and remained there till 2004. He held the same position from 1974 to 2004 at Delft University of Technology, and from 1984 to 2007 at Wageningen University. In 1994 he was awarded an honorary doctorate (honoris causa) in Engineering from the Northwestern University (Potchefstroom) in South Africa.

From 1983 to 2011, he was the chairman of the ChristenUnie (Christian Union) political party delegation in the First Chamber (Senate) of the Dutch Parliament. From 1983 to 1984, he participated in the international research team project concerning *Responsible Technology* in the U.S. From 1981 to 1983, he was a member of the so-called Broad DNA Committee, which, commissioned by the Dutch federal government, studied the social and ethical aspects of the use of human hereditary material. He edited two international magazines in the area of philosophy and technology. He was the member of the steering committee of the Dutch Royal Institute of Engineers which had the task of investigating *The Limits of Technical Applications.*

Schuurman has given guest lectures on subjects related to the philosophy of technology in Canada, the United States, England, Korea, Japan, South Africa, France, and Brazil. In 1994 he was awarded the *Templeton Award* for teaching about religion and science, granted to him in Berkeley, California, at the Center for Theology and Natural Sciences. In May 2002 he gave a concluding lecture at Delft University of Technology with the title: *Liberated from A Technical Worldview: A Challenge to Frame a New Kind of Ethics.* In 2003, he was named an Officer in the Order of Orange and Nassau. In September 2007, he gave a concluding lecture at Wageningen University with the title: *The Challenge of The Islamic Critique of Technology.*

Next to his doctoral dissertation, *Techniek en Toekomst,* translated into English *(Technology and the Future,* in its 2nd printing) and into Chinese, he has published, among other works, the following: *Techniek: middel of Moloch?* ("Technical applications: means to an end or Moloch?"), *Tussen technische overmacht en menselijke onmacht: Verantwoordelijkheid in een technische maatschappij* ("Caught between technical power and human powerlessness: responsibility in a technical society"), *Christenen in Babel* ("Christians in Babel"; in this edition, pp. 171ff.), *Het technische Paradijs* ("The technical paradise"), *Filosofie van de Technische Wetenschappen* ("The philosophy of technical sciences"), *Geloven in Wetenschap en Techniek—Hoop voor de Toekomst?* ("Faith and Hope in Technology," Toronto: Clements Publishing). The last book published in Dutch was *Tegendraads nadenken over techniek* [Thinking against the Grain about Technology] (Eburon, 2014). It was translated under the title: *Transformation of the Technological Society* (Dordt Press, 2022). Translations of that book have been published in Korean and Portuguese; a Spanish translation is in preparation.

www.ingramcontent.com/pod-product-compliance
Ingram Content Group UK Ltd.
Pitfield, Milton Keynes, MK11 3LW, UK
UKHW022029190726
13853UKWH00005B/2171